Summer Farms

Seasonal exploitation of the uplands from prehistory to the present

Edited by

John Collis, Mark Pearce and Franco Nicolis

Sheffield Archaeological Monographs 16

J.R. Collis Publications

Published by J.R. Collis Publications and distributed by Equinox Publishing Ltd

Editors: J. Collis, M. Pearce and F. Nicolis

UK: Office 415, The Workstation, 15 Paternoster Row, Sheffield, South Yorkshire S1 2BX
USA: ISD, 70 Enterprise Drive, Bristol, CT 06010

www.equinoxpub.com

First published 2016

© John Collis, Mark Pearce, Franco Nicolis and contributors 2016

All rights reserved. No part of this publication may be reproduced or transmitted in any form or by any means, electronic or mechanical, including photocopying, recording or any information storage or retrieval system, without prior permission in writing from the publishers.

British Library Cataloguing-in-Publication Data

A catalogue record of this book is available from the British Library.

ISBN 978-0-906090-55-8 (hardback)

Library of Congress Cataloging-in-Publication Data

Names: Collis, John, 1944 May 19- editor. | Pearce, Mark (Mark J.), editor. | Nicolis, Franco, editor.
Title: Summer farms : seasonal exploitation of the uplands from prehistory to the present / edited by John Collis, Mark Pearce, and Franco Nicolis.
Description: Sheffield, UK ; Bristol, CT : J.R. Collis Publications and Equinox Publishing, Ltd., 2016. Series: Sheffield archaeological monographs; 16 | Includes bibliographical references and index.
Identifiers: LCCN 2015031542 | ISBN 9780906090558 (hb)
Subjects: LCSH: Agriculture–History. | Hill farming–History. | Agriculture, Prehistoric. | Bronze age.
Classification: LCC S421 .S86 2016 | DDC 630.9–dc23
LC record available at http://lccn.loc.gov/2015031542

Typeset by J.R.Collis with the assistance of Queenston Publishing

Cover design: Mark Lee
Photograph: Brigitte Andres

Printed and bound in Great Britain by Lightning Source Inc. (La Vergne, TN), Lightning Source UK Ltd. (Milton Keynes), Lightning Source AU Pty. (Scoresby, Victoria).

Contents

1. Summer farms: An introduction 1
   John Collis

2. Pastoral exploitation of the Caspian and Don Steppes and the North Caucasus during the Bronze Age: Seasonality and isotopes 21
   N.I. Shishlina and Y.O. Larionova

3. 'Salaš': Summer farming and transhumance in the Czech Republic from a (pre)historic and environmental perspective 33
   Dagmar Dreslerová

4. Hard cheese: Upland pastoralism in the Italian Bronze and Iron Ages 47
   Mark Pearce

5. Shepherds and miners through time in the Veneto Highlands: Ethnoarchaeology and archaeology 57
   Mara Migliavacca

6. Seasonal settlements and husbandry resources in the Ligurian Apennines (17th–20th centuries) 73
   Anna Maria Stagno

7. The 'invisible' shepherd and the 'visible' dairyman: Ethnoarchaeology of alpine pastoral sites in the Val di Fiemme (eastern Italian Alps) 97
   Francesco Carrer

8. Going up the mountain! Exploitation of the Trentino Highlands as summer farms during the Bronze Age: The Dosso Rotondo site at Storo (northern Italy) 109
   Franco Nicolis, Elisabetta Mottes, Michele Bassetti, Elisabetta Castiglioni, Mauro Rottoli and Sara Ziggiotti

9. Pastoral land use and climate between the 17th and 19th century in the Italian Southern Alps (Pasubio Massif, Trento): A preliminary report 139
   Marco Avanzini and Isabella Salvador

10. Alpine huts, livestock and cheese in the Oberhasli region (Switzerland): Medieval and early modern building remains and their historical context 155
    Brigitte Andres

11. An historical ecology of the Neolithic to Medieval Periods in the southern French Alps: A reassessment of 'driving forces' 183
    Kevin Walsh and Florence Mocci

12. An archaeological approach to the *Brañas*: Summer farms in the pastures of the Cantabrian Mountains (northern Spain)   203
    David González Álvarez, Margarita Fernández Mier and Pablo López Gómez

13. Elusive *sel* sites: The geoarchaeological quest for Icelandic shielings and the case of Þorvaldsstaðasel, in northeast Iceland   221
    Patrycja Kupiec, Karen Milek, Guðrún Alda Gísladóttir and James Woollett

Index   237

List of Figures

1.1	The *buron* of Niercombe in the Cantal in central France.	2
1.2	The exceptionally ornate doorway to La Maison de Tailleur at La Courbatière.	3
1.3	Wooden vessels and implements used on the *seter* of Bjørge in Vestfold, Norway.	4
1.4	The traditional way of transporting the milk ('*gerle*') from the pasture to the *buron* in the Cantal.	5
1.5	The team at the *buron* of Espinasse Soubro.	5
1.6	Netta Karlsen Gravdal who was a *seterjente* on the Bjørge summer farm in Norway.	6
1.7	Visitors to the *seter* of Bjørge in Vestfold, Norway during the war, c. 1942.	6
1.8	The *buron* of Coulanjou, Vallée de l'Impradine and the '*bédélat*'.	8
1.9	Visitors to the *seter* of Bjørge, date unknown.	9
1.10	A platform constructed for one of the former buildings on the *seter* of Bjørge, 2011	10
1.11	A metal peg on the *seter* of Bjørge, 2011, used to anchor logs.	10
1.12	Sketch plan of a *buron* at Montagne de la Mouche B.	12
1.13	Sketch plan of a *buron* at Montagne de la Mouche A.	13
1.14	The 'comb' layout of rows of *burons* at Lac Servière in the Puy-de-Dôme.	14
1.15	The *seter* of Bjørge in Vestfold, Norway in 2011, used as a family summer house.	14
1.16	The restored *buron* of 'Belles Aigues' at Laveissière in the Cantal.	15
2.1	Location of sites: 1: Ulan IV; 2: Sukhaya Termista and Temrta III; 3: Mandjikiny 1; 4: Khar-Zukcha.	22
2.2	Kalmykia Steppes summer pastures.	23
2.3	Horses breaking the snow cover. Tyva region, 2013.	24
2.4	SukhayaTermista burial ground, kurgan 1; 1: grave 11; 2: grave 13.	27
2.5	Ulan IV burial ground, kurgan 1; 1: grave 14; 2: grave 15.	28
3.1	Geographical map of the Czech Republic and the position of the pollen profiles mentioned in the text.	34
3.2	Postcard of the Kleine Osser, Bavaria, showing the summer farm around 1920.	35
3.3	Ruins of the Scharfs Baude, the Krkonoše Mountains some 50 years after its abandonment.	36
3.4	Remains of the milk cellar of the Stará Bouda in the Krkonoše Mountains.	37
3.5	Summer farm and the sheep pen at the Radhošt' in the Beskydy Mountains, first half of the 20th century.	38
3.6	Drawing of the '*mraznica*' (sheep enclosure for overwintering).	38
3.7	Drawing of a traditional Valach shepherds' tool, '*obušek*' or '*valaška*'.	38
3.8	Settled area of the Neolithic in Bohemia (western part of the Czech Republic). The map shows only settlement sites.	39

3.9	Isolated neolithic finds (mostly stone axes) in Bohemia.	40
3.10	Settled area of the Iron Age (both Hallstatt and La Tène) in Bohemia.	41
3.11	Archaeological finds from the mountain regions (above 650 m.a.s.l.) in Bohemia. 1: Iron Age hillforts mentioned in the text; 2: The Prášily site.	42
4.1	Places mentioned in the text.	48
4.2	Apennine Bronze Age milk boilers. Pots with internal ledge and lids.	49
4.3	Apennine Bronze Age milk boiler.	50
4.4	Puglisi's reconstruction of how an Apennine milk boiler may have functioned.	50
4.5	Pot stand, Grotta di Pertosa (SA).	51
4.6	Whisk, Grotta di Pertosa (SA).	51
5.1	The area under study, with indication of the Lessini highlands (area 1) and of the Schio-Recoaro district (area 2).	58
5.2	The breeders' houses (*capanne*); the shelters (*ripari*); the sheepfolds (*ovili*) found in the Lessini highlands in 2010..	59
5.3	The remains of a wooden *casone* at Bagorno Nord (Lessini highlands) / at Magaello (neve) in the Lessini highlands.	60
5.4	The remains of a small building, connected with sheep farming, in the Lessini highlands.	61
5.5	A shelter at Campo Rotondo in the Lessini highlands.	62
5.6	A sheepfold at Campo Rotondo in the Lessini highlands.	63
5.7	Distribution of breeders' houses, shelters, sheepfolds and Bronze Age artefacts in the Lessini highlands; also Bronze and Iron Age fortified settlements are indicated.	64
5.8	Artefacts found on the top of Busimo mountain: three bronze daggers, one iron spearhead and one iron axe.	65
5.9	A bronze axehead found in the Fittanze mountain (Erbezzo).	65
5.10	A bronze axehead found in the Roccopiano mountain (Erbezzo).	66
5.11	A bronze axehead found in the Roccopiano mountain (Erbezzo).	66
5.12	A bronze axehead found in the Roccopiano mountain (Erbezzo).	67
5.13	A bronze dagger found at Montagna Coe Veronesi (Erbezzo).	67
5.14	A bronze dagger found at Montagna Modetto (Erbezzo).	67
5.15	A bronze dagger found at Montagna Modo (Erbezzo).	67
5.16	A bronze spearhead found at Montagna Gasparine di Mezzo (Boscochiesanuova).	68
5.17	The Schio–Recoaro mining district with the main archaeological sites quoted in the text.	68
5.18	The mines explored in the Schio-Recoaro mining district during the 2011–2012 field survey.	69
5.19	The entrance of Beata Maria Vergine mine (Torrebelvicino, Vicenza).	69
5.20	Fragments of First Iron Age pottery found at Monte Civillina.	70
6.1	Location of the cited sites.	74
6.2	Casoni of Lavaggi di Chiappozzo (820 m.a.s.l.), upper Graveglia valley.	75
6.3	Sketch of the location models of '*casoni*' before and after 1700–1750, derived from the Perlezzi case study (upper Sturla valley).	76

List of Figures

6.4	Domenico Carbonara map "Tipo geometrico delli Condotti, o Corse d'acqua fra Perleggi, Careggi e Caroso", 1752.	78
6.5	Localization on the present cartography of the features represented in the map of D. Carbonara.	80
6.6	Detail of the '*casoni*' represented in the '*Tipo geometrico*' of Domenico Carbonara (1752).	80
6.7	Hypothesis of terrace and water-works phases, and the chronology of *casoni*, derived from the comparison between historical cartography analysis and fieldwork results.	82
6.8	Casoni della Pietra. Sketch of the site.	84
6.9	Casoni della Pietra. Sketch of areas 100 and 200.	85
6.10	Historical cartography analysis. Transect Case delle Barche – Pian del Lago.	87
6.11	Maps of the surveys carried out in the upper Petronio valley.	89
6.12	Sketch of the standing wall stratigraphic analysis at Casone del Giazzo.	91
7.1	The location of Val di Fiemme within the alpine region.	98
7.2	The location of the selected seasonal pastoral sites within the territory of Val di Fiemme.	98
7.3	The stable of *Malga Cadinello Alta* (Val di Fiemme, Trentino, Italy).	100
7.4	The hut of *Malga Agnelezza*, where the herders live during the summer (Val di Fiemme, Trentino, Italy).	100
7.5	*Baito dei Ciocchi*, a former hay-makers' site (now a transhumant site) close to the Lusia pass, between Val di Fiemme and Val di Fassa (Trentino, Italy).	101
7.6	*Malga Lagorai*, a former dairying site, now a non-dairying pastoral site (Val di Fiemme, Trentino, Italy).	102
7.7	*Baito della Bassa*, a former hay-makers' site, now a non-dairying pastoral site (Val di Fiemme, Trentino, Italy).	103
8.1	Storo Dosso Rotondo (Trento). Geographical and geomorphological setting.	110
8.2	Storo Dosso Rotondo (Trento). Excavation area.	111
8.3	Storo Dosso Rotondo (Trento). Southern section.	111
8.4	Storo Dosso Rotondo (Trento). General view of the area subject to archaeological research	112
8.5	Storo Dosso Rotondo (Trento). Stratigraphic sequence of section 1.	113
8.6	Storo Dosso Rotondo (Trento). Synoptic table of microfacies (mf).	116
8.7	Storo Dosso Rotondo (Trento). Samples 252, 253b, 254.	118
8.8	Storo Dosso Rotondo (Trento). Samples 107–108.	119
8.9	Storo Dosso Rotondo (Trento). Building phases 1–4.	120
8.10	Storo Dosso Rotondo (Trento). Plan of hearth US 73 (mf_4).	121
8.11	Storo Dosso Rotondo (Trento). Depth of post-holes in the different building phases.	122
8.12	Storo Dosso Rotondo (Trento). Post-hole (US 98).	123
8.13	Storo Dosso Rotondo (Trento). Lithic industry: sickle blade.	128
8.14	Storo Dosso Rotondo (Trento). Lithic industry: arrowhead.	129
8.15	Storo Dosso Rotondo (Trento). Polish micro-wear on the sickle blade RR 136.	129

8.16	Storo Dosso Rotondo (Trento). Polish micro-wear on the sickle blade RR 329.	130
8.17	Storo Dosso Rotondo (Trento). Pottery.	133
8.18	Storo Dosso Rotondo (Trento). Fragments of strainers.	134
9.1	Comparison between the temperature anomaly calculated from the $\delta^{18}O_c$ of Giazzera Cave (GZ1) for the last 1000 years (tuned record) with other temperature reconstructions.	141
9.2	Chronology of pasture exploitation related to altitude.	141
9.3	The evolution of highland exploitation related to dairy production from historical documents.	142
9.4	Constructive scheme of a '*baito*' inferred from historical documents.	143
9.5	Construction scheme of the storage building; a: 18th century dairy store; b: 19th century '*casera*'.	143
9.6	Evidence of structures related to dairy production in Costoni pasture.	144
9.7	The study area in the Pasubio Massif. Pasture locations.	145
9.8	Stone alignment of casaria in Campobiso pasture; structure without wall remains in Costoni pasture; '*Baito*' remains in Corona pasture (Cor1, 1859); '*Baito*' remains in Campobiso pasture	146
9.9	Some coins found near structures useful for dating.	148
9.10	Iron nails of different size and functions.	148
9.11	Items of personal adornment and daily use objects.	149
9.12	Items related to a generic medieval presence.	149
9.13	Items related to horse presence.	150
9.14	Items related to cattle and goat presence.	150
9.15	Demographic curve, 'building density' curve and temperature oscillations of the study area.	151
9.16	A cross reference study of archive records and field data shows strong correlation between population pressure trends and environmental constraints.	152
10.1	View looking towards the Grimsel Pass. The landscape is punctuated by dams and electricity pylons.	157
10.2	The Oberhasli region, situated at the eastern end of the Bernese Oberland.	158
10.3	View of the Upper Gadmental Valley, looking towards the northeast, with the village of Gadmen and Mount Titlis in the centre of the picture.	159
10.4	View of the Gen Valley, looking northeast in the direction of the Engstlenalp. The pastures can be seen on the valley floor and on the mountain sides.	159
10.5	Permanent settlements above Meiringen on the Hasliberg Mountain with the alpine pastures in the background. View to the northeast.	160
10.6	Mountain passes link the Oberhasli region in every direction with other areas.	160
10.7	The three most important levels of the agricultural system exemplified in the Gadmen and Wenden Valleys.	162
10.8	Distribution of the features recorded during the surveys.	163
10.9	Overview of the categories of features. Half of all the features belong to the category of ground-plans.	164
10.10	Overview of altitude distribution. Most features lie above 1600 m.a.s.l.	165

List of Figures

10.11	An alpine building constructed in the shelter of a huge rock in Zum See, Innertkirchen.	166
10.12	The buildings of the Baumgarten Alp lie on a terrace on the steep northern slope of the Gental Valley.	166
10.13	The isolated features of Zum See, Innertkirchen, are distributed amongst the scree around the mountain lake.	167
10.14	Overview of the measurable inside areas. More than half of the ground-plans measure less than 30 m^2.	168
10.15	Ground-plan in Entlibüöch, Hasliberg, with a stone bench or shelf running along the walls.	169
10.16	A boulder integrated into the ground-plan at Zum See 1.	169
10.17	Present-day alpine hut with masonry base and timber super-structure in Mägisalp, Hasliberg.	169
10.18	The single-pitch roof of this building in Mettlenberg in the Gadmen Valley makes it less susceptible to avalanche damage.	170
10.19	The remains of walls built against an overhanging rock-face Hinder Tschuggi 9, Hasliberg.	171
10.20	A naturally-deposited lump of rock with a man-made masonry entrance and a completely roofed-over chamber in Zum See, Innertkirchen.	171
10.21	Purposely enlarged hollow under a naturally-deposited rock in the deserted settlement of Wendenboden in the Gadmen Valley.	171
10.22	The deserted settlement Gries 1 in the Gadmen Valley with a row of structures built on to one another, and some individual buildings.	172
10.23	The deserted settlement of Wendenläger 1 in the Gadmen Valley with several individual buildings and, beside the largest boulder, what is thought to have been an animal pen.	172
10.24	The animal enclosure system in the deserted settlement of Axalp Chüemad with the ground-plans of lean-to structures.	173
10.25	Ground-plan at Stäfelti 4 in the Gental Valley dug into the mountainside perhaps used as a cooling cellar.	174
10.26	A double cheese store in Mägisalp, Hasliberg, raised above the ground on a timber frame.	175
10.27	Milking hut in Axalp Litschentellti in Brienz, with its typical sheltered milking space.	176
10.28	Picture of a milking hut in Gsteigwiler, Breitlaunen, drawn in 1822 by G. Lory jr.	176
10.29	Existing alpine building in Hinder Tschuggi.	177
10.30	Outline of an alpine dairy with a cow-shed and a lean-to dung heap in Gadmen, Mettlenberg.	177
10.31	Ground-plan of a cow-shed in Hasliberg, Hääggen with a central passageway and cattle stalls to either side.	177
10.32	New tourist transport facilities pose a threat to archaeological sites.	178
11.1	Vertical ecological zonation and typical activities within those zones.	184
11.2	Location map of the study area.	185
11.3	Key pre- and proto- historic archaeological and palaeoecological sites in the Ecrins.	186
11.4	Bronze Age 'enclosure sites' excavated in the Ecrins.	187
11.5	The settlement complex at Serre de l'Homme.	188

11.6	Key (excavated) high altitude medieval and post medieval sites in the Ecrins.	195
11.7	The Roman and post-medieval enclosure of Col du Palastre in the Upper Champsaur.	196
12.1	Ruined hut in the summer *braña* of Saldepuesto (Cangas del Narcea, Asturias), mentioned in the medieval written sources from the Monastery of Corias as brannia de Soldepuesto.	204
12.2	The study area.	205
12.3	Ruined pastoral hut without a clearly defined chronology found in the summer *braña* of Los Cuartos (Somiedu, Asturias).	206
12.4	Summer *braña* of La Mesa (Somiedu, Asturias).	208
12.5	Equinoctial *braña* of La Pornacal (Somiedu, Asturias).	209
12.6	*Braña* of La Peral (Somiedu, Asturias) inhabited by *vaqueiros d'alzada* families.	210
12.7	Merino shepherds in the upland pastures of Braña Forada (Somiedu, Asturias).	211
12.8	Large pen for the merino shepherds' flocks in Sousas (Ḷḷaciana, León).	211
12.9	Braña of L'Estoupieḷḷu, used by the herders from Vigaña (Miranda, Asturias).	212
12.10	Braña of L'Estoupieḷḷu. Using a GPS navigator the test-pits were dug following a random stratified sample.	213
12.11	The village of Vigaña (Miranda, Asturias) in the Pigüeña valley.	214
12.12	Las Corvas site, in the agrarian surroundings of the village of Vigaña.	215
12.13	Annual cattle fair celebrated in El Puertu (Somiedu, Asturias).	217
13.1	Location of the Þorvaldsstaðasel site, northeast Iceland.	223
13.2	Plan of Þorvaldsstaðasel site showing the location of test trench B.	225
13.3	Section drawing showing the contexts visible in the test trench B, R1 at Þorvaldsstaðasel, and the location of the samples 44B and 45B.	227

LIST OF TABLES

2.1	Analysed cultures and their respective chronologies.	21
2.2	Climatic changes in the Caspian Steppes during time interval 3000–2000 cal BC.	23
2.3	Isotope ratios in bone collagen for two diet groups of steppe population.	25
2.4	$^{87}Sr/^{86}Sr$ isotope values in human teeth.	26
6.1.	Methods of investigation employed in the case studies. The grey colour highlights investigation based only on the literature.	77
6.2.	Comparison between the chronology of the aqueduct, terraces and *casoni* (with the indication of the sources for the chronology).	81
6.3.	List of the '*casoni*' of Perlezzi and of their owners located in private and permanent cultivated lands as listed in the '*Caratata*' of 1641, indicated in the *Tipo geometrico* of Domenico Carbonara (1752), and in the *Tipo geometrico* of Giuseppe Ferrretto (1789).	81
6.4	Synthesis of the archaeological investigation on *casoni* of Perlezzi.	83
6.5	Synthesis of the investigation at 'Casoni della Pietra'.	86
6.6	Synthesis of the investigation at Casoni di Bargone.	88
6.7	Historical cartography evidence for the *casoni* of 'Case delle Barche'.	90
6.8	Synthesis of Casone del Giazzo archaeological investigations.	91
6.9	Summary of the markers of *casoni* functions.	92
8.1.	Storo Dosso Rotondo (Trento). Profile description (section 1).	114
8.2	Storo Dosso Rotondo (Trento). List of the thin sections.	115
8.3	Storo Dosso Rotondo (Trento). Synthetic micromorphological description of microfacies anthropogenic components.	115
8.4	Storo Dosso Rotondo (Trento). Synthetic description of microfacies (mf).	117
8.5	Storo Dosso Rotondo (Trento). Anthracological remains.	124
8.6	Storo Dosso Rotondo (Trento). Carpological remains.	126
11.1	Radiocarbon dates for all excavated sites/structures from the Ecrins National Park.	190–93
13.1	Summary of features observed in thin sections 45B and 44B at the Þorvaldsstaðasel site.	226

LIST OF CONTRIBUTORS

Brigitte Andres	Viktoriastrasse 105, 3084 Wabern, Switzerland. Archaeological Service of the Canton Bern, Postfach 5233, 5001 Bern, Switzerland. Email: brigitte.andres@erz.be.ch
Marco Avanzini	Museo delle Scienze, Trento, Italy. Email: marco.avanzini@mtsn.tn.it
Michele Bassetti	Cora Società Archeologica S.r.l., Via Salisburgo 16, I–38121 Trento. Email: michele@coraricerche.com
J.-L. de Beaulieu	IMBE (UMR CNRS-IRD 7263), Aix-Marseille Université (AMU), Europôle méditerranéen de l'Arbois, BP 80, F-13545 Aix-en-Provence cedex 04, France. Email: jacques-louis.debeaulieu@imbe.fr
Francesco Carrer	Department of Archaeology, University of York, King's Manor, York, YO1 7EP, UK Email: francescokar@gmail.com
Elisabetta Castiglioni	Laboratorio di Archeobiologia dei Musei Civici di Como, Piazza Medaglie d'Oro, 1, I–22100 Como. Email: archeobotanica@alice.it
John Collis	9 Clifford Road, Sheffield S11 9AQ, UK. e-mail: j.r.collis@sheffield.ac.uk
M. Court-Picon	IMBE (UMR CNRS-IRD 7263), Aix-Marseille Université (AMU), Europôle méditerranéen de l'Arbois, BP 80, F-13545 Aix-en-Provence cedex 04, France Royal Belgian Institute of Natural Sciences, Department of Palaeontology, Vautier street 29, B-1000 Brussels, Belgium. Email: mona.courtpicon@naturalsciences.be
Dagmar Dreslerová	Institute of Archaeology of the CAS, Prague, Czech Republic, Letenská 4, 11801 Praha 1, Czech Republic. Email: dreslerova@arup.cas.cz
Margarita Fernández Mier	Lecturer in Medieval History. Department of History, Universidad de León. Email: margarita.mier@unileon.es
Guðrún Alda Gísladóttir	Fornleifastofnun Íslands, Bárugötu 3, 101 Reykjavík, Iceland. Email: fsi@instarch.is
David González Álvarez	FPU Researcher. Department of Prehistory, Universidad Complutense de Madrid. Email: davidgon@ucm.es
F. Guiter	IMBE (UMR CNRS-IRD 7263), Aix-Marseille Université (AMU), Europôle méditerranéen de l'Arbois, BP 80, F-13545 Aix-en-Provence cedex 04, France. Email: frederic.guiter@imbe.fr
Patrycja Kupiec	Department of Archaeology, School of Geosciences, University of Aberdeen, St. Mary's, Elphinstone Road, Aberdeen, AB24 3UF, UK. Email: r01pmk12@abdn.ac.uk

List of Contributors

Yuri Larionova	The Institute of Geology of Ore Deposits, Petrography, Mineralogy, and Geochemistry, Russian Academy of Sciences (IGEM RAS) Email: ukalarionova@gmail.com
Pablo López Gómez	Postgraduate student. Department of Prehistory and Archaeology, Universidad de Granada. Email: pirilopez@correo.ugr.es
Karen Milek	Department of Archaeology, School of Geosciences, University of Aberdeen, St. Mary's, Elphinstone Road, Aberdeen, AB24 3UF, UK. Email: k.milek@abdn.ac.uk
Mara Migliavacca	University of Padua, via Fazio 31/a, 36078 Valdagno (VI), Italy. e-mail: mara.migliavacca@unipd.it
Florence Mocci	Centre Camille Julian, CNRS, MMSH, Rue Chateau de l'Horloge, Aix-en-Provence. Email: Mocci@mmsh.univ-aix.fr
Elisabetta Mottes	Provincia autonoma di Trento, Soprintendenza per i beni culturali, Ufficio beni archeologici, Via Mantova 67, I–38122 Trento. Email: elisabetta.mottes@provincia.tn.it
Franco Nicolis	Provincia autonoma di Trento, Soprintendenza per i beni culturali, Ufficio beni archeologici, Via Mantova 67, I–38122 Trento. Email: franco.nicolis@provincia.tn.it
Mark Pearce	Department of Archaeology, University of Nottingham, Nottingham NG7 2RD, GB Email:mark.pearce@nottingham.ac.uk
S. Richer	Worcestershire Wildlife Trust, Hindlip, Worcestershire, WR3 8SZ, UK. Email: suzi@worcestershirewildlifetrust.org
Mauro Rottoli	Laboratorio di Archeobiologia dei Musei Civici di Como, Piazza Medaglie d'Oro, 1, I–22100 Como. Email: archeobotanica@alice.it
Isabella Salvador	Museo delle Scienze, Trento, Italy. Email: isabella.salvador@mtsn.tn.it
Natalia Shishlina	State Historical Museum, Moscow, Russia Email: nshishlina@mail.ru
Anna Maria Stagno	Laboratory of Environmental Archaeology and History, DAFIST, University of Genoa, Italy. Research Group on Heritage and Cultural Landscapes (GIPyPAC), University of the Basque Country (UPV/EHU). Email: anna.stagno@unige.it
B. Talon	IMBE (UMR CNRS-IRD 7263), Aix-Marseille Université (AMU), Europôle méditerranéen de l'Arbois, BP 80, F-13545 Aix-en-Provence cedex 04, France. Email: brigitte.talon@imbe.fr
Kevin Walsh	Dept of Archaeology, King's Manor, University of York, YORK, Y01 7EP, UK. Email: Kevin.walsh@york.ac.uk
James Woollett	Université Laval, Département d'histoire, 1030, avenue des Sciences-Humaines, Bureau 5309, Université Laval, Québec, G1V 0A6, Canada. Email: james.woollett@hst.ulaval.ca
Sara Ziggiotti	Via Matteotti 62/a I–35010 Villafranca Padovana (Padova). Email: sara.ziggiotti@gmail.com

1. Summer farms: An introduction

John Collis

The papers presented in this volume derive from two sessions held at the EAA conferences in Oslo (2011) and Helsinki (2012) on the topic of summer farms, settlements which are known by a wide variety of regional dialect names and languages (*malga, buron, Alm, hafod, shieling, seter, salaš, orry,* and *cayolar,* but these are only a few). The contributors are listed at the end of this article, though some had to drop out and/or have not been able to submit papers, or have published elsewhere, but where abstracts were submitted for the conference, I have included these as an appendix to this introduction to make readers aware of other projects and case studies not covered in the volume. Though I was the initiator of the sessions, it was mainly through the work of my co-organisers, Mark Pearce and Franco Nicolis, that we were able to get such a wide range of contributions, especially from the Alpine areas of Italy, Switzerland, France and Austria, a region which is very much an epicentre for transhumance studies. My interest derived especially from my experiences in central France where I was excavating in the rich agricultural plains of the valley of the Allier. I wondered how the highland areas of the Massif Central, whose peaks, clearly visible on fine days, and some still snow-covered when we were digging in July, were exploited in the later Iron Age contemporary with the lowland sites I was digging (still something of a mystery). I did manage some fieldwork in the Cantal looking at the evolution of the landscape around Pierrefort – field systems and boundaries as well as the *burons* themselves (Collis 2008). My other interest came from visiting friends in Norway while on holiday, including a *seter* which still functioned into the later 20th century as a summer farm, but which is now used as a summer house for family vacations. But I have also had to deal with archaeological questions, whether the Bronze Age remains on Dartmoor which I explored with Andrew Fleming in the 1970s, were permanently or seasonally occupied (Fleming and Collis 1973; Fleming 1988), and, when carrying out fieldwork in central Spain with Gonzalo Ruiz Zapatero and Jesús Álvarez Sanchís, I wondered what were the date and function of ruined stone structures in the highlands to the west of Ávila around the Iron Age hill-fort of Sanchorreja (still unresolved).

I have mainly been a lowland archaeologist, and often dealing with urban sites, but I was educated in the 1960s in the economic and environmental school of prehistory set up by Grahame Clark and Eric Higgs in Cambridge, but where at the same time a Peterhouse research fellow, David Clarke, was developing a more geographical, anthropological and socio-economic approach to prehistory, what became the 'New Archaeology'. Grahame Clark's preferred area of research had been the Mesolithic, but a major source of discoveries of Mesolithic flint assemblages in Britain was from below the peat on the bleak uplands of the Pennines which bisect northern England in a north–south line. The environmental approaches to hunter-gather sites was pioneered by Axel Steenstrup in the 1850s (Steenstrup 1859), using bones of animals and especially migrant birds to demonstrate seasonal occupation, coming up with a model under which the coastal fishers and hunters lived in the warmer lowlands in winter along the sea coast (the shell middens) and moved inland in the summer (hence the more inland distribution of the Danish megalithic tombs with their more elaborate polished stone axes); he argued that the tombs and middens were produced by the same population, in contrast to Jens Worsaae who argued (rightly) that the difference in stone technology was chronological. But Steenstrup had a major impact of Scandinavian archaeology leading to more economic and environmental approaches in Scandinavia than in the rest of Europe (Kristiansen 2002) and this had a major impact on the Cambridge school. In the case of the Pennines, Clark suggested that these were lightly wooded areas which provided good pasture for deer in the summer, and that the Mesolithic hunters followed them, returning to lowland sites such as Star Carr in the winter. It was thus an easy step to look at other periods and areas in terms of such regional economies, and arguing that no site could be interpreted without taking into account its regional context. Transhumance thus became a major theme for some archaeologists from Cambridge in the 1960s and 1970s, e.g. Graeme Barker (1985).

Transhumant or permanent?

My first direct encounter with the question of summer transhumance epitomised one of the problems for prehistorians: when there is no documentary evidence how can we recognise transhumance using only archaeological and environmental evidence, an especially difficult problem in highland situations where useful indicators like animal bones are usually not pre-

Figure 1.1   The buron of Niercombe in the Cantal in central France. Typically the burons in this part of the Cantal were built of stone from the 18th century onwards, and are sometimes dated on their ashlar door lintels or window frames. The roofs at this time are also typically quarried stone slates ('lauzes'). From Roc 1992:60, fig. 89.

served? When Andrew Fleming and I started our study of the Bronze Age reaves of Dartmoor (linear stone banks acting as boundaries) a major question was what of the surviving remains, if any, represented permanent year-round occupation sites and which were seasonal (Fleming and Collis 1973). Though relatively low-lying, climatic conditions and environmental change nowadays render these moorland areas marginal. So were all the surviving houses, settlement enclosures, field boundaries and linear boundaries also marginal in the Bronze Age, or was all, only part, or none of the area permanently occupied? In our first report Fleming and I agreed to disagree: he argued that some of the so-called 'contour reaves' marked a major division in land use, with the areas above the reaves mainly exploited seasonally and the lower areas with field systems and enclosures forming a part of the permanently occupied landscape; I argued that the environmental difference was not that great between the upper and lower areas and the contour reaves, like the other reaves, rather marked boundaries of a different land-use or were social, the boundaries of the territories of different residential groups on the moor. Later fieldwork by Sue O'Neill (1983) showed that the house structures on the higher ground were of generally lighter construction than those on the lower ground, so Fleming's interpretation is the more likely. In addition in nearby Cornwall there is evidence of seasonally occupied sites on the moorlands in the early medieval period, as evidenced by 'hafod' names, a term used in Wales for summer settlements up to relatively recent times, though transhumance had disappeared earlier in south-west England (paper given by Peter Herring at Oslo, see also Herring 2009, 2012).

Hunting or herding?

Over much of Europe transhumance was well established by historical times, primarily from the lowlands to the highlands, though other situations are known such as the use of salt marshes where herding could be combined with salt production in the warmer summer weather. But such situations cannot be simply projected back into the past and at some point in prehistory we have to identify the transition on highland sites from the hunting of wild animals to the pasturing of

Figure 1.2   The exceptionally ornate doorway to La Maison de Tailleur at La Courbatière. From Roc 1992:89, fig. 131.

domestic animals, though hunting might still continue as a subsidiary activity. One effect of transhumance is to remove livestock from the areas under cultivation to more distant areas where they can cause less damage to the growing crops and this might have been a primary motivation in temperate Europe in the initial stages of summer farming as much as the availability of high quality pasture and hayfields that might otherwise have gone to waste; in Mediterranean areas however it was probably seasonal drought in the home areas (Davies 1941) which led to the use of more moist areas for the summer pasture, as in the Spanish Mesta. In our sessions it was Klaus Oeggl who especially addressed the issue of seasonal hunting versus transhumance of domestic animals, in the case of 'Ötzi the Iceman'. The explanation for his presence on the high plateau originally presented by Konrad Spindler, and still the major theme in the museum in Bolzano, links him to the transhumance routes, especially for sheep; However Oeggl's team have looked at the diet of domestic animals from the adjacent farming settlements in the valleys using isotopic analysis suggesting that at that time the upland alpine pastures were not being exploited; the analysis of animal droppings found in the vicinity of the discovery of the body were from wild rather than domestic animals; also the pollen sequences from the area suggest that extensivepasture was not a feature of the landscape until the Middle Bronze Age, in the mid-2nd millennium BC, much later than the Iceman (Festi *et al.* 2014). 'Ötzi' now seems to have been a hunter rather than a herder.

Secondary products

A major development occurred in prehistory with what Andrew Sherratt (1981; 1983) termed the 'Secondary Products Revolution'. Though the cutting and storage of hay for winter fodder allowed an increase not only in herd sizes, and also in the availability of meat, hide and fleeces, the key product was cheese (and perhaps to a lesser extent butter) which could be stored for winter or for emergency use, or even as a trade item, and this was where the summer farms could provide a major change in agricultural potential. In part this change is marked by the appearance of more permanent structures on more static locations. Pure animal husbandry, as Francesco Carrer discusses in his contribution, can leave little trace for the archaeologist; the shepherd, cowherd or horse herder simply follow their herds and flocks around from one area of pasture to another, and only natural features such as caves or rock shelters might see regular re-occupation over periods of time. Cheese production, on the other hand, requires a permanent base for milking, for the processes of producing the cheeses and butter, and then for the maturing and storage of the finished product up to such time as it could be transported back to the permanent set-

Figure 1.3   Wooden vessels and implements used on the seter of Bjørge in Vestfold, Norway.

tlements in the lowlands; in cases where the distance between the summer farms and the permanent settlements was great this might not be until the end of the of the summer season. On the *casoni* of northern Italy (see Anna Stagno's contribution) and the *burons* in the Cantal in the southern Massif Central, the buildings were major investments of time and money (Fig. 1.1), and the stone lintels of the doors would proudly display the date when the building was constructed (Fig. 1.2) and the initials of the owner (Roc 1992).

There might also be other subsidiary buildings—a regular animal to be kept on the cheese producing sites was the pig which could be fattened up for the winter on the whey, the waste product from the cheese making, and though in some cases the pig would be allowed to wander around like a domestic pet, the provision of a sty was normal in some areas. Cheese making also demands specialist equipment; much of this would have been made out of wood and so only rarely survives (Fig. 1.3), but occasionally pottery and metal implements are found. In the 1950s Salvatore Puglisi was already interpreting some of the specialist pottery forms found on upland Bronze Age sites as being connected with cheese making, discussed by Mark Pearce in his contribution.

Who went to the shielings?

One thing which emerges from our overview is how varied practices could be, not only from one region to another, but even from one valley to another and from the different zones defined by height about sea level as in the Cantal or from variations in the quality of the soil (Bourdessoule 2002). In the Cantal the milk used for making cheese was transported from the pasture to the *buron* in large vats slung on a pole which could only transported by men (Fig. 1.4), and *buron* activities were considered too hard for women (Roc 1989); it is well illustrated in the film that Jean-Claude Roc made of La Croix Blanche which is shown at the restored *buron* of 'Belles Aigues' at Laveissière in the Cantal. Here the norm would be three men working together (Fig. 1.5), one in charge of the pasturing, one in charge of the milking and the *buron*, and one for the cheese making, (Bourdessoule 2002:141). In contrast in Scandinavia (Norway and northern Sweden) care of the *seter* was primarily a female activity, with older daughters of the family supplying the 'milkmaids') and for younger women it was considered a good training in household management before marriage (Daugstad 2006). The *seter* with which I am familiar was run for about fifty years by an unmarried lady, notable that she also had a wooden leg (Fig. 1.6), so clearly the methods of making cheese were very different from those in central France! The women were also expected to cut the hay and store it. In Scandinavia the men were engaged in other activities such as iron smelting and forestry. In one of the cases from Scotland described by Eve Boyle the whole community was obliged to move during the summer months, and individuals were liable to be

Figure 1.4   The traditional way of transporting the milk ('*gerle*') from the pasture to the buron in the Cantal, still in use in the 1980s. From Roc 1992:157, fig. 271.

Figure 1.5   The team at the buron of Espinasse Soubro, the '*boutilé*' Henri Valmier, the '*pâtre*' Elie Caumel, and the '*vacher*' Joseph Leron. From Roc 1992:166, fig. 294.

Figure 1.6   Netta Karlsen Gravdal who was a *seterjente* on the Bjørge summer farm in Norway for 32 years in the early 20th century (photographer unknown, c. 1918–1920).

Figure 1.7   Visitors to the *seter* of Bjørge in Vestfold, Norway during the war, c. 1942.

fined if caught by the landowner in the crop-growing area; however they did not move very far, just beyond the upper dyke which formed the boundary between the agricultural and pastoral areas.

Subsistence or commerce?

The reasons for exploiting summer pastures are also varied. Often it is linked in with the exploitation of other resources such as hunting and gathering, or the mining and smelting of metal ores such as copper, tin and iron, or the production of specialist minerals such as salt. Other situations might occur where pasture was in competition with other types of land use, especially for agriculture with the use of the best land reserved for crops during the summer months, or where there was a high population and the products of the lowlands were not sufficient to support the population especially over winter, or where there might be conflict between the use of land which might be under communal ownership and that of individual owners. The relationship between these different factors could be complex, as in Anna Stagno's study which shows where communal agreement was need to construct and maintain the leats which brought water to the meadows, or for the construction of terraces, but where buildings might be privately owned by individuals. In the case of the *seter* (summer farm) of Bjørge in Vestfold, Norway, different individuals might own the different resources – the land, the trees grown on it, or the cattle grazing on the vegetation, or the actual buildings and equipment used on the *seter* itself. In situations where the permanent farms and settlements were themselves marginal (e.g. on the northern limits of agriculture in Scandinavia, or on the higher parts of mountainous areas (e.g. of the Alps) the products of the summer farms might be an essential resource for survival in the winter, for instance in providing fodder for the over-wintering of livestock, or of cheeses which could be stored to supplement the diet in the lean months of winter. They could also be used as a buffer in cases of disaster like crop failure or in periods of warfare, e.g. the use of *seter* in Norway during the German occupation of Norway in the Second World War (Fig. 1.7), and where the summer farms could also provide hiding places for members of the resistance (the Massif Central was a major recruiting ground and base for the French *maquis)*.

But in other situations the products would provide a surplus which could then be traded. A good example is the Massif Central. We first hear of the cheese production in court cases where there was a dispute between the interested parties such as the land owners (often institutions such as monasteries as well as local magnates) and the farmers who were producing the cheese, and this is a clear indication of the value of the cheeses in monetary terms (Fournier & Fournier 1983). Under these conditions production was often tightly controlled. The normal production of Cantal cheese encompasses three varieties depending on the length of time the cheeses are stored and allowed to mature and ripen – *jeune, entre-deux* and *vieux* – but especially prized was Salers, produced by a special breed of cattle (also named Salers), which were on the highland pastures between April 15th and November 15th. The herbs and plants consumed by the cattle gave the cheese a special fragrance, and the small town of Salers grew rich on its trade. Salers and St Nectaire cheese (also from the Auvergne) reached the highest echelons of French society including the court of Louis XVI.

The decline of summer farms

For many communities in and adjacent to the major highland areas of Europe such as the Alps, the Pyrenees, the Massif Central or Scandinavia transhumance was an integral part of the culture, and the day when the cattle left for the summer pastures and the day when they returned were often fixed by convention and so became major days for festivities. But mechanisation and industrialisation not only of farming but also the expansion of rail and road networks opened up previously isolated rural communities to the outside world, not only flooding the market with agricultural goods produced more cheaply elsewhere, but also of farming life itself, with the arrival of mechanised forms of ploughing and harvesting, especially around the middle of the 20th century when tractors replaced the horse and ox. But there was also the exodus of substantial parts of the rural population to urban centres, if not emigration to expanding countries such as the United States. The small terraced fields typical of many of these highland communities were unsuited to mechanisation, and were abandoned unless the crop was one which was still marketable despite being labour intensive with high production costs, crops such as fruit trees, olives and grapes. Such crops were generally not viable alternatives to the production of cereals or hay in the higher regions.

A photographic exhibition in the beautiful and historical village of Blesle in the department of the Haute Loire in the Auvergne demonstrated the process graphically; the hills around the village may have only been a backdrop to the main focus of the photographs of village life and events such as the cattle markets, but they showed eloquently how, in the later 19th century the cultivation of the terraced fields reached to the top of the hills. With each successive photograph woodland regeneration, starting at the top of the hill, gradually descends lower and lower down the hill as the cultivation of the fields was abandoned. Many of the fields so accurately recorded by the early 19th century 'Napoleonic cadastre' can now only be found under dense woodland. With the decline of the population which had supported this intensive agricultural exploitation in the 18th and 19th centuries, farming (and the farms) was concentrated in fewer and fewer hands with larger and larger estates and less intensive methods of farming. With the decline of the home villages, transhumance

Figure 1.8 The *buron* of Coulanjou, Vallée de l'Impradine (right) and the '*bédélat*' used for overnight stalling of young cattle. The buildings have now been destroyed. From Roc 1992:117, fig. 204.

also declined, especially in the years after the Second World War. While many of the upland pastures were still exploited, the opening up of the highlands with tarmac roads revolutionised access, and especially the arrival of off-road 4×4 vehicles such as Land Rovers means that the cattle can be reached quickly without a long trek on foot up rough roads. Even the livestock started to be transported to the summer pastures and back by train and lorries. Where there has not been a shift from dairy to meat production, the summer farms have been replaced by milking parlours like the *laiteries* of the Massif Central, with investment in milking machines to replace hand milking. The milk can then be transported elsewhere to make the cheese and butter, usually in large and mechanised installations on lower ground and which can be used all the year round, and so are worth development with investment in machinery. The buildings of the summer farms mainly abandoned in the second half of the 20th century are a familiar sight in areas such as the Cantal (Fig. 1.8).

Distance travelled

One of the major variables in the use of summer farms is the distance between them and the permanent winter settlement, something which would impact in a number of ways. Who went to the summer quarters and who had to stay behind to carry out duties such as bringing in the harvest? To what extent did the two groups stay in contact – long distance would imply minimal contact whereas a walk of a few hours up the mountain would allow continuous contact and change of personnel (Fig. 1.9)? Where the distance was great what arrangements were needed to ensure the safety of the livestock and the drovers, and the control of the animals so that they did not cause damage to crops and property during the movements? How and when were the products, especially the dairy produce and hay sent back or marketed? The summer farms would normally have to be self-sufficient and so vegetable gardens would be a regular, if not essential, adjunct of the summer farm unless supplies could come from the home village.

A major factor is accessibility. In a deep and narrow valley, for instance in some mountain areas, the local farmers may simply be moving the animals up and down the valley sides according to the season, but access would be difficult for outsiders. In the case of islands such as the transhumance on the Isle of Man, this is more obvious, though examples of cattle being boated in are not unknown in western Britain. Secondly there is the size of the areas of pasture, and closely related to this the size of the local population in the immediate vicinity who could exploit it. If, as in the Massif Central or in Norwegian areas like the Hardanger Plateau the permanent population was low and the

Figure 1.9 Visitors to the *seter* of Bjørge, date unknown.

amount of summer pasture extensive and potentially under-exploited, outside groups living some distance away would be interested in bringing in livestock, but it had to be done on a sufficient scale to make the journey worthwhile and to cover the costs such as tolls that might be charged for the right of passage through territories on the route. Again, as Bordessoule (2002) describes for the southwestern part of the Massif Central, the quality of the pasture, and so the quantity and quality of the cheese produced, encouraged financial investment in buildings, etc., and it was outside institutions such as abbeys and urban financiers who were in a better position to invest than were the local landowners and especially the peasantry. Exploitation was also intense with the corralling of animals in moveable pens so that specific chosen areas would be well manured. In areas such as the Monts Dore where the fertility was less, exploitation would be more extensive, e.g. for meat rather than dairy products, or they could be utilized by the local peasantry, often acting communally, as part of a subsistence economy. Where the pasture was unable to support this continuous pasturing, more mobile regimes might operate, like the sheep flocks described by Francesco Carrer, but which could also be integrated with the more permanent exploitation by the cheese producers. But there were also the societies described by Natalia Shishlina and Yuri Larionova in eastern Europe, and the economic distinction between such groups and the purely mobile pastoralists of the steppes might not be particularly marked.

Long distance movement from outside meant that the herders could be ethnically different from the permanent population, a situation exemplified in the Balkans by the Vlachs or Wallachians from Romania who moved their livestock to the Carpathians and as far as Silesia, as described by Dagmar Dreslerová, or the merino shepherds who formed part of the Mesta bringing sheep from the south of Spain to the Cantabrian Mountains, or there could be a special group within the local society as in the Asturias, the *vaqueiros d'alzada*, both the latter described here by David González-Álvarez *et al.* But long distance transhumance is very reliant on a central authority to guarantee safety and access to the major routes which also depended on the size of the profits which could be made. This is best documented in Italy in the Roman era in such areas as Molise where the transhumant routes could be recorded by imperial inscriptions, and was also very much a feature of late medieval times up to the last century as in the case of the Mesta in Spain which was under royal patronage.

Methodology and teaching

Clearly, as is demonstrated in this volume, there has been an upsurge from the late 20th century in the study of summer farms and in transhumance. One cause of this is obvious, and that is the huge decline if not complete disappearance of transhumance, something which in some areas is considered an integral part of 'traditional' rural life and an important part of social iden-

Summer Farms: Seasonal exploitation of the uplands from prehistory to the present

Figure 1.10   A platform constructed for one of the former buildings on the *seter* of Bjørge, 2011.

Figure 1.11   A metal peg on the *seter* of Bjørge, 2011, used to anchor logs which were split using horse power to make shingles for the roof.

tity. Thus several articles in this volume (e.g. Carrer, Andres and González Álvarez *et al.*) have ethnoarchaeology and anthropology as a major component of the research with consultations with the last transhumant farmers and members of their families who as children were involved in the farming activities. This phase is gradually coming to an end as a generation dies off and much of the oral information dies with them. As a relatively experienced fieldworker it was a bit depressing walking round the *seter* at Bjørge with Astri Jahren Johansen as she pointed out where buildings had stood and the location of hearths in them which I had difficulty in seeing and interpreting (Fig. 1.10). I would never have understood the use of the metal peg hammered into the rock which held logs in position (Fig. 1.11) as they were split using horse power to make the shingles for the roofs of the buildings! Though there are many documents in official communal archives, there are also many in private family possession. This includes collections of artefacts with a description of their function, but there are also films such as that made by Jean-Claude Roc at La Croix Blanche in the Cantal, one of the last functioning *burons* in the Haute Auvergne, and which is shown at the restored summer farm of 'Belles Aigues' at Laveissière.

Several studies in this volume also use vital and informative historical sources, but these are heavily biased towards legal and fiscal matters: ownership of land, livestock and agricultural products; access to pasture and water; the construction and repair of buildings, field boundaries and terraces; and the ownership and exploitation of woodland and mineral deposits. They are the product of the literate elite rather than the illiterate peasant. But such written sources are rare or unknown for the late medieval period and generally non-existent before that (there are some authors and inscriptions from the Roman era). Historians still have written sources which have not yet been exploited, but in the last couple of decades it has increasingly been archaeologists who have taken the lead, not only with surface surveys as published here but also the excavation of the buildings, and test sampling the environs to understand better the nature and exploitation of the total landscape in which the sites lie. The increase in archaeology not only reflects the explosion of archaeology in the late 20th century, but also a shift in the paradigms and interests of archaeologists. Though in Scandinavia there has been a long history of environmental and economic interpretations of the past, the dominant archaeological paradigm of much of the late 19th and early 20th century has dealt with the origins of people, the so-called 'culture historical' paradigm with its emphasis on 'cultures', artefact typologies and chronology, to which summer farms with their general lack of artefacts and peripheral locations were of little interest, though there are cases such as the publication in 1961 by Lars Leinton, as noted by Kristoffer Dahle, where summer farms are linked with this paradigm, as a practice introduced by the 'Indo-Europeans', and an integral part of the culture of this invented 'race' of people. This was also true for some of the historically documented periods where archaeology was the 'handmaiden of history' and so primarily used to try to answer historical questions to which it was poorly adapted; even where useful methodologies were being developed such as systematic field survey and aerial photography (e.g. in Britain) they were little utilised except for historical interpretation (e.g. forts used for the reconstruction of Roman campaigns). But in many countries on the continent such as Italy, Spain and the former communist countries it was illegal as late as the 1990s for private individuals to take aerial photographs, but it is a technique which has literally 'taken off' in the last half century, now supplemented by laser-based LiDAR survey.

This historical emphasis was reflected in the training archaeologists received in universities, and in some countries such as France Archaeology was mainly taught as an ancillary of History or Art History rather than a subject in its own right. Archaeology courses taught general methodologies such as dating, typology and excavation, but mainly they specialised in a specific period such as the Palaeolithic, Prehistory, Roman or occasionally Medieval Archaeology. Though the advent of the 'New Archaeology' from the 1960s placed a new emphasis on topics such as the evolution of societies through time or 'least-effort' models of land use, in Britain it was mainly within Local History Studies and organisations such as the Royal Commission on Historical Monuments that landscape studies started to develop, influenced, for instance, by W.G. Hoskins' book *The Making of the English Landscape* (1955). This was not reflected in teaching in much of British archaeology until the 1980s with the development of Masters degrees in Landscape Archaeology, and early studies of landscape mainly occurred in evening classes run by specialists in Local History like Hoskins at Leicester University or Maurice Beresford at Leeds. In other countries archaeologists have generally been slow to develop an interest in the landscape and its history. In Sweden it was mainly the domain of historical geographers, and at the 2014 meeting of the EAA in Istanbul one participant from Finland in a session dealing with developments in EU legislation, suggested it was still difficult to find students who considered the history of the forest as an archaeological topic despite the increasing need for landscape archaeologists in national parks and in local and national government institutions.

Another problem is the complexity and the cost of work on rural landscapes. I abandoned my own research in the field both in the Cantal and around Ávila in part because of my retirement, my increasing age, and the loss of the student labour force, but also because I knew that the development of new methodologies such as GIS and LiDAR could produce in

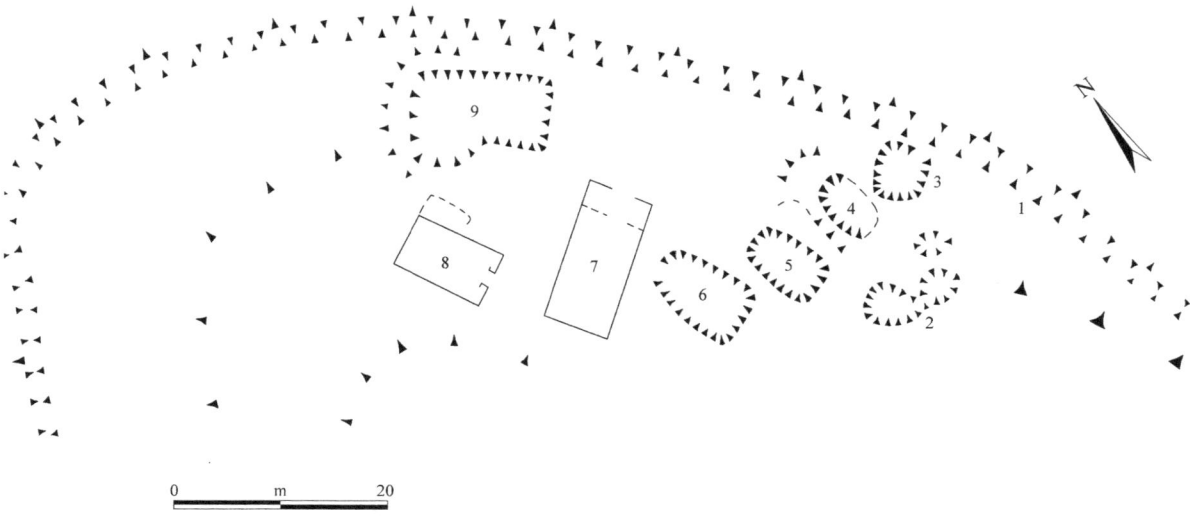

Figure 1.12 Sketch plan of a *buron* at Montagne de la Mouche B (X 638E Y 200.7N). 1: ditch or leat; 2–6, 9: *mazucs*; 7, 8: collapsed *burons*. From Collis 2008, fig. 3. One of the *burons* is photographed in Roc 1992:106, fig. 170).

a few hours much more accurate and detailed plans than I could achieve with years of fieldwork, though for me survey on foot is still an essential part of the interpretation process. No more spending a major part of the day carrying heavy equipment to inaccessible places! Locations are now plotted to the accuracy of a few centimetres whereas formerly I would have been contented in some cases with an accuracy of a few metres! Frédéric Surmely who has been working in the same area as I did in the Cantal has been carrying out LiDAR survey, proving the point, but such surveys are still very costly (Surmely *et al.* 2010).

The sites and finds also require a battery of scientific studies. This can include detailed pedological analysis (studies in this volume by the teams of Franco Nicolis and Patrycja Kupiec). Given the frequent lack of finds, dating can often only be achieved with a large number of radiocarbon dates (see the papers by Natalia Shishlina and Yuri Larionova, and by Kevin Walsh). Where there is skeletal material the use of isotopic analysis (again by Shishlina and Larionova), and by Klaus Oeggl in the case of the animals from the probable home valley of 'Ötzi the Iceman', which has completely changed the received interpretation for one of Europe's most iconic finds. But perhaps more than any other branch of Holocene archaeology the study of settlements in marginal areas occupied by summer farms, almost by definition, requires a careful study of climate and of environmental change, in this collection of essays best exemplified by the article by Marco Avanzini and Isabella Salvador. Such work may fundamentally change our knowledge, but does not come cheaply, and the acquisition of knowledge rather than finds that can be displayed in a museum makes raising funds a bit more difficult. It also raises the nature of museum displays, too often collections of artefacts rather than of the people who used them and their economies and environments.

Thus though there is a battery of techniques available for those studying summer farms (historical sources, cartographic evidence, place-names) ultimately there are certain questions which only archaeology can answer, for instance the origin and the development of the farms. This can in part be done by surface survey, and the identification of the earlier phases on sites and the building history of the structures as is done in several of the papers in this volume, for instance that by Brigitte Andres. As examples from my own experience I reproduce two surveys done on the Montagne de la Mouche north of Pierrefort in the Cantal (Fig. 1.12). What I have termed 'Site A' has the remains of a couple of *burons*, one of which Roc, on the evidence of his evolutionary typology of stone-built *burons*, supported by dated lintels on some of the sites, would date to the early 18th century (see his photograph in Roc 1992:106, Fig. 170). But there are what seem to be earlier structures, some of which are simple holes in the ground which he has suggested may be an early form of *buron* referred to in the medieval literature as *mazucs*. On many sites these earlier traces have been obliterated by later constructions, but a second site a few hundred metres away seems to represent a site undisturbed by

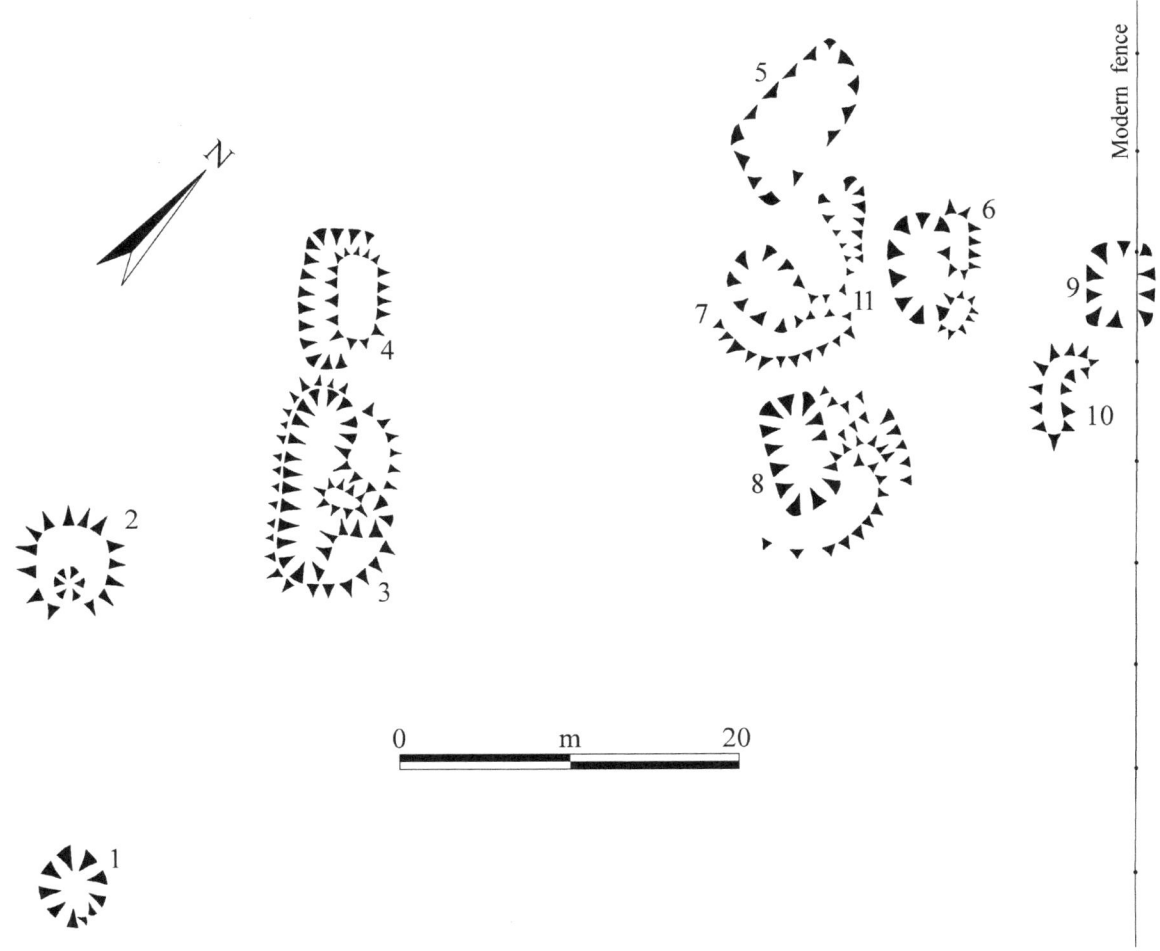

Figure 1.13 Sketch plan of a *buron* at Montagne de la Mouche A (X 637.5E Y 2003N). 1, 10: shallow depressions; 2: circular platform; 3: banked linear depression; 4–9: probable *mazucs*; 11: hollowed pathway. From Collis 2008, fig. 2.

later activity, and so represents a 'pre-stone *buron*' layout (Fig. 1.13). At present, with no excavation, we do not know the date or the function of these structures which could mark the beginning of summer farms in this region. Surmely's excavations, still in an early phase of research, suggests that the sites start in the medieval period, replacing an earlier medieval phase when there were more permanent structures on some of the areas which were abandoned as summer farming took off, perhaps in the 14th century (Surmely *et al.* 2010). So looking at the complexity of the sequence of structures on individual sites is an essential first stage of investigation, as in the papers by Walsh, Migliavacca, Andres, Avanzini, etc., but there may not always be surface indications which has led to systematic survey with test pits across the landscape as carried out by González-Álvarez. Thus an obvious question in many areas is simply the date when summer exploitation started, so one of the directions of research for archaeologists is to plot the presence of finds of artefacts in the highland areas, as in the articles by teams led Migliavacca, Dreslerová, Nicolis and Walsh which at least gives a temporal framework for research, even if some of the finds may relate to some other activity such a mining or hunting. Surface and aerial survey can also reveal the marked differences between different but adjacent areas. The sites I showed above are from the Cantal, but in another part of the Massif Central sites like the 'comb' settlements of rows of structures suggest a contemporary but very different from of social and economic organisation (Fig 1.14)

The future of summer farms

As is clear in the papers in this volume, summer farms are a phenomenon which has largely disappeared in Europe in the 20th century, overtaken by technological change (e.g. forms of transport) and also developments in the competitive nature of the world economy. In most cases too the physical structures are disap-

Figure 1.14 The 'comb' layout of rows of *burons* at Lac Servière in the Puy-de-Dôme which indicates a more communal organisation than the individual *burons* of the Cantal. Photographer and date unknown.

Figure 1.15 The *seter* of Bjørge in Vestfold, Norway in 2011; it is now used as a family summer house.

Figure 1.16 The restored *buron* of 'Belles Aigues' at Laveissière in the Cantal which has been opened to tourists to explain the history of transhumance and the methodology of cheese making. From Roc 1992:171, Fig. 301.

pearing as they are usually too far away from modern settlements to take on new uses, and for most of the contributors to the volume the relatively recent sites they may be studying are usually now without roofs, and with the walls in an advanced state of decay, or are already archaeological heaps of stones or levelled platforms. There are exceptions; in Scandinavia the tradition of having a summer home by the sea, in the woods or in the mountains has meant that some summer farms have taken over this function while losing their role as farms (Fig. 1.15), and are still used for family holidays. In areas such as the Massif Central a small number of *burons* have been kept in repair to act as overnight shelters for walkers, or, in the case of Belles Aigues (Fig. 1.16) to be used as a tourist attraction to explain the history of the *buron*. Though many sites are so far in the wild, they may remain relatively undisturbed, but development does affect some sites as major lines of communication such as motorways are constructed through highland areas, industrialisation through hydro-electric schemes, and even tourist developments such as skiing as described by Brigitte Andres. Sadly for the majority of summer farms the future is oblivion except as archaeological sites.

Acknowledgements

My thanks to all those who participated in our sessions or sent information, and especially to Mark and Franco for their involvement at busy times for both of them. I am thankful to Jean-Claude Roc for permission to reproduce illustrations from his beautiful volume *Burons de Haute Auvergne* and to Frédéric Surmely for information of the recent results of his archaeological surveys there. It was Sophie and Michel Couillaud at Lescure who invited me and my team to Pierrefort and helped me with the logistics of carrying out surveys there. In Norway Astri Jahren Johansen welcomed me twice, in 2002 and 2011 to the family *seter* and showed me round, and also gave me access to the family archives and photographs of life there in the 20th century. Eva Ek provided the contact and accompanied us, and Herman Berthelsen helped with the digitising of some of the photographs.

Appendix: the EAA sessions and abstracts given at Oslo and Helsinki

The following gives a list of the papers as advertised for the sessions at the meeting of the European Association of Archaeologists annual conferences, in Oslo in 2010 and Helsinki in 2011. For various reasons not all the papers were given or have been published here (some have been published elsewhere).

Oslo: Summer farms – upland economies through time.

Organisers: John Collis and Mark Pearce.

Lectures

Didier Galop (CNRS, France): *Origins and history of summer farming in the Pyrenees inferred from palaeoecological data.*

Klaus Oeggl (University of Innsbruck, Austria): *Was the Neolithic Iceman 'Ötzi' really a herdsman?*

Franco Nicolis, Elisabetta Mottes (Provincia autonoma di Trento, Italy): *Up to the mountain; exploitation of the highlands of Trentino as summer farms during the Bronze Age.*

Francesco Carrer (University of York, UK): *The 'invisible' shepherd and 'visible' dairyman; ethnoarchaeology of Alpine pastoral sites in Val di Fiemme (Trentino, Italy).*

Thomas Reitmaier (University of Zurich, Switzerland): *Prehistoric alpine animal husbandry; recent discoveries in the Silvretta range (Switzerland/Austria).*

Yolanda Alther (University of Zurich, Switzerland): *Last multi-level transhumance; archaeological insights into alpine settlement and economic systems; a case study from the Valtellina/Italy.*

Kristoffer Dahle: *Summer farming; constituent or conjuncture in Norse agriculture?*

Eve Boyle: *Not just 'some old huts in the hills'; shielings in the Scottish Highlands.*

Peter Herring (English Heritage, UK): *Early medieval Cornish transhumance; landscape* havos *and* hendre *names, huts, pens, lanes and landscape.*

Helsinki: *Malga, buron, Alm, shieling, seter, salaš, orry and cayolar: seasonal exploitation of uplands from prehistory to the modern day.*

Organisers: John Collis and Franco Nicolis.

Lectures

Mark Pearce (University of Nottingham; United Kingdom): *Why do people move to summer farms? Upland pastoralism in the Italian Bronze and Iron Ages.*

Michele Forte (University of Sheffield; United Kingdom): *The Cairo Massif; a changing upland economy and landscape between 1700 and 1970.*

Marco Avanzini and Isabella Salvador (Museo delle Scienze; Italy): *Mountain men under climate change; pastoral land use between the 17th and 19th centuries.*

Brigitte Andres (Archaeological Service of the Canton of Bern; Switzerland): *Hillside location with glacier view; Alpine settlement remains in the Bernese Alps (Switzerland) from medieval and modern times.*

Kevin Walsh (University of York; United Kingdom) and Florence Mocci (Centre Camille Julian, CNRS, MMSH; France): *Changing uses of a high altitude alpine landscape; an historical ecology of the Bronze Age to Medieval periods in the southern French Alps.*

David González-Álvarez (Universidad Complutense de Madrid; Spain), Margarita Fernández-Mier (Universidad de León; Spain) and Pablo López-Gómez (Universidad de Granada; Spain): *An archaeological approach to the 'brañas'; summer farms in the pastures of the Asturian mountains (northwest Spain).*

Andrew Johnson (Manx National Heritage; United Kingdom): *Shielings on the Isle of Man.*

Dagmar Dreslerová (Institute of Archaeology CAS, Prague; Czech Republic) and Mária Hajnalová (University of Constantine the Philosopher; Slovakia): *Summer farming in the Carpathians and the Bohemian Massif from a (pre)historic and ethnographic perspective.*

Natalia Shishlina (State Historical Museum; Russia): *Seasonality as the key principal in our understanding of Bronze Age nomadism.*

Posters

Mara Migliavacca (University of Padua; Italy): *Upland economies through time in the Italian Pre-Alps. Ethnoarchaeology and Archaeology: from the shepherds to the traces, from the traces to the shepherds.*

Anna-Maria Stagno (Genoa University; Italy): *Seasonal settlements and husbandry resources in Ligurian Apennines (17th–20th centuries).*

Abstracts

These are the abstracts of the papers (not all given) at Oslo and Helsinki, but which for various reasons have not been published in this volume.

Last Multi-Level Transhumance; archaeological insights into alpine settlement and economic systems; a case study from the Valtellina

Yolanda Alther, University of Zurich, Switzerland

Observing traditional economic systems which still exist in south alpine areas, is one possibility to interpret alpine and high-alpine settlement structures and their function. In the research area, the Valli del Bitto, in the Orobic Alps in Lombardy/Italy, one finds a seasonal transhumance which – in contrast to three-level transhumance of the northern Alps – is being carried out over a multi-level economic system. Dry-stone wall constructions on the alpine pasture level, so-called *calécc*, served as temporary settlements as well as for cheese production. While living in the *calécc*, the solid foundation walls were roofed over with organic material. The mobile roof, if demanded by food supply and the animals, could move on, be stripped down and erected once more at the next *calécc*. In this way the location can be changed up to twenty times during the alpine pasture time, which demands from each alpine dairy the same size of dry-stone foundation. While today the rearing of livestock in the alpine area since Neolithic times can be taken as proven, there are still a lot of open questions about economic and settlement systems. This contribution serves to enlarge the spectrum of interpretation of alpine settlements and their function.

Not just 'some old huts in the hills'; shielings in the Scottish Highlands

Eve Boyle, Royal Commission on Ancient and Historical Monuments of Scotland

There is abundant evidence that shielings were an important and integral part of the farming cycle in the Scottish Highlands before the advent of large-scale sheep farming in the late-18th century and early 19th centuries. Archaeologically the evidence lies in the footings of thousands of small huts of turf and stone, which are to be found clustered in small groups, dotted across the pastures in virtually every highland glen. There is also plentiful documentary evidence, particularly in the archives of many great estates, in which the workings of the shieling system from the 17th century onwards can be explored. Other sources include place names, oral tradition (preserved in stories and songs) and contemporary travellers' accounts. However, despite the huge amount of material available for study, it is only recently that shielings have come to be regarded as a worthwhile topic for archaeological research. In the late 19th century, the shielings of the Western Isles (some of which are still in use) were regarded by antiquarians as having value, not in themselves, but in their potential to aid the understanding of prehistoric structures excavated in other parts of Britain. It was not until the 1950s, as the study of traditional farming methods became more common, that shielings attracted any real attention but still the scope of fieldwork remained limited. From the 1980s the Royal Commission on Ancient and Historical Monuments of Scotland routinely mapped and recorded shielings in the course of their surveys, and in the same period several historians began to explore the documentary evidence, but it was only in the late 1990s that a major interdisciplinary project began to look at shielings in detail.

The Ben Lawers Historic Landscape Project (1996–2005) explored the history of a substantial area on the shores of Loch Tay, in the Central Highlands. The shielings were subjected to a programme of field survey, excavation, pollen analysis and documentary research, giving us a much better idea of how they were managed in the early modern period.

The paper offered a brief overview of the evidence for shielings across Scotland, and then moved on to focus on the results of the Ben Lawers project. It looked at the architecture of shieling huts, and their distribution on the mountainside. Evidence for changes in the use of shielings during the 17th and 18th centuries was explored, in particular the pressures on tenant farmers from their landlords' increasingly commercial approach to upland grazing, which led to the abandonment of the system in the early 18th century.

Summer farming; constituent or conjuncture in Norse agriculture?

Kristoffer Dahle

Summer farming or *seterbruk* played a significant role in Norse agriculture if not a constituent part. In 1961 the Norwegian historian Lars Reinton considered it to be part of the 'package' that was introduced through Indo-European migrations in the Neolithic. He argued it could have developed from nomadism but due to the apparent similarities all across Europe, this change must have taken place at an even earlier stage.

Now, 50 years later, archaeological evidence can shed new light upon the practice of summer farming in the North Atlantic, its origin and development. The idea of an introduced or self-grown '*seter*' is questioned. Comparisons also show great variations in time and space. Rather than one enduring and widely dispersed concept, could summer farming be seen as an intentional economic strategy, subordinate to conjunctures within Iron Age and Medieval economics? What factors caused summer farming to be adopted, sustained, modified and cease? And is it possible to return to any great synthesis? This paper was mainly based on the three different projects in Norway and Iceland, but other sites and regions were also taken into consideration.

The Cairo Massif: a changing upland economy and landscape between 1700 and 1970

Michele Forte, University of Sheffield; United Kingdom

For many centuries, the Cairo Massif in central Italy has provided local, and more distant, communities with important economic resources. This paper brought together archaeological, archival and oral historical sources to reconstruct the seasonal exploitation of the Massif's central uplands between the 18th and later 20th centuries, setting this within the context of broader developments in an upland economy based on agriculture, animal husbandry and woodland exploitation. It sets out the strategies which enabled households to exploit the uplands, primarily during the summer months, in the context of essential agro-pastoral activities; whilst short-term agricultural tasks could involve all household members, only males remained for prolonged periods, tending to livestock. The period saw significant changes, including variations in the character and scale of animal husbandry, the expansion of arable cultivation, the introduction of new crops, and the increasing protection and commercialisation of public woodland. This paper explored these changes and assessed their impact upon the seasonal exploitation of the central uplands. In demonstrating the mutable nature of this upland economy, and the extent of its connectivity with the wider world, it also highlighted the benefits of an integrated approach to understanding Mediterranean upland economies and their communities.

Origins and History of Summer Farming in the Pyrenees inferred from Palaeoecological Data

Didier Galop, Damien Rius, Carole Cugny, Florence Mazier and Laurent Carozza, CNRS and University of Toulouse II; France

In the Pyrenees the seasonal use of high altitude pastures is an essential component of the traditional agropastoral system, and short- or long-distance transhumance is strongly integrated into the socio-economic territory of mountain societies. Since the middle of the 1960s the great decrease in grazing pressure caused important transformations in highland zones such as abandonment or overexploitation according to the circumstances and the area, and threatens high mountain traditional cultural landscapes as well as biodiversity in alpine ecosystems. For many years, several interdisciplinary projects involving palaeoecological and archaeological research performed all over the Pyrenees have aimed to reconstruct the environmental history of these highland zones and the impact of summer grazing at high altitude in particular in a long-term perspective. Multi-proxy records (pollen, non-pollen palynomorphs and charcoal) and archaeological surveys show the first signs of small-scale grazing activities at the end of the Early Neolithic period. During Neolithic times the use of alpine pasture remained moderate and occasional while important changes occured during the Bronze and Iron Ages as suggested by deforestation, the increase of pollen grazing indicators and by numerous archaeological remains found at high altitude. The origin of such a change and the increase in grazing pressure can be correlated with the transformation in the agrarian system reflected by pollen and charcoal records from lower altitude. The Medieval period (*c.* 8th and 10th centuries AD) is characterized by a second step of intensification in summer pasture exploitation that involved massive deforestation and forest reduction. The maximum pressure was reached during the 19th century with a well-documented case of overgrazing.

Pre-eleventh century Cornish transhumance: suggesting meaning from havos *and* hendre *names, hut clusters, pens, lanes and landscape*

Peter Herring, English Heritage, United Kingdom

Transhumance (where the household or part of it, accompanied by summering livestock) ceased in Cornwall around AD 1000, but the practice had deep prehistoric roots, probably originating in the second millennium BC. Various sources and approaches, all supporting analytical landscape archaeology, enrich understanding of a practice of great importance in a mixed farming economy. It increased production, exploiting seasonally available resources, while reducing risks (removing livestock from growing hay and ripening crops).

Pre-Norman Cornish place name forms, some shared with the Welsh, confirm seasonal settlement was widespread: *havos* ('summer dwelling'); *hendre* ('old or winter farming estate'); *gwavos* ('winter dwelling'); and *kyniaf-vod* ('autumn dwelling'). Archaeological remains on Bodmin Moor, the largest of several upland areas, include clusters of small (single-person) sub-rectangular huts; associated pens; pounds for distraining trespassing stock; and driving lanes linking lowland and upland. Archaeological evidence also includes late prehistoric single-person huts and larger Bronze Age round houses that were probably summer houses for transhumants.

A simplified historic landscape characterisation (HLC) creates a spatial framework within which the archaeological sites and place-names can be placed, stimulating consideration of transhumance's impact on upland and coastal rough ground and on the lowland farmland whence the livestock came. Modelling transhumance activity includes predicting the overall herds or flocks (equivalent of *c.* 90,000 cows in Cornwall) that would achieve what the palaeo-ecological record shows: suppression of upland shrub and tree growth by grazing pressure. That scale fits with early medieval lowland settlement densities and tree growth and agricultural practice, and with a recent interpretation of later Bronze Age extension of 'commons' and their continued oversight in later prehistory as a response to prehistoric grazing pressure.

Analysis of the medieval hut groups places the individual household (with its owned and valuable animals and their products) in relation to the communal hamlet of which it was a part – shared pen and shared risk, company and pleasure. Consideration of why many or all households in a hamlet sent members to the hills suggests the animals were routinely processed, presumably milked for butters, cheeses, etc. to maximise outputs from the uplands. Social organisational implications of grazing pressure on resources are revealed by superimposing on to the HLC the nine large Cornish hundreds that probably policed rights and limits of transhumance.

More recent British ethnographic evidence, and selection on the roles of those remaining at the main farm, suggest that the household member who tended, milked and churned was probably (though not certainly) a young woman. We may reasonably imagine how the aspects of transhumance were experienced – the two great journeys up and down (perhaps around May Day and October's end) the partings and reunions, pleasures experienced, trust shared and responsibilities shouldered, etc.

Finally an understanding of the development of open field and convertible husbandry in the lowland farmlands, involving new ways of managing and milking livestock, probably late in the first millennium AD, provides a realistic context for the early ending of Cornish transhumance and a useful model for explaining why other areas of Britain and Europe did not undertake transhumance. See Herring 2009, 2012.

Shielings on the Isle of Man

Andrew Johnson, Manx National Heritage; United Kingdom

A quarter of the land on the Isle of Man – *c.* 150 km² – was historically common land. Most of it lay above the 180m contour, and from the medieval period to the early 19th century, it was communally exploited for seasonal grazing and also for peat-cutting and quarrying. Attempts were also made to enclose some of the better ground as permanent farmland. Within a relatively small island, effective management of the medieval rural landscape was important; the lowlands were intensively farmed wherever possible, and equitable land-division and stock-proof boundaries were an essential characteristic of a landscape which remained largely unchanged into the 20th century. Upland field survey is beginning to suggest that the same is true of the common land as well. Excavation in the late 1950s focussed on sample investigation at two out of 47 previously-identified groups of shieling huts, and suggested that stock control, dairying/cheese-making and corn-drying were significant activities. More recent fieldwork and survey have not only identified more sites, but imply a more nuanced range of sites, structures and associated activities, and the means of managing them. A range of sites was described and illustrated, together with other landscape, place name and documentary evidence, and folk life reminiscence.

Was the Neolithic Iceman 'Ötzi' really a Herdsman?

Klaus Oeggl, University of Innsbruck, Austria

In the course of the investigations on the Neolithic glacier mummy 'Ötzi' numerous questions were raised concerning his social status. This gave rise to several suggestions amongst which nowadays the assumption, that he was involved in an early form of alpine pastoral farming, has gained general acceptance. The existence of such a Neolithic pastoral economy in this inner Alpine region is based only on palynological studies conducted in the vicinity of the discovery site and is not confirmed as yet by archaeological finds. However, these palynological records require archaeological evidence because on closer examination this pollen analytical evidence for early alpine transhumance in the vicinity of Ötzi´s discovery site is inconclusive. The crucial thing is that the impact of highland pasture is recognized by indicator pollen-types which are of limited value due to the pollination mechanisms of the mother-plants, their regional or extra-regional provenance and consequently their presence in the pollen record. Finally, it has to be considered that transhumance establishes no 'new' habitats in the highland regions, but in contrast tillage does so in valley bottoms. Primarily, grazing in alpine regions causes an increase of plant species which are inedible for animals, or which are enhanced due to fertilization by animal dung. Thus pasturing creates an expansion of nutrient-rich plant communities in alpine mats. On the other hand the same effect causes an increase in precipitation, which induces a higher surface run-off, which leads also to a fertilisation of the alpine grass mats by the input of mineral matter. This may be reflected in increasing values of pollen-taxa including pasture indicators. Therefore raised values of these pollen-types do not imply exclusively grazing pressure. However, it is remarkable that the start of early transhumance in the uppermost Ötz valley coincides with climate changes (Haas *et al.* 1998; Magny and Haas 2004) and in this context the palynological results have to be scrutinized. Additional doubts on the existence of such an animal husbandry practice in the area during Ötzi's lifetime are posed by recent coprolite studies conducted on about a hundred caprine (sheep/goat or ibex/chamois) dung pellets recovered from the Iceman's find spot dated from 5400 to 2000 BC. Their results reveal that none of these dung pellets derive from domestic animals (sheep/goat) but all from game (ibex/chamois). Furthermore, palaeo-ethnobotanical and archaeozoological analyses of Neolithic dwelling sites are lacking for the Vinschgau and are also rare for the rest of the area. The recent discovery of Copper Age settlement layers near Latsch in the Vinschgau with well-preserved animal bones enables a comparative approach by Sr-isotope and palynological analyses combined with archaeological surveys along the traditional local transhumance route to evaluate the existence of alpine pasture in the area where the Tyrolean Iceman lived during the late Neolithic period. For a discussion of the pollen evidence see Festi *et al.* 2014.

Prehistoric Alpine Animal Husbandry; recent discoveries in the Silvretta range (Switzerland/Austria)

Thomas Reitmaier, University of Zurich, Switzerland

Few regions in Europe are so strongly associated with alpine animal husbandry and agriculture as the mountain regions of Switzerland and Austria. The seasonal use of high alpine pastures by sheep, goat and cattle herds and the immediate and local utilisation of animal products seems perfectly adapted to the alpine landscape, so much so that this tradition continues into the 3rd millennium AD. Perhaps this is part of the reason why origins and development of *Alpwirtschaft* in the central Alps are still so badly understood. However, this might also be due to the methodological difficulties of alpine archaeology.

An interdisciplinary research project was initiated in 2007 by the University of Zürich to study origins and development of alpine animal husbandry in the Silvretta range on the Swiss–Austrian border. Starting points of the surveys were a number of settlements on the valley floor dating to the Bronze Age and Iron Age. During four campaigns a large number of high alpine (over 2000 metres asl) sites dating between the earliest deglaciation and the modern age were discovered. These included Mesolithic, Neolithic and Bronze Age *abri* sites as well as unique structures from the 1st millennium BC, such as animal pens and huts. These are chronologically similar to and functionally complement the valley sites.

The results of the project show that the extensive alpine pastures were being used from at least the 2nd millennium BC for summer grazing. These archaeological results are supported by e.g. archaeobotany/palynology, archaeozoology, toponymy and dendrochronology. The presentation provided insight into current research results and discussed still open research questions and future research potential.

Bibliography

Barker, G. 1985. *Prehistoric Farming in Europe.* Cambridge: Cambridge University Press.

Bordessoule, E. 2002. La Grande "Montagne" des monts d'Auvergne. In D. Martin (ed.) *L'identité de l'Auvergne (Auvergne – Bourbonnais – Velay): mythe ou réalité historique?* Nonette: Créer. Pp. 131–141.

Collis, J.R. 2008. Unfinished business: fieldwork in the Massif Central, France. In P. Rainbird (ed.) *Monuments in the Landscape: Festschrift for Andrew Fleming.* Stroud: Tempus Publications, pp. 99–113.

Daugstad, K. 2006. *Kvinners Rolle i Seterbruket før og nå.* Trondheim: Norsk senter for bygdeforskning, Universitetssenteret Dragvoll, notat nr 6/06.

Davies, E. 1941. The patterns of transhumance in Europe. *Geography: Journal of the Geographical Association*, Sheffield. Pp. 155–168.

Festi, D., Putzer, A. and Oeggl, K. 2014. Mid and late Holocene land-use changes in the Ötztal Alps, territory of the Neolithic Iceman "Ötzi". *Quaternary International* 353:17–33.

Fleming, A.F. 1988. *The Dartmoor Reaves: exploring prehistoric land divisions.* London: Batsford.

Fleming, A.F. and Collis, J.R. 1973. A prehistoric reave system at Cholwich Town, Dartmoor. *Devon Archaeological Society* 31:1–21.

Fournier, G. and Fournier, P.-F. 1983. La vie pastorale dans les montagnes du Centre de la France: recherches historiques et archéologiques. *Bulletin historique et scientifique de l'Auvergne* 91:199–358.

Haas, J.N., Richoz, I., Tinner, W. and Wick, L. 1998. Synchronous Holocene climatic oscillations recorded on the Swiss Plateau and at timberline in the Alps. *The Holocene* 8: 301–309.

Herring, P. 2009. Early medieval transhumance in Cornwall, Great Britain. In J. Klapse (ed.) *Medieval Rural Settlement in Marginal Landscapes. Ruralia* 7:47–56. Turnhout, Belgium: Brepols.

Herring, P. 2012. Shadows of ghosts: early medieval transhumants in Cornwall. In S. Turner and R. Silvester *Life in Medieval Landscapes; people and places in in the Middle Ages; papers in memory of H.A.S. Fox.* Windgather Press: Oxford. Pp. 89–105.

Hoskins, W.G. 1955. *The Making of the English Landscape.* London, Hodder and Stoughton.

Kristiansen, K. 2002. The birth of ecological archaeology in Denmark: history and research environments 1850–2000. In A. Fischer and K. Kristiansen (eds.) 2002. *The Neolithisation of Denmark: 150 years of debate.* Sheffield, J.R Collis Publications, pp. 10–31.

Magny, M. and Haas, J.N. 2004. A major widespread climatic change around 5300 cal. yr BP at the time of the Alpine Iceman. *Journal of Quaternary Science* 19:423–430.

O'Neill, S.H. 1983. *The Bronze Age Settlement of the Plym Valley, Dartmoor.* University of Sheffield, Dept. of Prehistory and Archaeology, unpublished M.Phil. thesis.

Roc, J.-C. 1989. *Le Buron de la Croix Blanche.* Brioude: Éditions Watel.

Roc, J.-C. 1992. *Burons de Haute Auvergne.* Brioude: Éditions Watel.

Sherratt, A.G. 1981. Plough and pastoralism: aspects of the secondary products revolution. In I. Hodder, G. Isaac and N. Hammond (eds.) *Pattern of the Past: studies in honour of David Clarke.* Cambridge University Press: Cambridge. Pp. 261–305.

Sherratt, A.G. 1983. The Secondary Products Revolution of animals in the Old World. *World Archaeology* 15:90–104.

Steenstrup, J.J.S. 1859. Om Hr. Professor's tvedeling af Steenalderen. *Oversigten over Videnskabernes Selskabernes Forhandlinger 1859.* Copenhagen, Bianco Luno. English translation: On Professor Worsaae's division of the Stone Age. In A. Fischer and K. Kristiansen (eds.) 2002. *The Neolithisation of Denmark: 150 years of debate.* Sheffield, J.R Collis Publications, pp. 58–68.

Surmely, F., Nicolas, V., Tzortsis, S., Miras, Y., Savignat, A., Guenet, P., Servera, A. and Petit, S. 2010. Recherches sur l'histoire de l'occupation humaine sur la planèze du Plomb du Cantal. In S. Tzortsis and X. Delestre (eds.) *Archéologie de la montagne européenne; actes de la table ronde internationale de Gap, 29 septembre – 1er octobre 2008.* Bibliothèque d'Archéologie Méditerranéenne et Africaine 4. Aix-en-Provence, Editions Errances / Centre Camille Jullian. Pp. 235–251.

John Collis, 9 Clifford Road, Sheffield S11 9AQ, UK
Email: j.r.collis@sheffield.ac.uk

2. Pastoral Exploitation of the Caspian and Don Steppes and the North Caucasus during the Bronze Age: Seasonality and isotopes

N.I. Shishlina and Y.O. Larionova

The system of seasonal use of grassland has been one of the key principles of exploiting different ecological zones since the origin and subsequent evolution of the Eurasian steppes pastoral lifestyle during the Bronze Age. Seasonality data can therefore be used in identifying population mobility. Different values of ^{13}C and ^{15}N point to different diet systems, which have been reconstructed for various population groups. Therefore, the comparison between groups or even individuals can be used to identify the presence of immigrants who might have consumed a diet different from that of the population in a given area. $^{87}Sr/^{86}Sr$ ratio in human teeth enamel is another tool for the identification of human mobility and migration. With combination of all data available we are able to produce five models of population movements across the exploited Eurasian steppe areas during the Bronze Age and also gain insights into individual lives, so that personal stories begin to emerge.

Introduction

The Bronze Age Eurasian Steppes was the area of origin and development of a new lifestyle. Population groups began to occupy and settle on previously vacant lands. The exploitation of the new area was based on a multi-component mobile economy, which included the raising of domesticated animals, i.e. sheep/goats, cattle and, probably, horses; the fishing and gathering of wild plants as additional activities; and an exchange system. This lifestyle helped develop regular seasonal migrations, and was built on alternative use of different grasslands and water resources.

The aim of this paper is to show the development of population mobility based on the seasonal use of pastures and discuss the following questions:

i) What can the archaeology of the Steppe and the Caucasus during the Bronze Age tell us about the mobility of people?
ii) What are archaeological markers of population movement?

Objects and Methods

Data from several archaeological cultures, i.e. the Yamnaya, Steppe North Caucasus, Early Catacomb, Eastern Manych, have been used. A large-scale exploitation of different ecological niches of the Caspian and Don Steppes started at the beginning of the Yamnaya period dating back to 3000 calBC. Yamnaya population settled down not only on the already developed lands but also moved to new areas. The subsequent period saw the arrival of a new population group from the North Caucasus. This marked the launch of a new stage in the Caspian and Don Steppes exploitation. During a relatively short period several cultural groups, i.e. the Late Yamnaya, Early Catacomb, and Steppe North Caucasus groups must have coexisted. At the end of this period, i.e. around 2500–2450 calBC, the East Manych Catacomb population appeared (Table 2.1). Therefore populations of these cultures were mobile pastoralists who exploited different ecological areas of the Caspian Steppes and the nearby areas of the River Volga and the North Caucasus Steppes. The location of the sites is shown in Fig. 2.1.

The following approaches have been used in this study.

First, we analysed the environment. The study of the soil buried under Bronze Age kurgans, i.e. burial mounds of the local population, provided new data on the palaeo-environment in the areas under discussion. We studied geochemical characteristics of the soil as well as the preserved pollen and phytoliths. Therefore, climatic variations can be analysed within the range of 3000–2200 calBC. (Demkin *et al.* 2002; Shishlina 2008). To identify the productivity of pastures, ethnobotanical study of buried soils was conducted, using phytolith and pollen analyses as well.

Culture	Time Interval
Yamnaya	3000–2500 cal BC
Early Catacomb	2700–2400 cal BC
North Caucasus	2600–2400 cal BC
East Manych Catacomb	2500–2200 cal BC

Table 2.1 Analysed cultures and their respective chronologies

keywords: Eurasian Steppes Bronze Age, isotopes, diet, mobility, models of subsistence system.

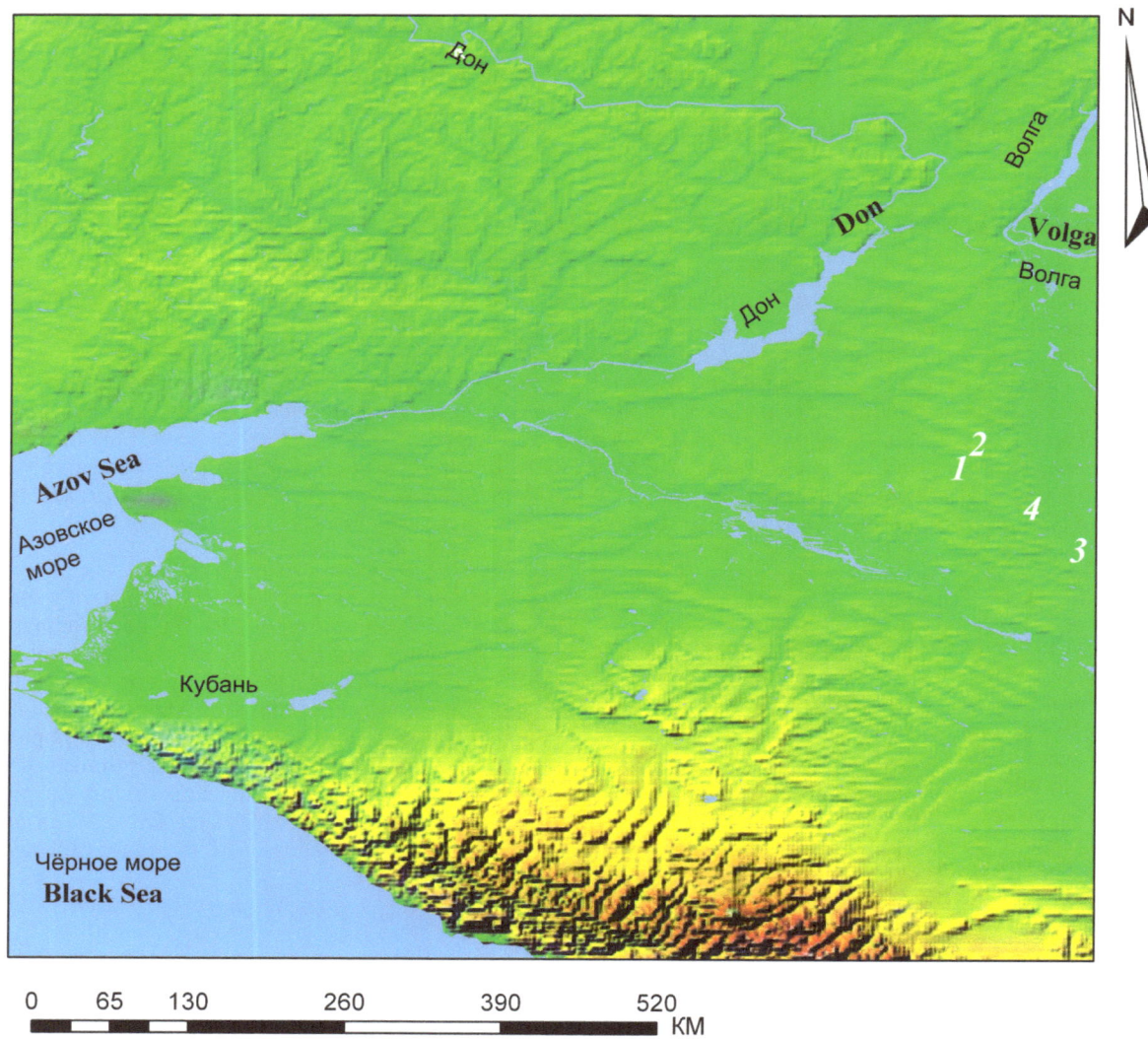

Figure 2.1   Location of sites: 1: Ulan IV; 2: Sukhaya Termista and Temrta III; 3: Mandjikiny 1; 4: Khar-Zukha.

The second thrust of the project was to study seasonality of the grave construction. We assume that seasonality of burials indicates that steppe grasslands located nearby were used. Each grave is a closed archaeological context which, like the surface of the earliest buried soil covered by the mound, contains important information on the season when the grave was constructed. The methodology used to determine the season of graves and kurgans is based on the following factors: 1) analyses of dentine and enamel of animal teeth found in burials and kurgans; 2) analyses of dentine and enamel of human teeth found in burials; 3) pollen analyses of a specific archaeological context, i.e. pillows under the skull, residue in vessels and the human stomach area, analyses of subsoil; 4) analyses of eggshell; 5) osteological analyses of animal bones from graves and kurgans (Kirillova *et al.* 2000; Klevezal and Shishlina 2001; Shishlina and Pakhomov 2002; Klevezal *et al.* 2006). Data obtained through several methods were cross-checked.

The third approach was based on the isotope study of human bone samples. The combination of carbon and nitrogen stable isotope ratios provides a direct measure of the diet of an individual. Isotope values help find out what main food products humans consumed during the last 10–20 years of their life. Different values of ^{13}C and ^{15}N point to different diet systems which were reconstructed for different population groups (Lanting and van der Plicht 1996; Thompson *et al.* 2005; Keenleyside *et al.* 2009).

Sr isotope variations in human enamel samples helped identify the possible place of birth of the individuals analysed (Price *et al.* 2012).

The correlation of data obtained made it possible to propose new models of population movements and link various exploited areas located in different zones of the Steppe and the Caucasus.

Time interval	Climatic characteristics	Culture
3000–2600 cal BC	A favourable climatic period which was humid and warm, an average annual precipitation rate of around 400mm, predominance of mixed grass steppes, at the end of the first half of the third millennium BC a process of gradual aridisation started.	Yamnaya culture
2600–2300 cal BC	An abrupt aridisation of the climate: summer temperature increased and winter temperature became lower, and the amount of precipitation decreased; dry steppes were replaced by semi-desert landscapes, characterised by wormwood and fescue, which correspond to a very dry climate; the forest area was reduced; the annual precipitation was 40–60mm lower than today, and 140–160mm lower than it was during the previous period, i.e. 300mm.	Late Yamnaya, Steppe North Caucasus, Early Catacomb and Early East Manych Catacomb cultures
2300–2000 cal BC	Continuation of the aridisation, predominance of semi-desert landscapes, further reduction of forest areas	East Manych Catacomb and Lola cultures

Table 2.2   Climatic changes in the Caspian Steppes during time interval 3000–2000 calBC.

Results and Discussion

Palaeoenvironment

The study of palaeoclimate based on analyses of soils buried under Bronze Age kurgans (Kremenetsky 1997:32–33; Ivanov and Demkin 1999; Demkin *et al.* 2002: 118–122; Shishlina 2008) resulted in identification of distinct climatic changes during time intervals 3000–2000 calBC. Periods of relatively humid and dry climate could last for several centuries and even slight variances in the ground-water level may have caused changes in the main landscape components of vegetation, soil and fauna. The climatic changes are presented in Table 2.2.

Steppe ecological niches offer a variety of landscapes. The productivity of different steppe niches with their specific landscapes and climatic features changed due to season and drought. Pastures with steppe vegetation are more productive in spring and have less vegetation in early winter. Saltwort desert pastures are more productive in autumn, but they cannot be used in spring and summer. Grasslands with ephemeral vegetation are most productive during the rainy summer (Fig. 2.2) and autumn seasons, while wormwood grasslands are more productive in autumn after the first frost. Saline soil pastures with saltwort vegetation are used in autumn and winter. Sand grasslands are productive throughout the year, but they are in use only when other pastures are inaccessible. The depth of snow cover provides evidence as to whether it is possible to use various areas as winter pastures. Sheep and cattle cannot get fodder from under deep snow. But horses have an ability to break the snow cover for smaller animals (Fig. 2.3). On the contrary, if the snow cover is not deep or if there is no snow at all, and the temperature drops down significantly below zero, the soil gets thoroughly frozen and the fodder becomes inaccessible for the cattle as well. Different vegetation conditions in pasturing were taken into account.

Knowledge of the location of water resources is another key landscape characteristic. All water resources were used: rivers and lakes throughout the year; juicy (rich) grasslands in spring and snow in winter. Animal stock raising using pastures during the whole year is the main characteristic of a nomadic pastoral economy (Masanov 1995).

Ethnobotanical studies indicate that there are pastures of different types of vegetation on the Caspian and Don Steppes: pastures located on sands, alkaline soils, and watershed plateaux, in river and lake valleys, flood plains, along the coastline.

Figure 2.2. Kalmykia Steppes summer pastures.

Figure 2.3   Horses breaking the snow cover. Tyva region, 2013.

Basic seasonality concepts of economic exploitation of the Caspian and Don Steppes during the Bronze Age

Several seasonality models of economic population movements have been proposed (Shishlina 2008):

Model 1 supports the Yamnaya culture economic cycle. The spatial location of Yamnaya sites points to the inconsistency between the potential (optimal) exploitation of grasslands and a predominant number of Yamnaya kurgans, which are located in flood plains of rivers, along lake coasts, and on the plateaux of watersheds closest to rivers. No large permanent settlements on the Caspian and Don Steppes that would date back to the Yamnaya culture have been found. The season when Yamnaya burials and kurgans were made in several landscape zones demonstrates how these areas were exploited: herders of small family groups migrated across flood plains of the river in spring and summer. The migration pattern can be characterised as a river flood plain → a watershed. People moved within a small area not more than 10–20km away from the river where they lived in winter towards the nearest watersheds where they moved around in summer and autumn. They selected sites where reed and cane stood high above the snow cover. As they lived near a water source, they did not have to arrange artificial water places. This system resulted in overgrazing of grasslands within the limited region of exploitation. A lack of permanent settlements indicates that such moves within the region were quite regular.

Model 2 characterises the economy of the subsequent period when newcomers from the North Caucasus areas settled down in the vast steppe areas. Open southern boundaries and developed river routes (the east–west (latitude) direction: Kalaus – East Manych – West Manych – Lower Don Rivers; the south–north (meridian) direction: Terek – Kuma – lakes of the Western Caspian Plain – the Sarpa Lakes) are linked with the spreading of this population. Newcomers began to organise year-round movements across the entire Caspian and Don Steppe area.

Data suggest that some independent Catacomb groups exploited river valley areas only in spring and summer (the first part of summer; early spring–summer) but spent other seasons somewhere else. Other groups moved to watershed plateaux during summer and to unknown winter pastures when the colder season set in.

Seasonal data also point to two variants of summer migrations: 1) migrations of the entire family; 2) migrations of only adult males (and, possibly, adolescents).

It has been found out that two camps of the Early Iron Age, i.e. Gashun-Sala and Manych, were visited in summer. The occupation layers of these camps contain Catacomb ceramics that are few in number. It implies that some watershed plateaux were used predominantly in summer (Shishlina *et al.* 2008).

Therefore the picture of seasonal moves based on the data from graves indicated a wide range of possible population movements. But the proposed models have one significant flaw, as we know the exact place where people were buried and when (having in mind that they were accompanied by their family groups), but we do not know where they spent the major part of their lives. There are still more questions that need an answer:

i) Is it possible to identify long and short-distance seasonal movements at the level of individual herders and families?

ii) Is it possible to identify more precisely the areas visited by an individual during his/her life?

iii) Is it possible to identify the place where the individual might have been born?

The database needs to be expanded.

Isotope study

Identification of the diet system based on the ^{13}C and ^{15}N values of human bone collagen

Data obtained for 110 humans of the cultures under discussion with a mixed and multicomponent diet sys-

Model	δ^{13}C, ‰	δ^{15}N, ‰
Model 1 (Steppe Model): Flesh/milk of herbivore animals, river and lake aquatic products, wild C_3 plants: *Majkop, Yamnaya, Catacomb, North Caucasus*	from -21 to -17	from +11 to +15
Model 2 (Marine Model): marine products are predominant: *Yamnaya, Early Catacomb, East Manych Catacomb cultures*	from -17 to -15	from +15 to +18

Table 2.3 Isotope ratios in bone collagen for two diet groups of steppe population

tem show large variations (most data are published in Shishlina *et al.* 2012) and form the basis for discussing two diet system models (Table 2.3).

Steppe Model 1 is consistent with most data, and is characterised by average values of δ^{13}C from -21 to -17 ‰ and δ^{15}N from +11 to +15 ‰. It is based on consumption of large quantities of aquatic resources (fish, molluscs, water plants) and wild C_3 plants. We assume that this model corresponds to the steppe environment. Stress caused by the struggle to survive and the aridity of climate made values of the isotope ratio of nitrogen in human collagen higher. Population groups of the Yamnaya and Catacomb cultures made use of all food resources in the exploited areas of the Volga, Don, the Caspian Steppes and North Caucasus Steppes. This model is supported by archaeological finds of freshwater (from both lakes and rivers) fish bones and scales, edible molluscs and C_3 plants uncovered in the Bronze Age graves obtained through phytolith and pollen analyses (Shishlina *et al.* 2009).

Marine Model 2 is characterised by average values of δ^{13}C from -17 to -15 ‰ and δ^{15}N from +15 to +18 ‰. Such values might indicate the consumption of seafood by some individuals of the Yamnaya, the Early Catacomb and the East Manych Catacomb cultures.

An old man from a Chilguir Caspian Plain grave and three females of 14–17 and 45–50 years old from the Middle and South Yergueni Hills (Mandjikiny-2, Khar-Zukha and Mu-Sharet-4) refer to this group. The diet system of these Yamnaya individuals was based on marine food. Seasonality data obtained for these graves indicate that these people were seasonal pastoralists and must have moved from one pasture to another. Maybe, two young women of 14–17 years old were born somewhere near the coastlines of the Caspian or Black Seas and later got married and moved to the steppe. As newcomers, they lived in a new steppe environment for a short period and died. At least the isotope data confirm this fact, i.e. the isotope values of two females differ from those of other local people, whose diet system was typical for Steppe Model 1.

Several individuals of the Early Catacomb culture, for example, *senilis* female from Khar-Zukha have a very high value of ^{13}C, i.e. -15‰, and a very high value of ^{15}N, i.e.+18‰. The aforesaid woman consumed marine food. Maybe, she moved to the open steppe from a coastline area. She died after having lived among people whose diet system was different.

Maybe, East Manych Catacomb individuals buried at Temrta III, kurgan 1, grave 6 (male 20–30 years old), Mandjikiny-1, kurgan 10, grave 2 (female of 45–50) and Mandjikiny-1, kurgan 14, grave 1 (male of 30–35), were members of family groups who migrated during winter seasons to the coastal areas of the Azov and the eastern Black Sea. These three individuals have the highest values of nitrogen and carbon. These values indicate the marine food consumption.

Therefore, a group of people who consumed quite a lot of marine food was identified among the steppe population. Anyway, having these data, we are not able to say which areas those people came from or visited.

Identification of the possible place of origin based on Sr isotope values in human enamel

The strontium isotope ratio in human tooth enamel is another tool for the identification of human mobility and migration (Buzon *et al.* 2007). Individuals growing up in different regions can be differentiated by their ^{87}Sr/^{86}Sr because the strontium isotope ratios vary depending on the environment (Bentley *et al.* 2004; Bentley and Knipper 2005; Keenleyside *et al.* 2009).

^{87}Sr/^{86}Sr ratio in human enamel reflects type of diet at the time of their formation (Bentley *et al.* 2004). The local geochemical environment of different regions is characterised by specific ^{87}Sr/^{86}Sr ratios obtained for rock, groundwater, soil, plants and animals and through components of diet, is incorporated into human tooth enamel during the time of formation. It provides data for identifying an individual's possible place of origin (Buzon *et al.* 2007; Eckardt *et al.* 2009).

Therefore, strontium reflects local geology, and Sr isotopes are fixed in enamel biogenic phosphate at the time of tooth formation. We took 25 teeth from 17 human skeletons buried in four kurgans located on the Don and Caspian Steppes areas in order to differentiate local from non-local individuals (Figs. 2.4 and 2.5). The samples represent a broad cross-section of the variables of age, sex, burial style and grave offerings as well as diet and seasonality (Table 2.4).

The Sr isotope data were obtained at the Laboratory of Isotope Geochemistry and Geochronology at the IGEM RAS, Moscow. The Sr isotopic composi-

No.	Samples/analyses done	Tooth sample	Sr, ppm	$^{87}Sr/^{86}Sr$	^{13}C	^{15}N
Ulan IV, kurgan 4						
1	grave 17, sheep tooth, Catacomb Culture	M1, enamel	649	0.70920	-18.33	+7.61
2	grave 8, female 40–45, Yamnaya Culture, model 1, summer	M1, enamel	150	0.70901	-18.37	+12.67
3	grave 8, male 35, Yamnaya Culture, model 1, summer	M1, enamel	532	0.70936	-18.03	+13.0
4	grave 10, male 40–45, Catacomb Culture, model 1	M1, enamel	275	0.70903	-18.13	+13.17
6	grave 14, female 40–50, Catacomb Culture, model 1	M1, enamel	284	0.70888	-18.49	+13.80
Ulan IV, kurgan 3						
7	grave 15, female 30–40, Yamnaya, model 2	M1, enamel	180	0.70917	-17.58	+15.88
8	grave 15, female 30–40, Catacomb Culture, model 3	M2, enamel	144	0.70910	-17.58	+15.88
9	grave 5, male 40–59, model 2	M1, enamel	263	0.70921	-16.89	+16.89
10	grave 14, female 25–30, model 2, spring	M1, enamel	375	0.70915	-17.31	+14.93
Sukhaya Termista						
11	grave 11(wife) female 20–25, model 2	M2, enamel	156	0.70918	-15.70	+16.40
12	grave 13 (husband), male 55, model 2 autumn? winter?	M1, enamel	165	0.70908	-16.80	+15.20
13	grave 10, female 45–35, model 3, summer	M1, enamel	193	0.70904	-14.37	+15.37
14	grave 14 ,Yamnaya? Culture, female 18–25 M2, spring	M1, enamel	271	0.70914	–	–
15	grave 9, adult, early Iron Age	M1, enamel	150	0.70901	–	–
Mandjikiny 1, kurgan 14						
16	grave 13, male 50, Majkop Culture, model 1, summer	M1, dentine	967	0.70908	-18.80	+11.64
17	grave 10, male 20–25, Yamnaya Culture, model 1, winter	M2, enamel	317	0.70900	-18.80	+11.64
18	grave 12, male 17–25, Yamnaya Culture, model 1, late summer	M1, enamel	297	0.70906	-18.60	+14.10
19	grave 12, male 17–25, Yamnaya Culture, model 1, late summer	M1, dentine	–	0.70909	-18.02	+14.31
20	grave 12, male 17–25, Yamnaya Culture, model 1, late summer	M3, enamel	302	0.70908	-18.02	+14.31
21	grave 12, male 17–25, Yamnaya Culture, model 1, late summer	M3, dentine	850	0.70908	-18.02	+14.31
22	grave 1, male 40, Catacomb Culture, model 2, late summer–early autumn	M1 enamel	232	0.70924	-16.50	+17.60
23	grave1, male 40, Catacomb Culture, model 2, late summer–early autumn	M1, dentine	834	0.70915	-16.50	+17.60
24	grave 6, male 55–60, Early Catacomb culture, model 1	M2, enamel	230	0.70899	-17.50	+15.40
25	grave 6, male 55–60, Early Catacomb Culture, model 1	M2, dentine	912	0.70915	-17.50	+15.40

Table 2.4 $^{87}Sr/^{86}Sr$ isotope values in human teeth.

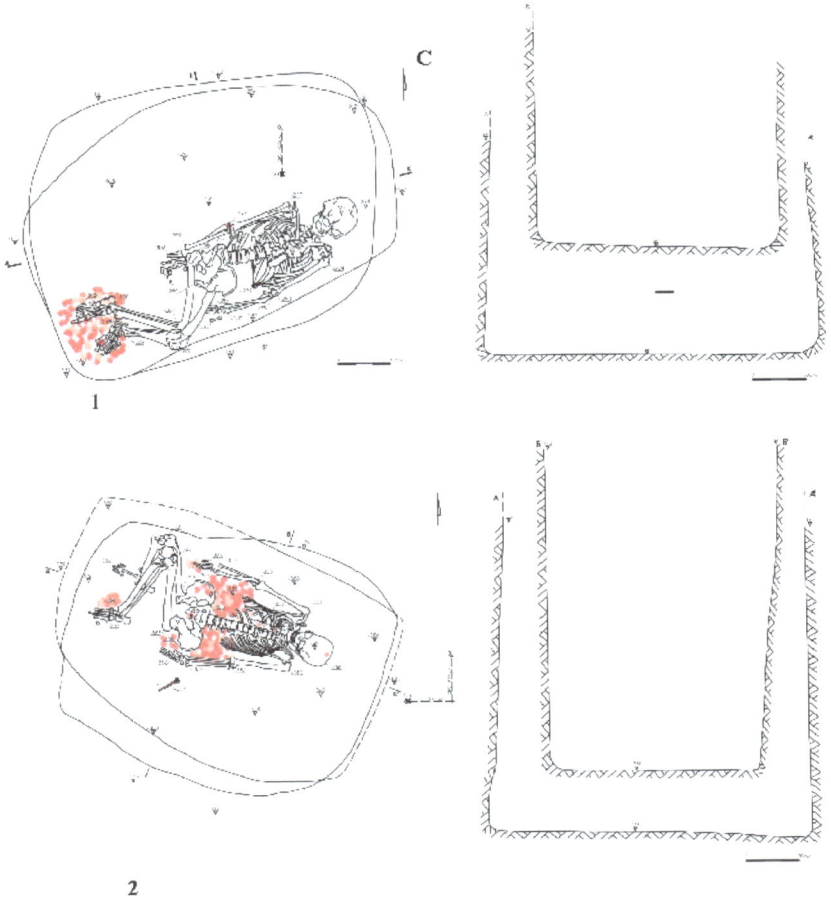

Figure 2.4 Sukhaya Termista burial ground, kurgan 1; 1: grave 11; 2: grave 13.

tions were measured on the Faraday collector VG Sector-54 thermal ionization mass-spectrometer in multi-dynamic mode. Corrections for isotope mass-fractionation were applied by normalising to $^{87}Sr/^{86}Sr=0.1194$ using an exponential law. The accuracy and precision of the obtained Sr isotope compositions were monitored by replicate analyses of SRM-987 standard. The average $^{87}Sr/^{86}Sr$ value for this standard obtained in the course of this study was 0.710250 ± 8 (2stdev, N=60).

As is clear from Table 2.4, isotope data obtained make it possible to assess the mobility level at the individual level. The people under analysis were buried on the western slope of the Middle Yergueni Hills or on the Southern Yergueni Hills. With regard to their diet system, they are referred to two models, i.e. the Steppe model and the Marine model. So it turns out, that at least nine out of 17 individuals must have spent a rather long period of time in the areas close to the marine coast, and local seafood evidently predominated in their diet so that the isotope signal in their bone collagen was made different from the isotope signal in the bone collagen of people who, as we presume, lived predominantly in steppe ecological niches.

Variations of the Sr isotope ratio in the tooth enamel of these people were compared with local Sr variations obtained by analysing soils, water and contemporary snails (Shishlina and Larionova 2013).

In reconstructing migration routes of ancient people with the help of the Sr isotope composition recorded in their tooth enamel, we must be sure that after burial the Sr isotope system of the teeth we have been studying remained closed, while the measured isotope composition is indeed consistent with the one that appeared *in vivo*. To check the possibility of the isotope composition transformation due to the process of Sr absorption on the phosphate matrix of the enamel post-mortem, in parallel to the study of the enamel isotope composition, we studied water and acid (HCl) soil leaches that the body and the teeth had contacts with after burial. In our view, when the enamel isotope composition and the soil water leach composition coincide within an error, the enamel isotope composition has to be excluded from the sample due to a high probability of the Sr isotope in this composition not being primary. As a rule, in case the isotope compositions of the enamel and soil-water leaches coincide, we observed a

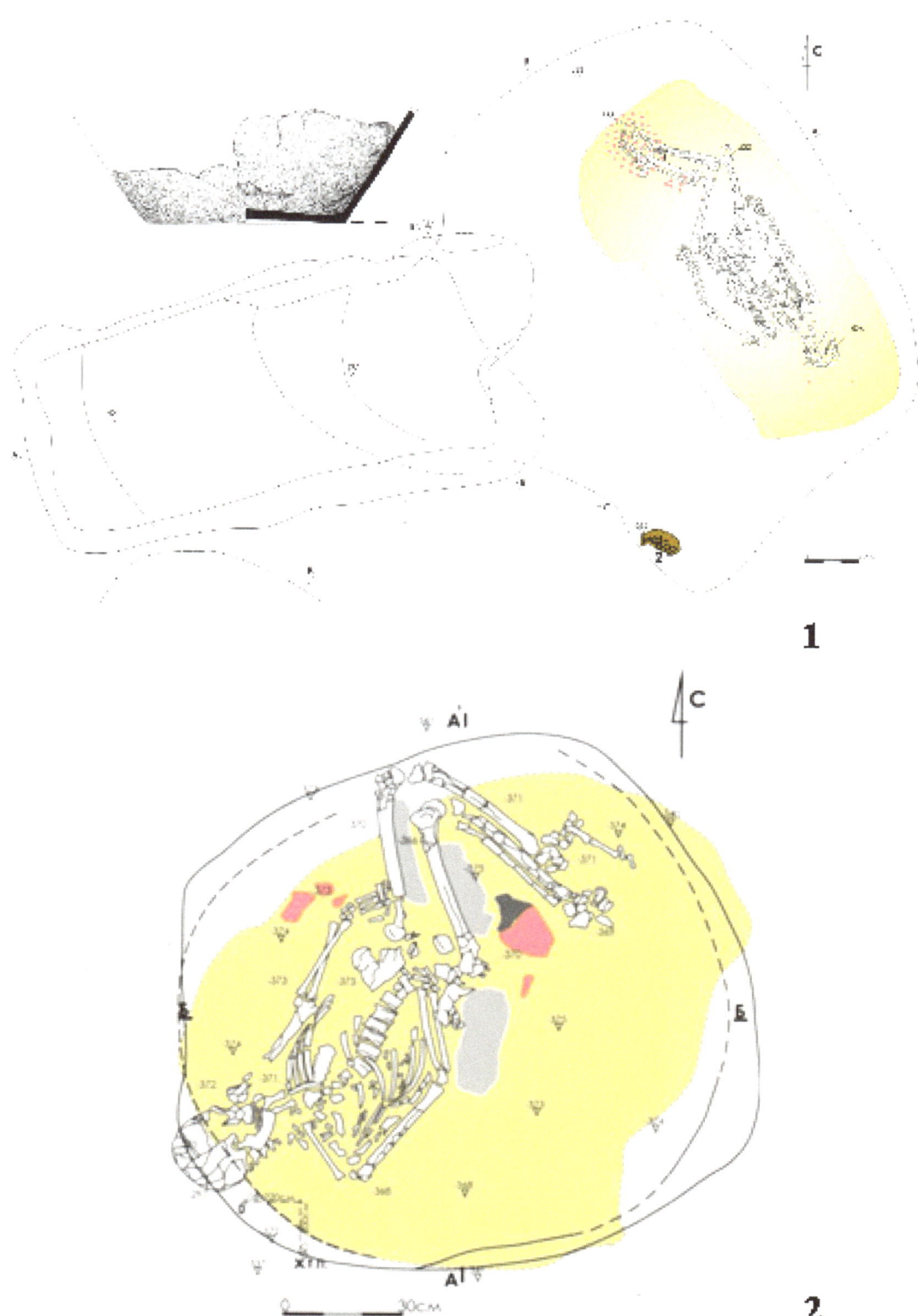

Figure 2.5   Ulan IV burial ground, kurgan 1: 1: grave 14; 2: grave 15.

higher content of Sr in the enamel (ppm>350). Apparently, a warm climate combined with calcareous soils and rainwater created favourable conditions for the absorption of 'external' Sr by the tooth enamel and led to overbalancing of the isotope composition in the enamel. Too high concentrations of Sr in the dentine (800–1000 ppm), which are 2–3 times higher than the concentration of this element in the enamel, also point to post-mortem contamination, as in such cases the Sr concentration in dentine *in vivo* is always lower than in the enamel. Apparently, a tubular and more porous structure of the dentine contributed to a more efficient concentration of the 'external' strontium. Hence, $^{87}Sr/^{86}Sr$ isotope values in human dentine should also be treated as post-mortem contamination and taken as a local signal (samples 16, 19, 21, 23 and 25). This fact should be taken into consideration before proposing reconstructions.

Strontium isotope ratios for humans analysed varied from 0.7088 to 0.7093 (Table 2.4). We compare those ranges with local signals obtained from contemporary snails (Shishlina and Larionova 2013).

Due to the $^{87}Sr/^{86}Sr$ isotope values, we may consider that a female from Ulan IV, kurgan 4, grave 14 (sample 6), was born in the local area, near the Dzhurak-Sal River, and she spent all her life on the steppe areas consuming local food. She died and was buried on the steppe together with her three children, one of which was 2 years old and according to the isotope values in his bone collagen was being breastfed. The same life story may be reconstructed for old and young males from Mandjikiny, kurgan 14, grave 6 and 12 (samples 18–21, 24–25), as these people were born, lived and were buried on the steppe. Their $^{87}Sr/^{86}Sr$ isotope values vary from 0.70899 to 0.70915.

Five individuals from Ulan IV, kurgans 3 and 4, Sukhaya Termista and Mandjikiny according to $^{87}Sr/^{86}Sr$ isotope values, which are 0.70901/0.70908 (samples 2, 4, 13, 15 and 12), were also probably locally born, but some humans did not move far and exploited food resources only in the steppe niches. But an old female from Sukhaya Termista, kurgan 1, grave 10, was probably locally born, but she spent the last 10–20 years of her life in other locations, i.e. in the steppe area with a different signal of vegetation or near the coastline where she consumed marine food.

In the case of Ulan IV, grave 8 (samples 2 and 3), we assume that two individuals were probably a couple and spent life together; their ^{15}N and ^{13}C values are similar, so they consumed the same food from the same area. But Sr composition differs significantly (0.70901 and 0.70936). We compare the Sr ratio of individuals with Sr ratio from the sheep tooth and leaches of buried soils obtained from the same kurgan. We assume that the enamel of the male was contaminated after the funeral.

$^{87}Sr/^{86}Sr$ isotope values of two teeth (M_1 and M_2) of another female from Ulan IV, kurgan 3, grave 15 (samples 7–8), show that this woman was born somewhere else on the steppe (with $^{87}Sr/^{86}Sr$ isotope values=0.70917), probably, in another part of the steppe but with different local $^{87}Sr/^{86}Sr$ isotope values and during her childhood (M_1=1–4 years, M_2=5–8 years) she did not move around much. She consumed a lot of marine food and died at the age of 30–40 years. It is possible that she was born somewhere near the Azov Sea or the Lower Don region (due to the local isotope signal) and did not move around much during her childhood as well as during her lifetime. But there might have been some reasons why she moved far from the place of origin at the end of her life and died among people with a different steppe diet.

A couple from Sukhaya Termista, kurgan 1, grave 11 (wife, 20–25 years old; husband, 55 years old), spent their life consuming marine food not far from the Lower Don or some areas near the Azov or Black Sea coastlines, but they were born in different places (samples 11 and 12; $^{87}Sr/^{86}Sr$ isotope values 0.70918 and 0.70908), although they were both buried under the same kurgan quite far (probably, more that 400–500 km) away from that place on the steppe.

Isotope values of M_1 (=1–4 years) and M_3 (=9–12 years) teeth of a young male from Mandjikiny-1, kurgan 14, grave 12, indicate that he was born and spent his childhood in the same place and did not travel much (samples 18–21, $^{87}Sr/^{86}Sr$ isotope values 0.70908).

Outsiders buried in Ulan IV, kurgan 3, grave 5 (sample 9), and Sukhaya Termista in kurgan 1, graves 2 and 14 (samples 11 and 14), as well as in Mandjikiny-1, kurgan 14, grave 1 (samples 22–23) (two females and two males), were born in the area with non-local $^{87}Sr/^{86}Sr$ isotope values, i.e. 0.7092, probably near the coastline of the Azov Sea, or somewhere in the North Caucasus near the coastline of the Black Sea, where they spent quite a lot of time according to isotope values in their bone collagen. Two females of age 20–25 and 18–25 probably got married and died on the steppe among people with different steppe cuisine. Another case is a male from Ulan IV, kurgan 3, grave 5, who was buried accompanied by extraordinary goods including river pearls. He was non-locally born and was non-locally fed. He probably moved to the steppe from the North Caucasus marine coast area or the Azov Sea and died quite old, but the isotope signal in his bone collagen was still typical for marine food consumers.

Conclusions

Thus, if we look at the mobile pattern, we may assume that two groups of people were buried in the steppe kurgans: locally born and non-locally born. This helps us reconstruct the following level of individual mobility based on five more detailed models:

i) Seasonal small movements across the exploited steppe areas where people were, probably, born and spent their whole lives, some of them as the Yamnaya 'couple' from Ulan IV, grave 8, who, according to their isotope values, consumed food from the same pot (economic reasons).

ii) Possible far-away seasonal movements of people who were probably locally born on the steppe across the vast exploited steppe areas with different vegetation. Such movements might have exceeded 100km. During their seasonal movement those small families might have consumed local steppe food with slightly different isotope signals (economic reasons).

iii) Possible large-scale individual movements of young females 500–600 km from the place of origin (near the coastline of the Azov Sea or the Lower Don region) to the occupation place of their husband on the steppe (personal or economic reasons).

iv) Possible seasonal large-scale individual movements 500–600 km from the steppe to the south with a long stay near the coastlines of the Azov Sea and the west coast of the Black Sea areas (economic or personal reasons).

v) Possible large scale individual adult male and female movements 500–600 km from their place of origin (near the coastline of the Azov Sea or the lower Don region) to the steppe due to some business (trade, visiting relatives or personal reasons).

Seasonality data indicate that all movements occurred in different seasons.

Therefore, seasonality already resulted in a high level of population movements across the exploited area, i.e. the steppe and the North Caucasus. With the combination of seasonal and isotope data, we are able to confirm that during the Bronze Age people exploited different pastures located sometimes far from each other. It means that during the year small population groups, including the new born and little children, moved with the herds from one pasture to another and they consumed local food with different strontium isotope ratios. We assume that some of the exploited lands might have been located near the coastlines of the Azov and Black Seas and the adjacent areas in the foothills of the North Caucasus Mountains. New data also give insights into individual lives. Once developed, the model of seasonal use of pastures gradually became incorporated into the everyday life of subsequent nomadic population groups.

We assume that it will be very difficult to identify the locality of the possible area of individual origin. Only in cases where the child was born and spent several years of his/her life in one locality with a very distinctive strontium isotope signal, the area of his/her origin may be identified more precisely. At the moment the first version of the map with local $^{87}Sr/^{86}Sr$ isotope values for the southern Russian steppe areas, foothills and mountain areas of the North Caucasus as well as western coastlines of the Azov, Black and Caspian Seas helps identify only the possible locations of some individual places of origin.

Seasonality of archaeological sites and isotope values of human skeletons are considered to be markers of population movement.

Acknowledgement.

This work was supported by RFFI grant 13–06–12003 ofi_m.

Bibliography

Bentley, R.A., Price, T.D. and Stephan, E. 2004. Determining the local $^{87}Sr/^{86}Sr$ range for archaeological skeletons: a case study from Neolithic Europe. *Journal of Archaeological Sciences* 31:365–375.

Bentley, R.A. and Knipper, C. 2005. Geological patterns in biologically available strontium, carbon and oxygen isotope signatures in prehistoric SW Germany. *Archaeometry* 47/3:629–644.

Buzon, M.R., Simonetti, A. and Creaser, R.A. 2007. Migration in the Nile Valley during the New Kingdom period: a preliminary strontium isotope study. *Journal of Archaeological Sciences* 34:1391–1401.

Demkin, V.A., Borisov, A.V., Demkina, T.S., Yeltsov, M.V., Borisova, M.A., Klepikov, V.M., Sergatskov, I.V., Dyachenko, A.N., Shishlina, N.I. and Tsutskin, E.V. 2002. The Yergueni Hills soil and environment development during the Eneolithic and Bronze Ages (Razvitiye pochv i prirodnoy sredy Ergeninskoy vozvyshennosty v epochi eneolita i bronzy). In N.I. Shishlina and E.V. Tsutskin (eds.) *Ostrovnoy Kurgan Burial Ground: results of interdisciplinary investigation of northwestern Caspian archaeological sites*: Moscow, Elista: State Historical Museum. Pp. 107–131.

Eckardt, H., Chenery, C., Booth, P., Evans, J.A., Lamb, A. and Muldner, G. 2009. Oxygen and strontium isotope evidence for mobility in Roman Winchester. *Journal of Archaeological Sciences* 36:2816–2825.

Evans, J., Stoodley, N. and Chenery, C. 2006. A strontium and oxygen isotope assessment of a possible fourth century immigrant population in a Hampshire cemetery, southern England. *Journal of Archaeological Sciences* 33:265–272.

Ivanov, I.V. and Vitaly, A.D. 1999. Soil sciences and archaeology. *Soil Sciences* 1:106–113.

Jay, M. and Richards, M.P. 2006. Diet of the Iron Age cemetery population at Wetwang Slack, East Yorkshire, UK: carbon and nitrogen stable isotope evidence. *Journal of Archaeological Sciences* 33:653–662.

Kirillova, I.V., Klevezal, G.A., Mikhailov, K.E., Golyeva, A.A., Trunova, Y.E. and Shishlina, N.I. 2000. Complex method for determination of the season of the Bronze Age graves in Kalmykia (Komplexny metod opredeleniya sezona soversheniya pogrebeniy). In N. Shishlina (ed.) *Seasonality Studies of the Bronze Age Northwest Caspian Steppes*. Papers of the State Historical Museum 120. Moscow: State Historical Museum. Pp. 30–42.

Keenleyside, A., Schwarcz, H.P, Stirling, L., Lazreg, N.B., 2009. Stable isotopic evidence for diet in a Roman and Late Roman population from Leptiminus, Tunisia. *Journal of Archaeological Sciences* 36:51–63.

Klevezal, G.A. and Shishlina, N.I. 2001. Assessment of the season of death of ancient human from cementum annual layers. *Journal of Archaeological Sciences* 28:481–486.

Klevezal, G.A., Shishlina, N.I., Pakhomov, M.M. and Khokhlov, A.A. 2006. Establishing of season mortality of the Bronze Age man according to dental cement layers (Identifikatsiya sezona smerti cheloveka po sloyam v tsemente zubov (epokha bronzy). *Russian Archaeology* 2:15–23.

Kremenetsky, K.V. 1997. Holocene Environment of the Lower Don and the Kalmyk area (Prirodnaya obstanovka golotsena na Nizhnem Donui v Kalmykii). In P. Kozhin (ed.) *Steppe and the Caucasus (cultural traditions)*. Papers of the State Historical Museum 97. Moscow: State Historical Museum. Pp. 30–45.

Lanting J.N. and van der Plicht, J. 1998. Reservoir effects and apparent ^{14}C ages. *Journal of Irish Archaeology* 9:151–165.

Masanov, N. 1995. *Nomadic Civilization of the Kazaks (the base of the vital functions of nomadic society) (Kochevaya tsivilizatsiya kazakhov) (osnovy zhiznedeyatelnosti nomadnogo obschestva)*. Almaty-Moscow: Sotsinvest-Gorizont.

Nafplioti, A. 2012. Late Minoan IB destruction and cultural upheaval on Crete: a bioarchaeological perspective. In E. Kaiser, I. Burger and W. Schier (eds.) *Population Dynamics in Prehistory and Early History: new approaches using stable isotopes and genetics*. Berlin/Boston: Walter de Gruyter GmbH & Co. KG. Pp. 241–263.

Price, T.D., Frei, K.M., Tiesler, V. and Gestsdottir, H. 2012. Isotopes and mobility: case studies with large samples. In E. Kaiser, I. Burger and W. Schier (eds.) *Population Dynamics in Prehistory and Early History; new approaches using stable isotopes and genetics*. Berlin/Boston: Walter de Gruyter GmbH & Co. KG. Pp. 311–321.

Shishlina, N.I., 2008. *Reconstruction of the Bronze Age of the Caspian Steppes: life styles and life ways of pastoral nomads*. Oxford: BAR International Series 1876.

Shishlina, N.I. and Pakhomov, M.M. 2002. Pollen studies of soil samples from the Ostrovnoy burial ground in Kalmykia (Sporo-pyltsevoye issledovaniye pochvennikh obraztsov iz mogilnika Ostrovnoy in Kalmykia). In N.I. Shishlina and E.V. Tsutskin (eds.) *Ostrovnoy Kurgan Burial Ground: results of interdisciplinary investigation of northwestern Caspian archaeological sites*. Moscow, Elista: State Historical Museum. Pp. 186–195.

Shishlina, N.I., Gak, E.I. and Borisov, A.V. 2008. Nomadic Sites of the South Yerguieni Hills on the Eurasian Steppes: models of seasonal occupation and production. In H. Barnard and W. Wenrich (eds.) *The Archaeology of Mobility: Old World and New World nomadism*. Los Angeles: Cotsen Institute of Archaeology. Pp. 230–249.

Shishlina, N.I., Zazovskaya, E.P., van der Plicht, J., Hedges, R.E.M., Sevastyanov, V.S. and Chichagova, O.A. 2009. Paleoecology, subsistence and ^{14}C chronology of the Eurasian Caspian Steppe Bronze Age. *Radiocarbon* 51:481–499.

Shishlina, N., Sevastyanov, V. and Hedges, R. 2012. Isotope ratio study of Bronze Age samples from the Eurasian Caspian Steppes. In E. Kaiser, I. Burger and W. Schier (eds.) *Population Dynamics in Prehistory and Early History. new approaches using stable isotopes and genetics*. Berlin/Boston: Walter de Gruyter GmbH & Co. KG. Pp. 481–499.

Shishlina, N. and Larionova, Y.O. 2013. $^{87}Sr/^{86}Sr$ isotope values variation in contemporary snail samples obtained from the southern part of Russia: preliminary results. In A.B. Belinsky (ed.): *Material of the investigation of the cultural-historical heritage of the North Caucasus*. Moscow: *Pamyatniki istoricheskoy mysly* 9:159–168.

Thompson, A.H., Richards, M.P., Shortland, A. and Zakrzewski, S.R., 2005. Isotopic paleodiet studies of Ancient Egyptian fauna and humans. *Journal of Archaeological Sciences* 32:451–463.

N.I. Shishlina, (corresponding author), State Historical Museum, Moscow, Russia
Email: nshishlina@mail.ru

Y.O. Larionova, The Institute of Geology of Ore Deposits, Petrography, Mineralogy, and Geochemistry, Russian Academy of Sciences (IGEM RAS)
Email: ukalarionova@gmail.com

3. 'Salaš': Summer farming and transhumance in the Czech Republic from a (pre)historic and environmental perspective

Dagmar Dreslerová

*Mountain summer grazing has formed a significant part of the economy in many regions of Europe. In the modern history of the Czech Republic such a system was practiced in the eastern part of the country, East Moravia in the Outer Western Carpathians. It started with the arrival of nomadic shepherds (the Wallachians) in the 15th and 16th centuries and ceased to exist at the beginning of the 20th century. In contrast, in the western part of the country (Bohemia) transhumance has been almost unknown despite the fact that the whole of Bohemia is surrounded by mountain ranges. The only exceptions were so called 'mountain cabin farming' (*Baudenwirtschaft*) in the Krkonoše Mountains, introduced in this region by Alpine woodcutters in the late 17th–19th centuries, and the insufficiently recognised animal husbandry in the Šumava Mountains taking place at the same time. The reason why summer farming was not practiced in Bohemia on a bigger scale has not been fully understood. Environmental rather than cultural factors may be behind it. In this context the possibilities of prehistoric summer farming/transhumance are discussed as well as the limitations of its detection in archaeological and palynological records.*

Introduction

Transhumance pastoralism has been a significant part of the economy in many European regions. The report of the EU project concerning transhumance and biodiversity in European mountains published in 2004 (Bunce *et al.* eds.) reviewed the past and present state of transhumance in many countries all over the Europe, with an exception of the Czech Republic (CR). The question arises what was the cause of the exclusion of this country from a community of 'herding' countries; perhaps our frontier mountain ranges surrounding almost the whole country except southern Moravia (Fig. 3.1) were not taken as 'mountains' or, more probably, a lack of evidence of the present day transhumance in the CR lead to the presumption that such practices perhaps never existed. But is this the truth and if yes, why it is so?

What is transhumance?

The definitions of a seasonal movement of animals are numerous as well as their forms (long or short distance transhumance, alp-system, etc.). For the purpose of this study, transhumance is understood as a seasonal vertical movement of livestock from a valley to higher altitudes and summer farming is understood as a specific form of vertical transhumance connected with dairying activities (milking and cheese production, cf. Spindler 2003). There is a vast amount of local/regional variations of transhumance/summer farming. The differences lie in the selection of animals, in distances from the base settlement, the gender- and age-based division of work, etc. Herzog and Bunce (2004:303) define transhumance as 'the seasonal oscillatory movement of livestock (which) includes many diverse practices depending upon local situations and traditions'. The common features of transhumance are flexibility, complexity and the utilisation of complementarities in space (between habitat/landscape) and time (between seasons).

Festi and Oegl (poster) described conditions for transhumance as follows: 'transhumance is practiced in the environment with the stable valley settlement based on an agro-pastoral economy, where there is a limited space for agriculture in the valley. Moving of animals thus saves valley/lowland pastures and the sources of hay. Travelled distance might go from a few kilometres up to 40–50km.

To sum up, the conditions for practicing transhumance/seasonal highland exploitation are: the scarcity of pastures near the home village (mostly due to limited space in the valley); and the simultaneous availability of pastures in the uplands/other regions nearby. Suitable conditions are enhanced where there are natural meadows/plant communities. In central Europe, it is mostly the case of high elevations above the tree line.

Natural pre-conditions for mountainous summer farming/transhumance in the area of the present day Czech Republic

There are obvious differences between Bohemia (the western part of the CR) and Moravia and Silesia (the eastern part of the CR), and the countries with known

KEYWORDS: transhumance, summer farming, Czech Republic, prehistory, montane farming system, pollen analysis, Wallachians

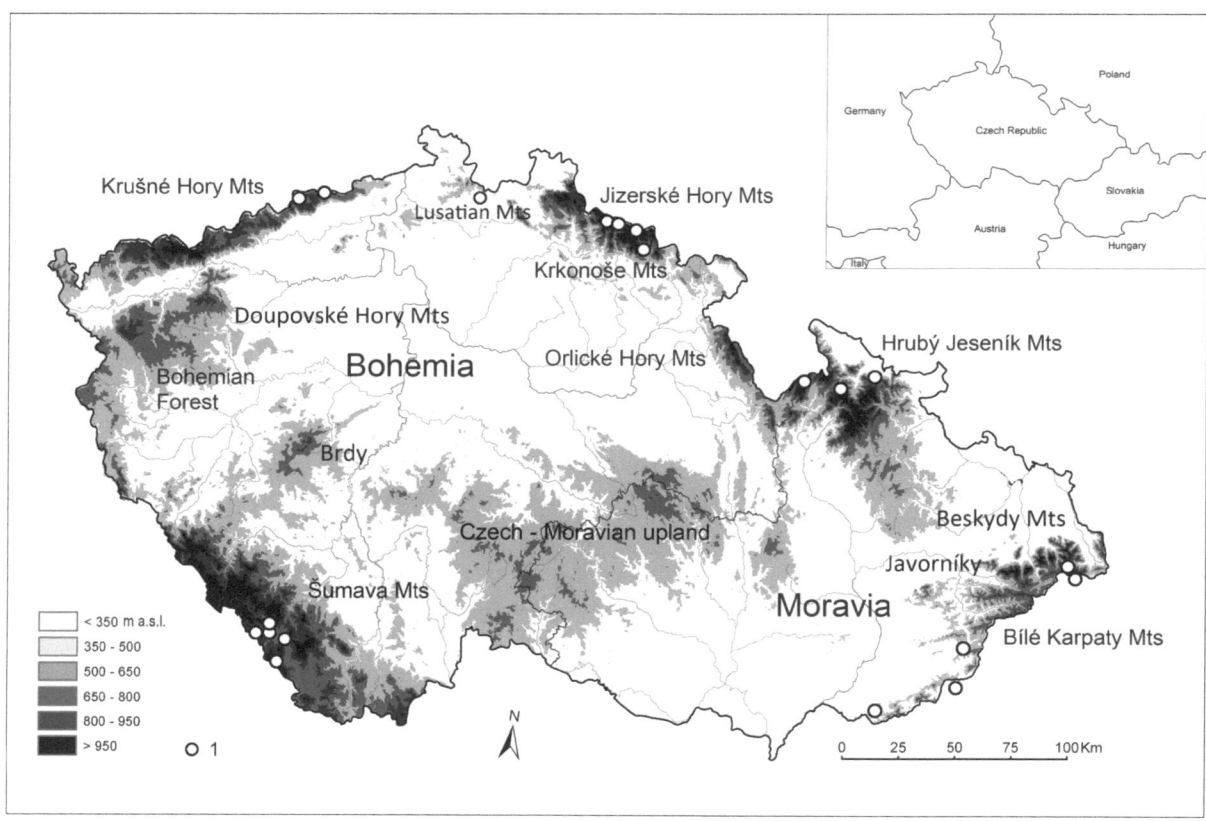

Figure 3.1   Geographical map of the Czech Republic and (1) the position of the pollen profiles mentioned in the text.

(pre)historic transhumance/summer pasturing (like Switzerland, Bosnia, Slovakia, Romania).

First it is necessary to take into consideration the natural habitat of the landscape. Bohemia and the western part of Moravia are created by the Bohemian massif (*Hercynicum, Panonicum*), dissected plateaux bordered by the mountain ridges (for the purpose of this study areas over 600 m.a.s.l. are taken as mountains, as they host only an inconsiderable number of archaeological sites. Uplands are indicated areas between 400 – 600 m.a.s.l.). The geological substrate of uplands and mountains consists mostly of crystalline rock (especially acidic gneiss and granite), avoided by prehistoric settlement even in the lower, climatically more favourable regions. Most of southern and central European summer farming systems 'operate' in calcareous regions with a dynamic relief (e.g. short distances between lowlands/valleys and upland pastures).

The lowland Bohemian Cretaceous basin is formed by sea sediments. The eastern part of Moravia belongs to the Outer Western Carpathians (Carpaticum), mostly formed by Cretaceous and Palaeogenic sea sediments. Excluding the latter, these areas are more suitable for arable farming than for extended animal husbandry.

The Czech Republic belongs to the central European phytogeographical province. Natural grasslands in this province are relatively rare and due to the natural conditions, their area must have been quite limited here for the substantial part of the Holocene (Hejcman *et al.* 2013). Also the productivity of the semi-natural hay meadows is generally lower than in, for example, the karst limestone terrains of central and southern Europe (the Dinarides – Bosnia-Herzegovina, the Dolomites, the Carpathians, the Schwabian Jura, the French Alps), where, according to our observations, past transhumance mainly operated.

Last but not least, since prehistory, in the Czech lands sheep husbandry as the main animal of transhumance has been less significant than cattle husbandry, unlike some countries of southern Europe. Sheep husbandry started growing as late as the 16th century specifically at feudal manor farms in upland regions. Its importance continued throughout the 17th and 18th centuries due to the development of textile manufacturing. Since the mid-19th century sheep husbandry has lost its significance through the decline of the open-air system as well as the import of cheap wool from overseas (Beranová and Kubačák 2010).

Historical evidence of mountain pasturing

No evidence of mountain pasturing is known so far from the Krušné Hory (the Ore Mountains, the Erzge-

Figure 3.2   Postcard of the Kleine Osser, Bavaria, showing the summer farm, around 1920.

birge) and Lužické Hory (Lusatian Mountains), which form the border between Saxony and Bohemia, and the Jizerské Hory (the Jizera Mountains, the Isergebirge) on the border between Bohemia and Poland.

The Šumava Mountains (Bohemian Forest, Böhmerwald), southwest Bohemia are the largest range rising to 1400 m.a.s.l. Nowadays forests cover more than a third of the range. The beginnings of specialised summer pasturing of cattle and, to a lesser extent, of sheep as well, are not precisely known. The first colonists came to the Šumava region in the 13th–14th century and in subsequent centuries settlement gradually moved into higher elevations. The second colonisation in the 17th and 18th centuries related to glass making and logging. These activities caused the appearance of large-scale clearings and openings, which attracted pasturing. However, according to K. Klostermann (2012), a popular novelist born in the Šumava in 1848, around the mid-19th century the summer pasturing of livestock (mostly for meat) took place in the rest of the primeval forests. Herds of cattle came from the foothills in thousands; a herd consisted of 800–1200 head. Shepherds stayed in wooden cabins in the forest; they protected animals against predators and also thieves, usually coming from Bavaria. Forest pasture naturally caused tensions between foresters and herders; the more profitable solution usually won. In 1870 a strong gale destroyed most of the Šumava forests and the subsequent wood extraction of large upland plains encouraged summer pasturing of cattle.

Little is known about sheep pasturing. Very limited evidence exists from the Bohemian side of the mountains, but across the border the tradition is documented, among other sources, by the photograph of a summer farm (*Almhütte*) situated in the saddle of the Kleine Osser mountain at the height of 1293 m. a.s.l. (Fig. 3.2). An archaeological surface survey in 2012 did not find any traces of the building except for a charcoal layer in the lower part of the slope on which the structure was situated.

The Krkonoše Mountains (Giant Mountains, Riesengebirge) create the northern fringe of the Bohemian Massif and reach 1603 m.a.s.l. The history of (known) transhumance goes back to 17th century. The knowledge of summer farming was brought here from the Alps with the woodcutters, who decided to settle down permanently after they had cleared a substantial part of the Krkonoše forests (Hartmanová 2004). A specific form of local highland farming (*Baudewirtschaft*) consisted of the gathering of hay and pasturing of livestock (cows and goats) both for meat production and milk, butter and cheese. In 1804 there were 2600 farms (*Bauden*) on the Czech side of the Krkonoše, some of them being used all year. Permanent farms were usually solid buildings composed of a living room, black kitchen, hall and stables. Semi-permanent buildings were smaller, less stable, more primitive structures. As a rule the so-called grass garden, e.g. meadow, free of stones and manured which was meant for a high hay production, used to be situated on a slope

Figure 3.3 Ruins of the Scharfs Baude, the Krkonoše Mountains some 50 years after its abandonment. After Hartmanová 2004.

immediately below the farm building/*Baude* (Hartmanová 2004). It is estimated that during the biggest boom of the *Baudewirtschaft* about 18–21,000 people from the foothills took part in pasturing, hay collection and milk and cheese production (Hoser 1804; Lokvenc 1978). Forest protection and the development of tourism at the end of the 19th century contributed to the restriction of the transhumant economy; its definitive end, however, came with the resettlement of the German population after the Second World War (Lokvenc, 1978; Bartoš and Nováková, 1997; Hartmanová 2004).

An archaeological survey of the area of Čertova louka, central Krkonoše Mountains, revealed remains of several mountain farms in the form of stone foundations, cellars or cold store, stone fences and pathways (Fig. 3.3). Excavation of (1136 m.a.s.l., Fig. 3.4), which ceased to exist in 1893,was also interesting methodologically as it shows what can be found more than one hundred years after such a type of dwelling has disappeared. The most numerous finds were over 500 fragments of pottery (both vessels and stove tiles), then glass (beer bottles and window-panes), various iron fittings, spikes and nails, a piece of a scythe, pieces of grindstone and whetstones, six iron spoons and a knife, two copper coins, a brass sheep/goat bell, the porcelain head of a pipe, a leather shoe sole and a fragment of a felt hat (Hartmanová 2004).

The Outer Western Carpathians, Beskydy Mountains, Vsetínské vrchy (Vsetín Mountains, Wsetiner Berge), Javorníky and the Bílé Karpaty Mountains, form the eastern part of Moravia and the border between Moravia and Slovakia (Fig. 3.1). Though some parts of the Outer Western Carpathians are lower, reaching only c. 600 m.a.s.l., they give the impression of a truly montane landscape. Due to their position between the valleys of two large rivers, the River Morava in the west and the River Váh in the east, prehistoric occupation reached the foothills of the Bílé Karpaty Mountains as early as the Neolithic.

In the late 15th–16th centuries, a new unique grazing method was introduced to the region with the arrival of the new colonists – Wallachians (Valachs, Vołosi, Valahians) during so called Wallachian colonisation. Originally, from the 14th century, nomadic shepherds from the region of the present day Romania were gradually moving towards the north and northwest in order to find new pastures. They spread across the western Ukraine (Sub-Carpathian Rus), the mountain areas of Slovakia, and the Carpathian regions of southern Poland. At the end of the 15th century they reached the westernmost edge of their expansion by hitting eastern Moravia and Silesia. The colonists gradually merged with the local population and their name was preserved in the term 'valachian husbandry' (= mountain dairy farming) or in the term for a local shepherd, '*valach*'. Another specific feature of the assimilation process in Moravia and Silesia was their symbiosis with the so-called '*pasekarska*' (clearance) colonisation of suitable areas in the mountains by peasants from the valleys below (Štika 2001).

Figure 3.4 Remains of the milk cellar of the Stará Bouda in the Krkonoše Mountains. After Hartmanová 2004.

Wallachians settled even the highest parts of the mountains above the limit of a sub-montane farming colonisation. It was enabled by special breed of 'walachian' sheep resistent to harsh conditions, kept for milk, production of cheese, and of wool for making special heavy blankets. Sheep were pastured in the forest or de-forested mountain clearings, while sub-montane sheep were pastured on fields and fallow ground (Štika 2001; Jongepierová ed. 2008; Mróz and Olszańska 2004). In some regions (namely the eastern part of the Lower Tatra Mountains, Slovakia) animals were kept in the high pastures all year round. They were confined in a special sheep pen called '*mraznica*' a robust rounded wooden enclosure sloping inwards (Fig. 3.5). Animals were fed hay, conifer brushwood and leaf fodder (Podolák 1982). Shepherds stayed in various types of huts called '*salaš*' or '*koliba*' ranging from simple open wooden shelters to more solid half-timbered or timbered, usually one storey, constructions (Fig. 3.6, Kunz 2005).

A typical multifunctional instrument of shepherds (valachs) was an '*obušek*' or '*valaška*', a light axe on a long stick (Fig. 3.7). It served as tool for cutting leaf fodder, as a weapon, a walking stick, an attribute of a young man, and also an attribute of a special dance (Kunz 2005). With the help of this tool shepherds also kept pastures treeless by cutting off the young shoots of trees while walking through pastures.

During the 17th and especially in the 18th centuries the mountain husbandry gradually merged with lowland sheep husbandry. It reached its peak in the 1780s. In the 19th century the decline of mountain farming came with the need of wood for developing industries. The mountain pastures were quickly replaced by spruce monocultures. By the middle of the 20th century, the last remnants of traditional sheep farming had disappeared (Štika 2001).

Traces of transhumance in the archaeological record

The oldest permanently settled land, at least since the Neolithic covers the area of the Bohemian and Moravian lowlands up to 350 m.a.s.l. (Fig. 3.8). Higher elevated large hilly areas (uplands up to 550–600m) were populated later, usually from the Bronze Age. Mountain foothills (highlands) were primarily settled in the Iron Age (Early Sub-Atlantic) when settlement reached its largest spatial distribution in prehistory (Fig. 3.10). In comparison to the lowlands the density of population in the uplands seems significantly lower. The colonisation of the highlands appeared in the High Medieval period and was usually associated with the activities of colonists brought in from abroad.

Seasonal movement of animals covers both pure movement of herds and shepherds into summer pastures and their stay without any solid dwelling structures or animal pen/enclosure. Another form, probably later, is represented by so called *Almwirtschaft* (summer farming). It consists of milk production, building of specific farmhouses (*salaš*) and usually also by the creation of enclosures. Naturally these activities leave more material

Figure 3.5 Summer farm and the sheep pen at the Radhošť, in the Beskydy Mountains, first half of the 20th century. After Kunz 2005.

traces and therefore they are archaeologically more easily detected compared to the previous case.

Archaeological finds supported by archaeobotanical evidence place the beginnings of transhumance practices in the Neolithic/Late Neolithic period in the Pyrenees (Gedes 1983), French Alps (Walsh and Mocci 2011), Bosnia-Hercegovina (Mlekuž 2003; Müller-Scheessel *et al.* 2010), Switzerland (Akeret and Jacomet 1997). The latest evidence is furthermore supported by the archaeozoological research of the LBK settlement site at Vaihingen, southwest Germany. The ratio of isotopic strontium ($87Sr/86Sr$) in cattle teeth confirmed that transhumance was practiced, and domestic animals were pastured in regional uplands (Bentley and Knipper 2005). Pollen analyses show indicators of human impact in profiles from the sub-montane/montane areas also in the Neolithic, but most authors interpret them as a result of the long-distance transport of pollen; they place the first indisputable human influence in the Late Sub-Boreal/ Early Sub-Atlantic – the Bronze/ Early Iron Ages (e.g. Rösch 2009; Röpke *et al.* 2011; Pokorný 2004; Jankovská 2006).

The oldest archaeological finds in the mountain areas of the CR come from the Upper Palaeolithic and Mesolithic in the Šumava Mountains. They are usually connected with the seasonal exploitation of food resources, for example fishing in the upper stream basins (Vencl 2006; Čuláková *et al.* 2012). The finds of the Neolithic stone tools, both flaked and polished industry, are found relatively often, although not in such quantities as in the Black Forest, Germany, where

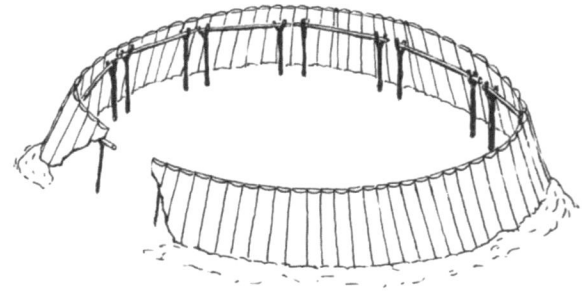

Figure 3.6 Drawing of the '*mraznica*' (sheep enclosure for overwintering). After Botík and Slavkovský (eds.) 1995.

Figure 3.7 Drawing of a traditional Valach shepherds' tool, '*obušek*' or '*valaška*'. After Kunz 2005.

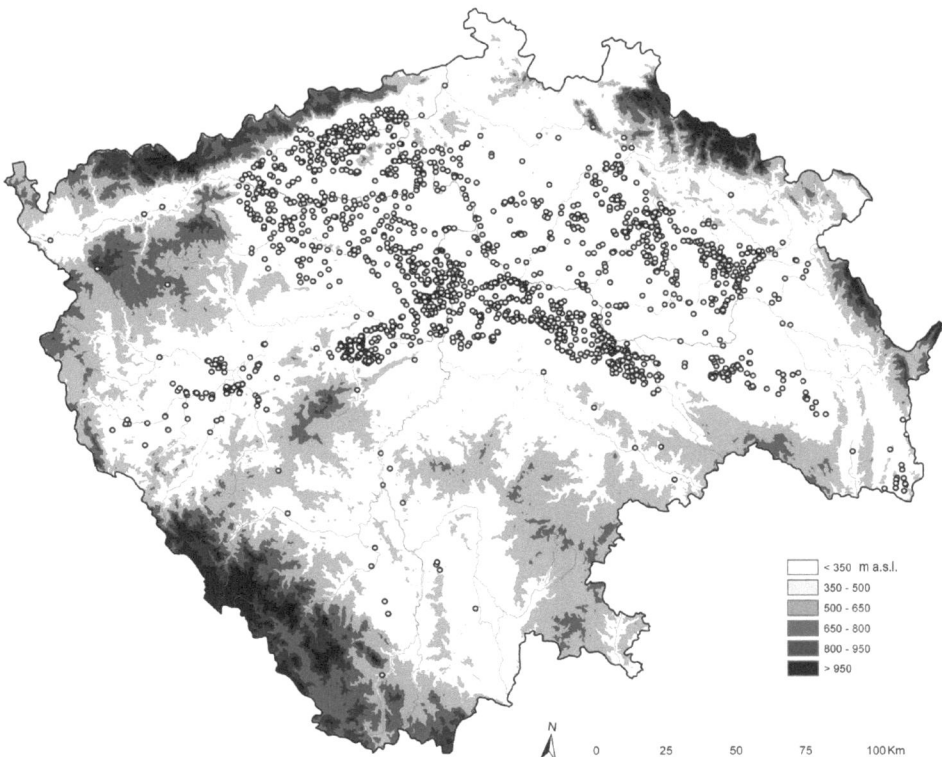

Figure 3.8   Settled area of the Neolithic in Bohemia (western part of the Czech Republic). The map shows only settlement sites. Archive of the Institute of Archaeology of the CAS, Prague.

Kienlin and Valde-Nowak (2004) connect such finds with transhumance activities. Also in Ecrins, France, numerous finds of isolated scatters or individual pieces of flint are seen as the evidence of Neolithic transhumance. The finds of the stone industry are furthermore supplemented by a stone sub-circular enclosure and an adjacent small structure, interpreted as pastoral structures. They were found at the Late Neolithic site Chichin III, dated to the mid and latter half of the third millennium BC (Walsh and Mocci 2011). Valde-Nowak (2008) argues that also in the Polish West Carpathians (the Beskydy Mountains) a Neolithic system of stable settlement on the mountain edges supplemented by mobile settlement deeper into the mountains may have existed. The map of the Neolithic settlement sites in Bohemia shows rather densely populated lowlands up to 350 m.a.s.l. with only limited settlement evidence in the higher elevations up to 500m (Fig. 3.8). In contrast, the map of isolated Neolithic finds (Fig. 3.9), usually a single stone axe, shows a plentiful presence even at heights of 500–650 m.a.s.l., especially at the Českomoravská vrchovina upland on sites where finds from later periods are not to be found. The rather surprising appearance of these isolated finds within forested areas avoided by settlement till the medieval colonisation has not been discussed and its connection with any form of pasture is yet to be investigated.

Sparse Neolithic finds from high ranges have been, due to their often unique position in the landscape, considered as traces of a cult of hills or specific geomorphological structures, and as witnesses of prospector activities, and possibly as the markers of trade routes (Fröhlich 2009). Archaeological evidence from the Bronze Age has a similar character of single finds. In addition, several Late Bronze Age settlement sites (some of them enclosed) are known from the Krušné Hory Mountains (Fig. 5.10). Their existence is linked with ore mining, especially iron (Koutecký and Bouzek 2009). For the same reason (ore mining) complemented by glass production settlement is found on the higher reaches of the Krušné Hory Mountains on both sides of the border, settled from the mid-12th century AD (Kenzler 2009). The nearest Bronze Age transhumance is recorded in the Dachstein region, Austria (Mandl 1996).

The relatively massive growth in the settlement of the uplands and highlands took place in the Iron Age, especially in the Šumava Mountains and its foothills (Fig. 3.10). The appearance of a number of hillforts and hilltop settlements at heights above 700 m.a.s.l. situated on the margins or outside the regular settlement zone is a characteristic feature of the period. A number of sites cover an area of several hectares, e.g. Sedlo u Albrechtic 902 m.a.s.l. and 3ha, Obří Hrad u Studence, 980 m.a.s.l. and 2.5ha, Věnec u Lčovic, 765 m.a.s.l. and 7.1ha (Fig.

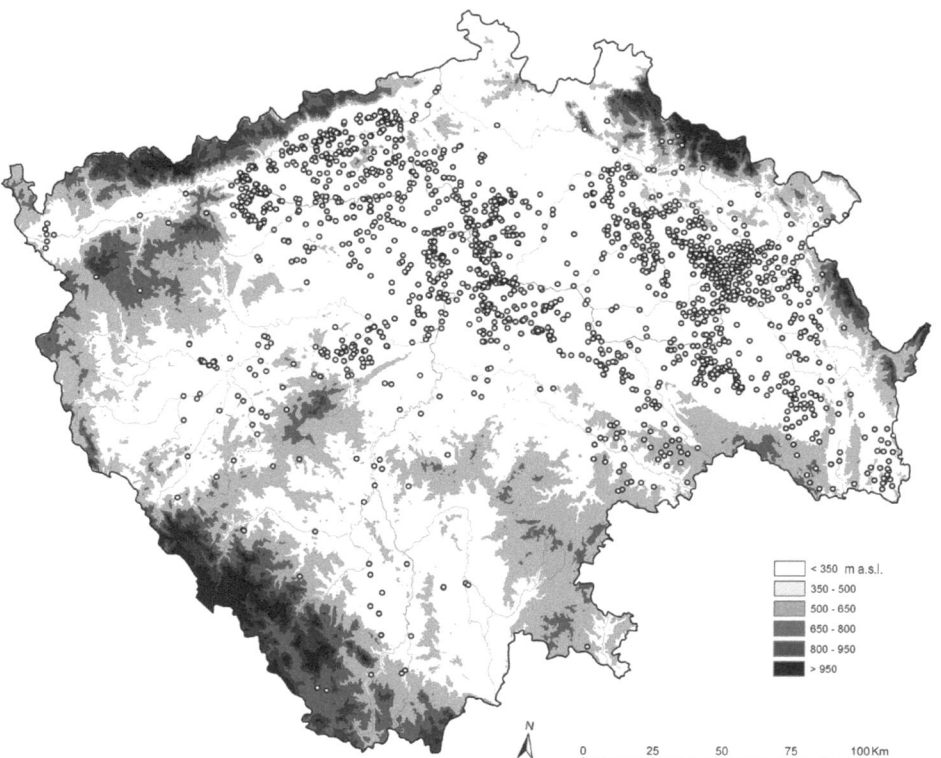

Figure 3.9　Isolated neolithic finds (mostly stone axes) in Bohemia. Archive of the Institute of Archaeology of the CAS, Prague.

3.11); others are much smaller. Fortifications are usually made of dry stone walls often connecting with natural rock formations. However these 'mountain fortresses' do not have any direct connection to the settlements in the foothills and the reason for them being built there has not so far been clarified. In several cases there is speculation about the connection with exploiting local sources of graphite and iron ore (Dreslerová and Hrubý 2004), and additionally also gold. There is also speculation concerning the exploitation of this kind of enclosure for the benefit of mountain animal husbandry. However, the research of these hillforts has been rather limited and the ability to prove the connection between hillforts and seasonal pasturing is further complicated by the fact that bone material, which could significantly clarify this problem, does not survive in the acid mountain and sub-mountain environment.

The Iron Age hillfort of Klisura-Kadića Brdo on the Glasinac Plateau, Eastern Bosnia, represents a similar type of the site located in an agriculturally marginal highland region. The site consists of a small plateau protected on one side by a steep rocky hillside, while the other three sides were fortified by a massive stone wall. The zooarchaeological study of animal remains revealed the production strategies and forms of animal husbandry, which varied according to animal species. The cementum analysis showed that caprines were probably moved in a transhumant pattern. All of the caprines in the cementum sample were killed during the warmer half of the year. This would indicate that the animals were probably moved out of the region during the winter (Greenfield and Arnold 2005).

During the research project focused on the Stubai Alps (southeast of Innsbruck, Austria), two stone structures, probably the remains of houses dating back to the Hallstatt period, were discovered in Wörgetal at an altitude of 2170m. These findings are clearly associated with summer grazing and are supported by evidence of human activity in the pollen spectra. The same is true of another stone structure at an altitude of 2265m in Wörgetal dating back to the Roman period (Weishäupl 2014). There is no evidence, however, for pasturing or other activities from the Bohemian or Moravian mountains in the Roman or Early Medieval periods up until the 12th–13th centuries.

The evidence of mountain exploitation in pollen records

The results of pollen analyses belong to the usual evidence often employed by archaeologists, for transhumance and highland use. At present, a comparatively large number of pollen spectra exist from the Czech and Moravian frontier mountain ridges (Kuneš

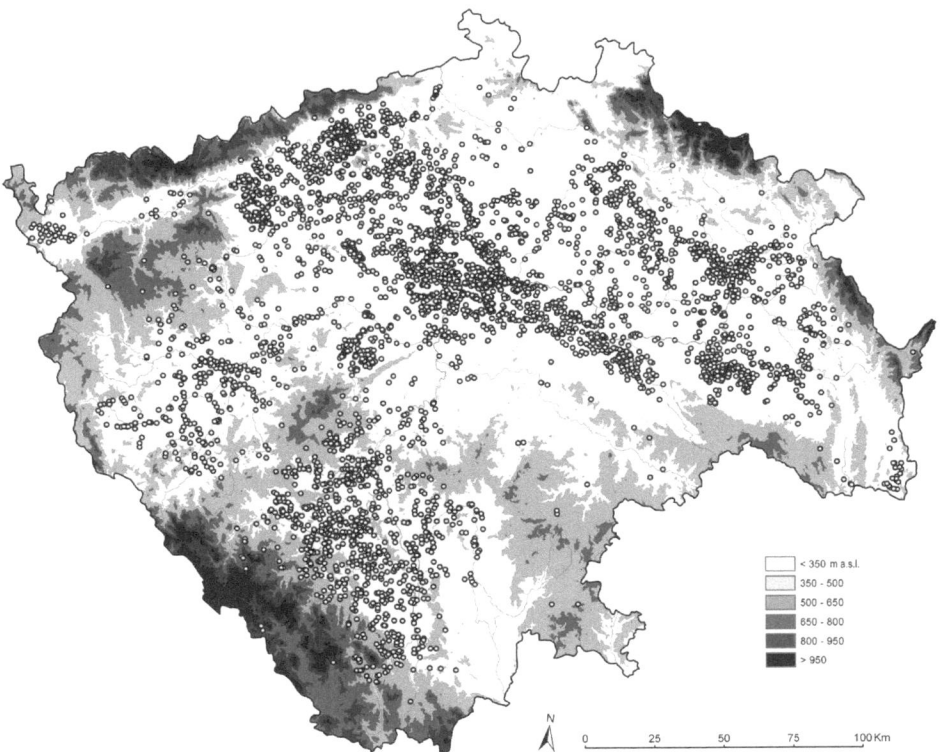

Figure 3.10 Settled area of the Iron Age (both Hallstatt and La Tène) in Bohemia. Archive of the Institute of Archaeology of the CAS, Prague.

et al. 2009, see www.palycz.cz for the map and the extended data). However, the reading and interpretation of the pollen records still yield some uncertainties.

First of all, it is difficult to decide whether the occurrence of so-called secondary anthropogenic indicators (pasture indicators in particular, *Plantago lanceolata* and *major*) are a result of the presence of naturally forest-free areas or the consequence of human activities. Another still poorly understood feature is whether the indicators of human activities come from the close vicinity of the pollen profiles (as is usually claimed at least for small sources e.g. mires, kettle holes, bogs several metres/several tens of metres in diameter), or whether pollen grains could reach mountain locations of pollen sampling from their foothills, as various authors presume, at least for the period prior to the later Sub-Atlantic (Jankovská 2004, 2006; Speranza *et al.* 2000; Rybníček and Rybníčková 2004, 2008). This hypothesis seems to be based on the obvious discrepancy between finds of some pollen, especially *Cerealia,* in the profiles situated in high elevations in which, according to presumptions, cereals cannot be planted naturally. *Cerealia* appear sporadically in the mountain profiles almost exclusively in the form of individual grains from the Neolithic without direct relation to evidence of settlement or economic activity. The problem, however, is more complex: except rye (*Secale cereale*), which is anemogamous and has a high pollen production capacity and good dispersion, the other cereals – wheats, barley and oat (*Triticum, Hordeum vulgare, and Avena sativa*) are autogamous. Most pollen grains remain in husks and are therefore poorly spread. Moreover, pollen productivity of the prehistoric wheats einkorn and emmer is poorly understood (Behre 1981). Pollen grains of the aforementioned cereals can come from threshing, can be transported together with husks remains, and their remains can be found along paths of transport. Another odd fact is that the oldest pollen grains, in mountain pollen profiles, belong to rye. It became intentionally cultivated very late, in the La Tène period. Until then, it is believed to have been a weed admixture in other cultivated cereals, although Kočár and Dreslerová (2010) do not consider this possibility probable.

Last, but not least, it is virtually impossible to standardise certain combinations of taxa (for example, the precise combination of an increase of *Poaceae, Betula, Pinus,* etc.) in the pollen spectrum which would decidedly document human activity. One of the reasons is that natural vegetation cover varies from region to region (for the critical overview see Court-Picon *et al.* 2006). It is always the result of the individual decision of an author and therefore the results can vary substantially. In many cases the direct human activities are admit-

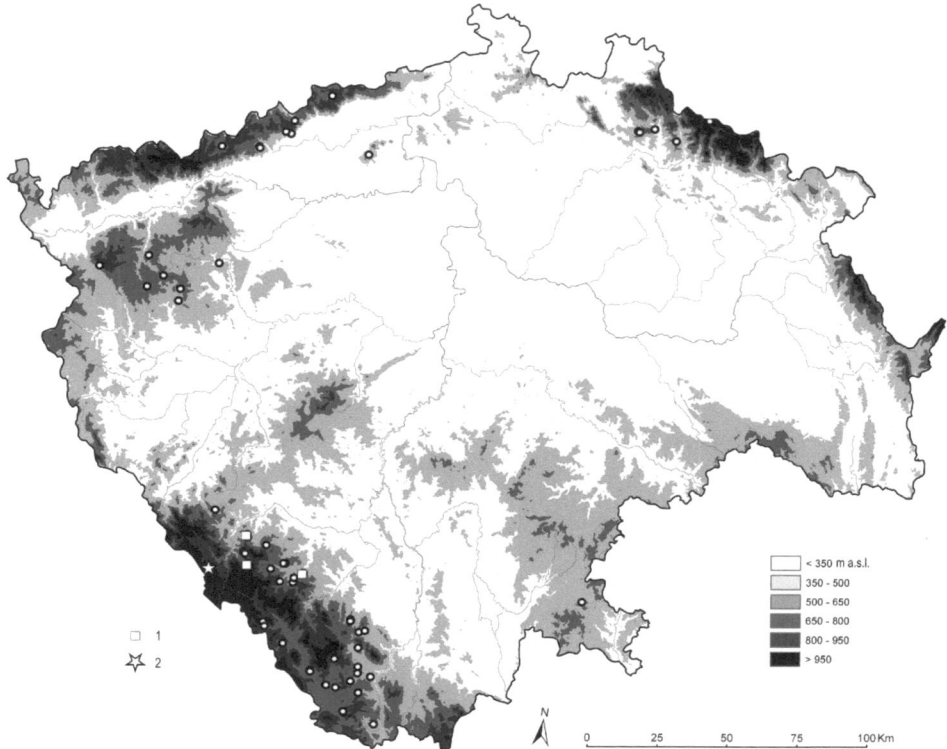

Figure 3.11 Archaeological finds from the mountain regions (above 650 m.a.s.l.) in Bohemia. Archive of the Institute of Archaeology of the CAS, Prague. 1: Iron Age hillforts mentioned in the text; 2: The Prášily site.

ted only when this fact has already been supported by written sources. All the aforementioned critical observations also hold true for the evaluation of profiles from the Czech border mountains.

Summarizing the previous remarks, it can be stated that the presence of pollen grains of anthropogenic indicators in the mountain profiles can be explained as:

1. remote spread of pollen from foothill areas;
2. direct transport by people or domestic animals;
3. proof of local mountain farming.

The largest amount of pollen data is known from the Šumava Mountains (Svobodová et al. 2001; Svobodová 2002, 2004; Jankovská 2006). About ten pollen profiles were taken from the mires situated in the elevations between 750–1200 m.a.s.l., e.g. above the elevation limit for cereal cultivation. Anthropogenic indicators, namely *Plantago lanceolata, P. media/major, Cerealia* type and *Secale cereale* appeared in some profiles (Strážkovská slať, Velká niva-Volary, see Svobodová et al. 2001) quite early, already around the Neolithic. Very early occurrences of the so-called pastoral indicators may be attributed to the specific local conditions; although the Šumava Mountains lie below the level of the natural sub-alpine forest-free area, numerous forest disconnections appeared through frequent occurrence of climate and edaphic extremes blocking the forest succession (Jeník 1961).

The first cereal pollen grains (Cereal type and *Secale cereale*) also appear very early (in the Late Atlantic) in the Šumava Mountains pollen samples, although the regular occurrence of cereals emerges in the Early Sub-Atlantic (Svobodová et al. 2001:196). The earlier case may be the result of a (very) long-distance transport of pollen grains, as even the Šumava Mountains foothills were not inhabited then. This idea is supported by the persistent occurrence of hornbeam (*Carpinus*) pollen in all pollen profiles for practically the whole of the Holocene. Hornbeam has never grown naturally in the Šumava and it cannot be seen there even today outside of plantation.

The later appearance of cereals and other anthropogenic indicators in the Early Sub-Atlantic pollen record may correspond to reality. From the Hallstatt period settlements spread into the elevation of above 600m (e.g. the Šumava foothills). The pollen record of human impact is supported by a new evidence of the La Tène settlement activities (four or five settlement sites) from the Šumava uplands themselves. Pollen spectra around the highest site Prášily (at 802 m.a.s.l., Fig. 3.11) contain an increasing amount of anthropogenic indicators (*Plantaginaceae*) which may be caused by pasturing (Čuláková et al. 2012).

The most distinctive Krkonoše Mountains pollen profiles are represented by the Labský důl (Jankovská

2004) and the Černá hora bog sites (Speranza *et al.* 2000). As Jankovská (2004) states, although some of the anthropogenic indicators occur already in the early Sub-Atlantic (e.g. *Plantago lanceolata, Rumex acetosella, Urtica* and even cereals) they cannot be taken as evidence of anthropogenic activities directly in the mountains. Pollen grains could reach here from the foothills; possibly the source vegetation of pollen could grow there naturally. Conclusive pollen evidence of direct anthropogenic activities in the Krkonoše Mountains comes from the 14th and especially from the 16th centuries. Speranza *et al.* (2000) found proof of earlier openings in the sub-mountain and mountain forest, at approximately 1380 ± 45 BP (7th to 8th centuries AD). They were created, possibly either for forest pastures or for the first exploitation of ores. Concerning the pollen of cereals, the authors suppose that they probably originated from the lowlands.

Pollen spectra are known from the Krušné Hory Mountains from both sides of the Czech/German border. At Fláje/Kiefern (CR, Jankovská *et al.* 2007) anthropogenic indicators show the first human impact in the foothills in the Sub-Boreal. The first cereal pollen and what is suggested to be a strong human impact on vegetation, presumably also at lower elevations, is detectable in the Late Sub-Atlantic. The decline of forest is attributed to ore mining, building, and glass production. At Georgenfelder Hochmoor (Germany, Stebich and Litt 1997) strong anthropogenic influence on vegetation can be seen from the medieval period onwards. The lack of anthropogenic indicators in earlier periods at the Krušné Hory Mountains cores is rather surprising especially when compared to those from the Šumava Mountains. Unlike the latter, settlement activities in the Krušné Hory Mountains foothills were distinctive already since the Neolithic.

Four pollen diagrams at the Hrubý Jeseník Mountains come from small mires situated in high elevations above 1300 m.a.s.l. They document the vegetation history since Sub-Boreal/Early Sub-Atlantic respectively. At the same time the continuous increase of cereal pollen begins (*Secale cereale* predominates). However, arable agriculture was limited to altitudes below about 600m, and therefore it has been concluded that cereal pollen must have originated in the foothills (Rybníček and Rybníčková 2004). According to written sources exploitation of the mountain summits appeared later, at the beginning of the 17th century.

Three of the five pollen diagrams from the Beskydy Mountains are situated at lower elevations below 615 m.a.s.l. (Rybníček and Rybníčková 2008). It may be the reason why the anthropogenic indicators are present already from the Sub-Boreal and their amount increases from the Early Sub-Atlantic, including the pollen of cereals (*Secale cereale* and *Triticum* type). Nevertheless, according to the authors of the study, human impact started rather late, in the course of the Middle Ages between the 13th and 14th centuries at the lowest elevations and only later was it followed by the establishment of mountain farming. The appearance of anthropogenic indicators in earlier periods thus remains unclear.

The evidence of a long land-use history was discovered in the Bílé Karpaty Mountains (the Flysch belt of the Outer Western Carpathians). At present, the area is well-known for extremely species-rich calcareous semi-dry grasslands classified as *Brachypodio pinnnati – Molinietum arundinaceae* association (Chytrý *et al.* 2007). The history of occupation of the foothills dates back to the Neolithic. In the late Neolithic, settlement penetrated deeper into the mountains and during the Bronze Age it covered the whole area of the present day grasslands (Hájková *et al.* 2011:191). The distribution of species-rich semi-dry grasslands largely overlaps with the distribution of prehistoric settlement and the density of settlement coincides with the richness of species. It is suggested that the extreme species richness of semi-dry grasslands may be partly attributed to the long history and continuous usage of these grasslands maintained by human activities such as grazing, mowing and burning. Macrofossil and pollen data indicate the existence of a cultivated landscape with a mosaic of forests and open habitats in the Roman era at the latest, but its history may even go back much further into prehistory (Hájková *et al.* 2011).

The Lusatian Mountains in northern Bohemia only reach an altitude of between 420 and 760 m.a.s.l. but, like the Bílé Karpaty Mountains, retain the character of mountains. The only pollen profile from the region, the Mokrá žába (400 m.a.s.l.) illustrates the historical development of the local vegetation (Kozáková *et al.* 2015). It was probably permanently influenced by man since the Bronze Age when rye first appears together with a continuous occurrence of *Plantago lanceolata*, whose representation increases in the Iron Age. Depending on changes and types of anthropogenic indicators, and changing composition of forest vegetation, it can be stated that during the Hallstatt and La Tène periods small-scale deforestation / logging, forest grazing and grazing on open sites existed; the existence of roads is likely. The intensified regime of the local landscape exploitation probably began in the Early Medieval period (Kozáková *et al.* 2015).

To sum up, human activities in the foothills/lower parts of the mountains have been documented in pollen records by the 'trace amount' of anthropogenic indicators since the Sub-Boreal (Late Bronze Age), and in a larger quantity since the Early Sub-Atlantic (Late Hallstatt and La Tène periods). However, archaeological evidence does not usually support this statement. The so-called pasture indicators (especially *Plantago lanceoloata*), usually appear in the pollen spectra together with the pollen grains of *Secale cereale* or *Cerealia* sp. Moreover *Plantago lanceolata* can also be the indicator of early agriculture and also *Rumex* and *Artemisia* can grow in various ecological conditions including cultivated fields (Court-Picon *et al.*

2006). Therefore the direct evidence that sub-mountain/mountain land was used for pasturing/summer farming only is missing. It is more likely, that a special form of sub-montane/montane farming management, based on animal husbandry and cereal production was applied. Rye (*Secale cereale*) appears to have been used in the above-mentioned system though in the lowlands it started being cultivated intentionally from the Late Iron Age (Dreslerová and Kočár 2013). Such an hypothesis needs to be proven both archaeologically and with the help of macro-remain analysis. However the largely forested terrain of the Bohemian mountains makes this task extremely difficult.

Conclusion

The attempt to answer the question asked at the beginning of this paper, 'Did transhumance and seasonal summer pasture exist in the mountain regions of the present day Czech Republic in the past?' was confronted with the fact that this subject has been largely avoided both by archaeologists and historians. The only exception proved to be Walachian pasture in the Outer Western Carpathians, which was a significant integral part of local farming and economic history. From the archaeological point of view it is fully understandable as the finds coming from mountains are scarce; they are extremely hard to find in forested mountain terrains. As indicated by pollen records 'something' specific must have happened in the mountains long before medieval colonisation, possibly a specific type of agropastoral management. Today it is considered that the profitable boundary contour of cereal production lies at 600 m.a.s.l. Historical sources, however, record fairly succesful cereal cultivation in the Šumava Mountains at the beginning of the 20th century in the altitudes around 1000 m.a.s.l. (Zeithammer 1902).

If sub-montane/montane settlement really existed, it cannot be decided whether there was any connection with the lowland settlement; for the whole period of prehistory (perhaps with the exception of the Late Iron Age in southern Bohemia), the distance between lowland settlement and mountain regions was further than the presupposed maximum distance for transhumance, e.g. 40–50km.

The reason for the lack of transhumance seems to be environmental rather than cultural. In large areas of the lowlands and uplands natural conditions were more favourable for crop than animal husbandry. Moreover, vast and relatively flat upland areas (under 600 m.a.s.l., the limit for cereal production) seem to have been sparsely populated and they may provide enough space for summer pasture and fodder for winter during prehistory/Early Medieval period. If transhumance truly existed it would be logical to expect that the livestock from the lowlands moved to surrounding hilly areas instead of remote mountains. In this case the traces of seasonal pasture in uplands would have been destroyed by later agricultural cultivation.

Of all Czech and Moravian regions only the Bílé Karpaty Mountains meet the requirements of an 'ideal' transhumance landscape: a bedrock rich in calcium carbonate; and a dense prehistoric settlement of warm foothills and lower part of mountains. But even in this specific case it is not clear whether the settlement of grasslands was permanent or human activities took place in the form of transhumance from the foothills only.

Acknowledgements

The author thanks Č. Čišecký for the graphics and J. Fröhlich, O. Hartmanová-Hájková and M. Hajnalová for valuable discussions and help.

Bibliography

Akeret, Ö. and Jacomet, S. 1997. Analysis of plant macrofossils in goat/sheep faeces from the Neolithic lake shore settlement of Horgen Schelle: an indication of prehistoric transhumance? *Vegetation History and Archaeobotany* 6:235–239.

Bartoš, M. and Nováková, Z. 1997. *Nejstarší obrazová mapa Krkonoš kronikáře Šimona Hüttela.* Trutnov.

Bentley, R.A. and Knipper, C. 2005. Transhumance at the early Neolithic settlement at Vaihingen (Germany). *Antiquity* 79, No. 306, December 2005, online project gallery.

Beranová, M. and Kubačák, A. 2010. *Dějiny zemědělství v Čechách a na Moravě.* Prague, Libri.

Botík, J. and Slavkovský, P. (eds.) 1995. *Encyklopédia ľudovej kultúry Slovenska 1*. Bratislava, Veda.

Bunce, R.G.H., Pérez-Soba, M., Jongman, R.H.G., Gómez Sal, A., Herzog, F. and Austad, I. (eds.) 2004. *Transhumance and Biodiversity in European Mountains; report of the EU-FP5 project Transhumount (EVK2-CT-2002–80017).* IALE publication series no. 1. Wageningen, Alterra.

Chytrý, M., Hoffmann, A. and Novák, J. 2007. Suché trávníky (Dry grasslands). In M. Chytrý (ed.), Vegetace České republiky 1. Travinná a keříčková vegetace (*Vegetation of the Czech Republic. 1: grassland and heathland vegetation*). Prague, Academia, pp. 371–470.

Court-Picon, M., Buttler, A. and de Beaulieu, J.-L. 2006. Modern pollen/vegetation/land-use relationships in mountain environ-ments: an example from the Champsaur valley (French Alps). *Vegetational History and Archaeobotany* 15:151–168. DOI 10.1007/s00334-005-0008-8

Čuláková, K., Eigner, J., Fröhlich, J., Metlička, M. and Řezáč, M. 2012. Horské laténské sídliště na Šumavě: Prášily, Sklářské údolí, okr. Klatovy. *Archeologické výzkumy v jižních Čechách* 25:97–115.

Dreslerová, D. and Hrubý, P. 2004. Halštatské výšinné lokality v jižních Čechách: nové výzkumy dvou hradišť. (Hallstattzeitliche Höhensiedlungen in Südtschechien: neue Grabungen auf zwei Burgwällen). *Študijné zvesti AÚ SAV Nitra* 36:105–129.

Dreslerová, D. and Kočár, P. 2013. Trends in cereal cultivation in the Czech Republic from the Neolithic to the Migration Period (5500 b.c.–a.d. 580). *Vegetation History and Archaeobotany* 22:257–268.

Festi, D. and Oeggl, K. T*ranshumance and Alpine Summer Farming as Potential Subsistence Strategy in the Alps during Prehistory.* Poster published in http://www.uibk.ac.at/himat/sfb-himat/pps/pp11/transhumance/05-p-v-7_festi---oeggl.jpg

Fröhlich, J. 2009. Pravěké osídlení v horských a dalších vysokých polohách v jižních Čechách. (Prehistoric settlement in the mountainous and higher-altitude locations of southern Bohemia). *Časopis Společnosti přátel starožitností* 117/3: 150–156.

Gedes, D.S. 1983: Neolithic transhumance in the Mediterranean Pyrenees. *World Archaeology* 15:52–66.

Greenfield, H.J. and Arnold, E. 2005. Production strategies and transhumance: the zooarchaeological remains from Early Iron Age hill-top fortress at Klisura-Kadića Brdo, eastern Bosnia: taphonomic analysis. *Godišnjak, Centar za balkanološka ispitivanja* (Annual of the Centre for Balkan Studies, Academy of Sciences and Arts, Sarajevo) 334/32:107–150.

Hájková, P., Roleček, J., Hájek, M., Horsák, M., Fajmon, K., Polák, M. and Jamrichová, E. 2011. Prehistoric origin of the extremely species-rich semi-dry grasslands in the Bílé Karpaty Mountains (Czech Republic and Slovakia). *Preslia* 83:185–204.

Hartmanová, O. 2004. Budní hospodářství v Krkonoších z pohledu archeologie (Die Baudenwirtschaft im Riesengebirge aus archäologischer Sicht). *Památky Archeologické* 96:165–204.

Hejcman, M., Hejcmanová, P., Pavlů, V. and Beneš, J. 2013. Origin and history of grasslands in central Europe: a review. *Grass and Forage Science*. DOI:10.1111/gfs.12066.

Herzog, F. and Bunce, R.G.H. 2004. Conclusions from the Policy Workshop. In R.H.G. Bunce et al. (eds.) *Transhumance and Biodiversity in European Mountains: report of the EU-FP5 project Transhumount (EVK2-CT-2002–80017)*. IALE Publication Series, Wageningen, Alterra, pp. 303–306.

Hoser, J.K.E. 1804. *Das Riesengebirge in einer statistisch-topografischen und pittoresken Uebersicht*. Wien.

Jankovská, V. 2004. Krkonoše v době poledové: vegetace a krajina (The Giant Mountains in the Postglacial: vegetation and landscape). In J. Štursa, K.R. Mazurski, A. Palucki and J. Potocka (eds.) Geoekologické problémy Krkonoš. Sborn. Mez. Věd. Konf., Listopad 2003, Szklarska Poręba. *Opera Corcontica* 41:111–123.

Jankovská, V. 2006. Late Glacial and Holocene history of Plešné Lake and its surrounding landscape based on pollen and palaeo-algological analyses. *Biologia* 61/Supplement 20:371–385.

Jankovská, V., Kuneš, P. and van der Knaap, W.O. 2007. Fláje/Kiefern, (Krušné Hory Mountains): late Glacial and Holocene vegetation development. *Grana* 46:214–216.

Jeník, J. 1961. *Alpinská vegetace Krkonoš, Králického Sněžníku a Hrubého Jeseníku*. Prague, NČSAV.

Jongepierová, I. (ed.) 2008. *Louky Bílých Karpat (Grasslands of the White Carpathian Mountains)*. Veselí nad Moravou, ZO ČSOP Bílé Karpaty.

Kenzler, H. 2009. The medieval settlement of the Ore Mountains: the development of the settlement structure. In J. Klápště and P. Sommer (eds.) *Ruralia VII: medieval rural settlement in marginal landscapes*. Seventh Ruralia conference, 8th–14th September 2007, Cardiff, Wales, UK. Turnhout, Brepols, pp. 379–392.

Kienlin, T. and Valde-Nowak, P. 2004. Neolithic transhumance in the Black Forest Mountains, SW Germany. *Journal of Field Archaeology* 29:29–44.

Klostermann, K. 2012. *Vzpomínky na Šumavu II: sbírka rozptýlených pamětí*. Strakonice, Nakladatelství Hrad Strakonice.

Kočár, P. and Dreslerová, D. 2010. Archeobotanické nálezy pěstovaných rostlin v pravěku České republiky; archeobotanical finds of cultivated plants in the prehistory of the Czech Republic. *Památky archeologické* 101:203–242.

Koutecký, D. and Bouzek, J. 2009. Horská sídliště v Krušných horách. *Archeologie ve středních Čechách* 13:213–282.

Kuneš, P., Abraham, V., Kovařík, O. and PALYCZ contributors, 2009. Czech Quaternary Palynological Database (PALYCZ): review and basis statistics of the data. *Preslia* 81:209–238.

Kunz, L. 2005. *Rolnický chov ovcí a koz. Svazek 2: Rožnov pod Radhoštěm*. Valašské muzeum v přírodě v Rožnově pod Radhoštěm.

Kozáková, R., Pokorný, P., Peša, V., Danielisová, A., Čuláková, K. and Svitavská-Svobodová, H. 2015. Prehistoric human impact in the mountains of Bohemia. Do pollen and archaeological data support the traditional scenario of a prehistoric 'wilderness'? *Review of Palaeobotany and Palynology* 2015:29–43.

Lokvenc, T. 1978. *Toulky krkonošskou minulostí*. Hradec Králové.

Mandl, F. 1996. Dachstein: vier Jahrtausende Almen im Hoch-gebirge. Bd. Das östliche Dachsteinplateau. *Mitteilungen der ANISA* 17, H. 2/3. Gröbming.

Mlekuž, D. 2003. Early herders of the eastern Adriatic. *Documenta Praehistorica* 30 (Neolithic Studies 10):139–51.

Müller-Scheesel, N., Hofmann, R., Muller, J. and Rassmann, K. 2010. The socio-political development of the Late Neolithic settlement of Okoliste/Bosnia-Hercegowina: devolution by transhumance? In Kiel Graduate School 'Human Development in Landscapes' (eds.) *Landscapes and Human Development: the contribution of European archaeology*. Proceedings of the International Workshop 'Socio-environmental dynamics over the last 12,000 Years: the creation of landscapes', 1st–4th April 2009, Bonn, pp. 181–191.

Mróz, W. and Olszańska, A., 2004. Poland: traditional pastoralism and biodiversity in the western and eastern Carpathians. In R.G.H. Bunce et al. (eds.) *Transhumance and Biodiversity in European Mountains. Report from the EU-FP5 project Transhumount (EVK2-CT-2002–80017)*. IALE publication series no. 1. Wageningen, Alterra, pp. 171–182.

Pelisiak, A. 2013. Man and mountains: settlement and economy of Neolithic communities in the eastern Polish Carpathians. In S. Kadrow and P. Włodarczak (eds.) *Environment and Subsistence: forty years after Janusz Kruk's 'Settlement studies'*. Studien zur Archäologie in Ostmitteleuropa 11, Rzeszów: Institute of Archaeology Rzeszów University. Pp. 225–244.

Podolák, J., 1982. Tradičné ovčiarstvo na Slovensku *(Traditionelle Schafzucht in der Slowakei)*. Bratislava, Vydavateľstvo SAV.

Pokorný, P., 2004. The effect of local human-impact histories on the development of Holocene vegetation: case studies from central Bohemia. In M. Gojda (ed.) *Ancient Landscape, Settlement Dynamics and Non-Destructive Archaeology*. Prague, Academia, pp. 171–185.

Röpke, A., Stobbe, A., Oeggl, K., Kalis, A.J. and Tinner, W. 2011. Late-Holocene land-use history and environmental changes at the high altitudes of St. Antönien (Switzerland, northern Alps): combined evidence from pollen, soil and tree-ring analyses. *Holocene* 21:485–498.

Rösch, M., 2009. Botanical evidence for prehistoric and medieval land use in the Black Forest. In J. Klápště and P. Sommer (eds.) *Ruralia VII: medieval rural settlement in marginal landscapes.* Seventh Ruralia conference, 8th–14th September 2007, Cardiff, Wales, UK. Turnhout: Brepols, pp. 379–392.

Rybníček, K. and Rybníčková, E., 2004. Pollen analyses of sediments from the summit of the Praděd range in the Hrubý Jeseník Mountains (Eastern Sudetes). *Preslia* 76:331–347.

Rybníček, K. and Rybníčková, E., 2008. Upper Holocene dry land vegetation in the Moravian–Slovakian borderland (Czech and Slovak Republics). *Vegetation History and Archaeobotany* 17:701–711. DOI 10.1007/s00334-008-0160-z

Speranza, A., Hanke, J., van Geel, B. and Fanta, J. 2000. Late-Holocene human impact and peat development in the Černá Hora bog, Krkonoše Mountains, Czech Republic. *The Holocene* 10/5:575–585.

Spindler, K. 2003. Transhumanz. *Preistoria Alpina* 39:219–225.

Stebich, M. and Litt, T. 1997. Das Georgenfelder Hochmoor: ein Archiv für Vegetations-, Siedlungs- und Bergbaugeschichte. *Leipziger Geowissenchaften* 5:209–216.

Štika, J. 2001. Moving with the flock. *Central European Review* 3/14:23. April 2001. http://www.ce-review.org/01/14/stika14.html

Svobodová, H. 2002. Preliminary results of the vegetation history in the Giant Mountains (Úpská rašelina mire and Černohorská rašelina bog). *Opera Corcontica* 39:5–15.

Svobodová, H. 2004. Migrace klimaxových dřevin na Šumavu v holocénu. *Bulletin Slovenskej botanickej spoločnosti,* Supplement 11, Bratislava, pp. 207–216.

Svobodová, H., Reille, M. and Goeury, C. 2001. Past vegetation dynamics of Vltavský luh: upper Vltava river valley in the Šumava mountains, Czech Republic. *Vegetation History and Archaeobotany* 10/4:185–199.

Valde-Nowak, P. 2008. Neolithic in the European mid-mountains: case study from the Polish Carpathians. In S. Grimaldi, T. Perrin, and J. Guilaine, (eds.) *Mountain Environments in Prehistoric Europe: settlement and mobility strategies from the Palaeolithic to the Early Bronze Age.* Oxford, Archaeopress, pp. 131–135.

Weishäupl, B. 2014. Anthropogene Strukturen in den nördlichen Stubaier Alpen: Bericht über die Prospektionen von 2008 bis 2011. *Forschungsberichte der ANISA für das Internet.* 10 (ANISA FB I. 10, 2014). Pp. 1–58. http://www.anisa.at/Stubaier_Alpen_B_Weishaeupl_2014.pdf

Vencl, S. (ed.) 2006. *Nejstarší osídlení jižních Čech: paleolit a mesolit.* Prague, Archeologický ústav AVČR.

Zeithammer, L.M. 1902. *Šumava, kraj a lid.* Nákladem vlastním. Tiskem J. Přibyla v Český Budějovicích.

Dagmar Dreslerová, Institute of Archaeology of the CAS, Prague, Czech Republic, Letenská 4, 11801 Praha 1, Czech Republic.
Email: dreslerova@arup.cas.cz

4. Hard cheese: Upland pastoralism in the Italian Bronze and Iron Ages

Mark Pearce

By moving livestock to summer farms, fodder at the home base is saved but the milk and other animal products produced during the animals' absence are no longer immediately available to those left at the home base. In this paper I shall explore the economic implications of the use of summer farms, in particular the effect on carrying capacity, on the number of livestock which can be over-wintered, and on the use of the milk produced while the animals are at the summer grazing lands. I then explore archaeological evidence from the Bronze and Iron Ages of the Italian uplands (Apennines and Alps). I argue that the production of hard cheese, which converts milk into an easily conservable and transportable commodity, is key to the expansion of summer farms in the Bronze Age of Italy. Cheese production is an essential part of models for the pastoral use of Mediterranean uplands in prehistory but it is commonly held that in the Alps the production of hard cheese only begins in the Middle Ages. I examine the literary and archaeological evidence for the prehistoric production of hard cheese and argue that its production in prehistory is the most parsimonious explanation for the summer use of high mountain pastures and thus for the origins of the Alpwirtschaft *economy in the southern Alps.*

"I'm going a milking, kind sir", she answered me,
"For roving in the dew makes the milk maids fair".
Traditional Sussex Folk Song

Why do people move to summer farms?

It may be useful to begin this discussion by reflecting on the purpose of seasonal movements of stock in the Alps and other high mountain environments. The first point to make is that grass does not grow in the winter months. This is well known, but often forgotten. Furthermore, any available vegetation may be covered by deep snow during the winter (particularly at higher altitudes) and so livestock, whether sheep, goats or cattle, will have to be stabled indoors and fed on fodder collected during the growing season. Indeed, we know that this was precisely the practice at the Bronze Age lake village of Fiavé (TN), which is situated at 648m above sea level (Karg 1998).

If livestock are moved away from the home farm, then available fodder is spared and may be harvested and stored for the winter – thus, if livestock are away for one third of the year, then one third of the available fodder is saved. Because animals have legs and can walk to grazing, while hay must be cut and then transported to be stored, it is clearly more efficient to grow fodder as close as possible to the farm or outlying barn, moving the animals away to graze elsewhere.

In simple terms, the more land that can be exploited as pasture for grazing or as meadows for hay production, the more animals that can be kept during the summer months. But the more animals that are kept, the more critical the availability of winter fodder becomes (at Early / Middle Bronze Age Fiavé herb and grass fodder was supplemented in early spring with leafless and foliating twigs of hazel, birch, alder, beech and other trees – Karg 1998). The use of upland summer pasture away from the farm is an efficient way of raising carrying capacity, as well as exploiting the excellent grazing of high Alpine pastures, because it spares fodder growing round the farm for use in the winter months.

Another issue that needs to be considered is that sheep, goats and cattle all produce milk in the conventional Alpine economy; as we shall see, Alpine dairy production is documented from at least the Bronze Age. However, this secondary product, unlike wool, *cannot* be stored for long periods, and in the absence of refrigeration, milk deteriorates rapidly. There are two answers to this problem: either the milk may be used to feed young animals, kids, lambs and calves, rather than humans, or alternatively the milk may be transformed into a product than can be stored and transported more easily, such as cheese. Writing in the 1st century AD, Columella (*De Re Rustica* 7, 8, 1) advises precisely this solution: "Casei quoque faciendi non erit omittenda cura, utique longinquis regionibus, ubi mulctram devehere non expedit" ("It will be necessary too not to forget the task of cheese-making, especially in distant parts of the country, where it is not convenient to take milk to the market in pails"; translation: Forster and Heffner 1954:285). It is perhaps superfluous to note that hard cheese is much easier to store and to transport than soft cheese. However, the conventional view, typified by Margarita Primas (1999:3), is that in the Alps the production of hard cheese only begins in the

Keywords: *Alpwirtschaft*, pastoralism, milk, cheese, Bronze Age

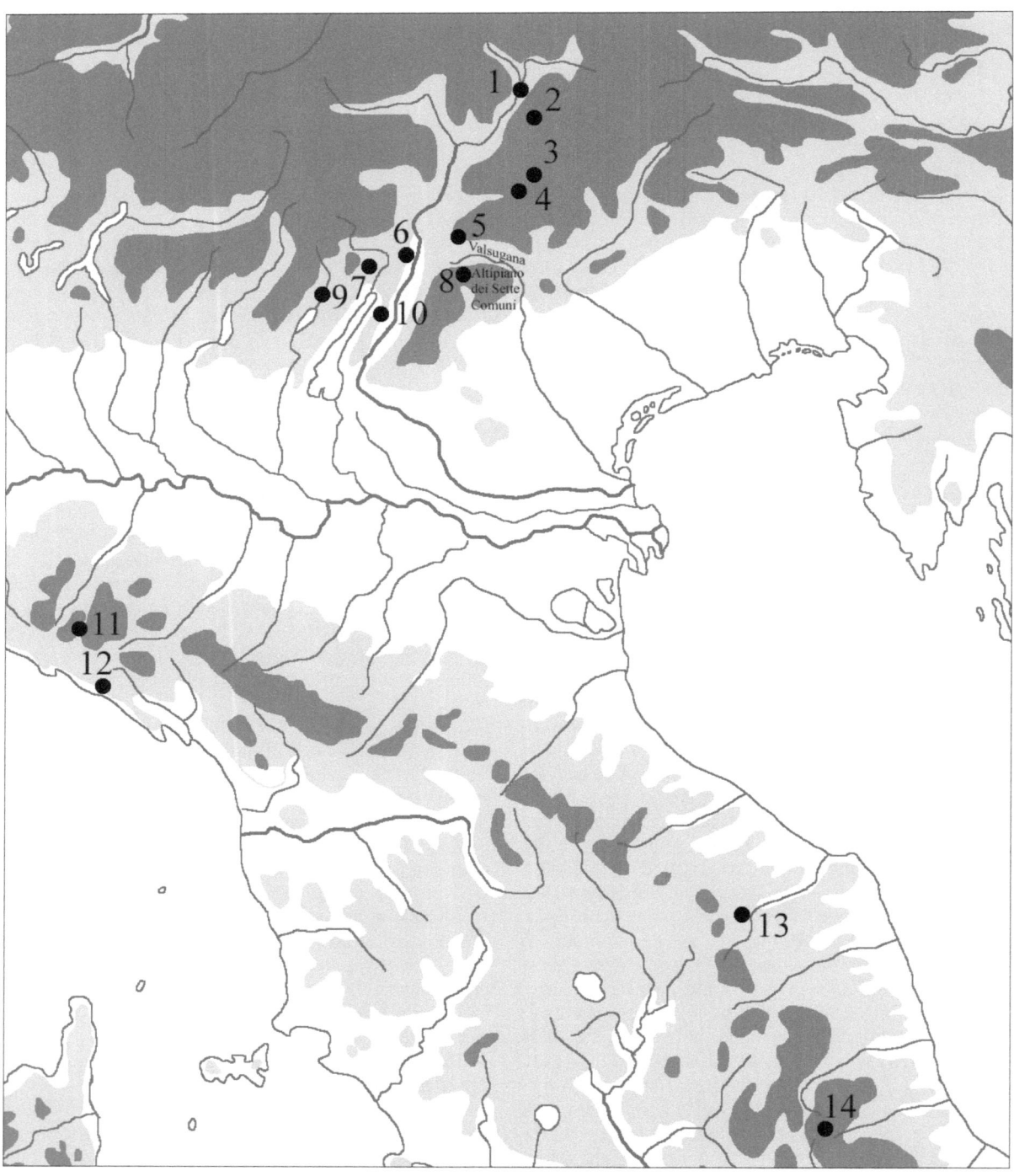

Figure 4.1 Places mentioned in the text. 1: Albanbühel (Bressanone / Brixen BZ); 2: Castelir di Bellamonte (Predazzo TN); 3: Sotćiastel (Badia / Abtei BZ); 4: Selva di Val Gardena / Wolkenstein in Gröden BZ) and Lech Sant (Santa Cristina Valgardena / Sankt Christina in Gröden BZ); 5: Acqua Fredda (Bedollo TN); 6: Dos Grum di Cadine (Trento TN); 7: Fiavé (TN); 8: Malga Principi (Luserna TN); 9: Dosso Rotondo and Malga Vacil (Storo TN); 10: Mandrom de Camp (Brentonico TN); 11: U Puzzu di Ertola (Rezzoaglio GE); 12: Monte Loreto (Castiglione Chiavarese GE); 13: Sentino Gorge (AN); 14: Campo Pericoli (Pietracamela TE) and Grotta a Male (L'Aquila AQ).

Middle Ages and that the traditional Alpine economy was not practised in prehistoric times.

Another reason for moving livestock to mountain pastures is the quality of the grazing – indeed some of the most prized matured cheeses made in Italy today are produced in high Alpine pastures, such as Vézzena or Bagós (Viviani 1993). It should also be borne in mind that the lowlands are hot in summer: it is generally considered to be too hot in the Po plain to make good cheese in the summer and so, for example, connoisseurs avoid eating Gorgonzola cheese made in the summer months.

Prehistoric production of hard cheese?

Having established these basic principles, let us now consider whether, despite the conventional view, it may be shown that the Alpine economy and the production of hard cheese can be dated to the Bronze Age in the southern Alps (Fig. 4.1).

I shall start by examining the Italian peninsula. In 1959 Salvatore Puglisi posited a transhumant pastoralist economy for the Apennine Bronze Age. His argument was that the Italian peninsula is ideal for such an economy as the high mountain pastures of the Apennine chain can be reached easily from the coastal lowlands via well-watered river valleys (Puglisi 1959:18). He also noted faunal samples with high percentages of sheep and goats (Puglisi 1959:31–33), and posited an ethnographic analogy between the material culture of the shepherds of the Abruzzo in the 1950s and the so-called milk boilers that characterise the Bronze Age inventory of peninsular Italy (Figs 4.2, 4.3 and 4.4; Puglisi 1959:33–37). Finally, he adduced the presence of other artefacts that could be interpreted by analogy as having been used for cheese-making, such as whisks, skimmers, perforated strainers and pot-stands (Figs. 4.5, 4.6; Puglisi 1959:38–41). Indeed, many Apennine Bronze Age sites seem best adapted to a pastoral-

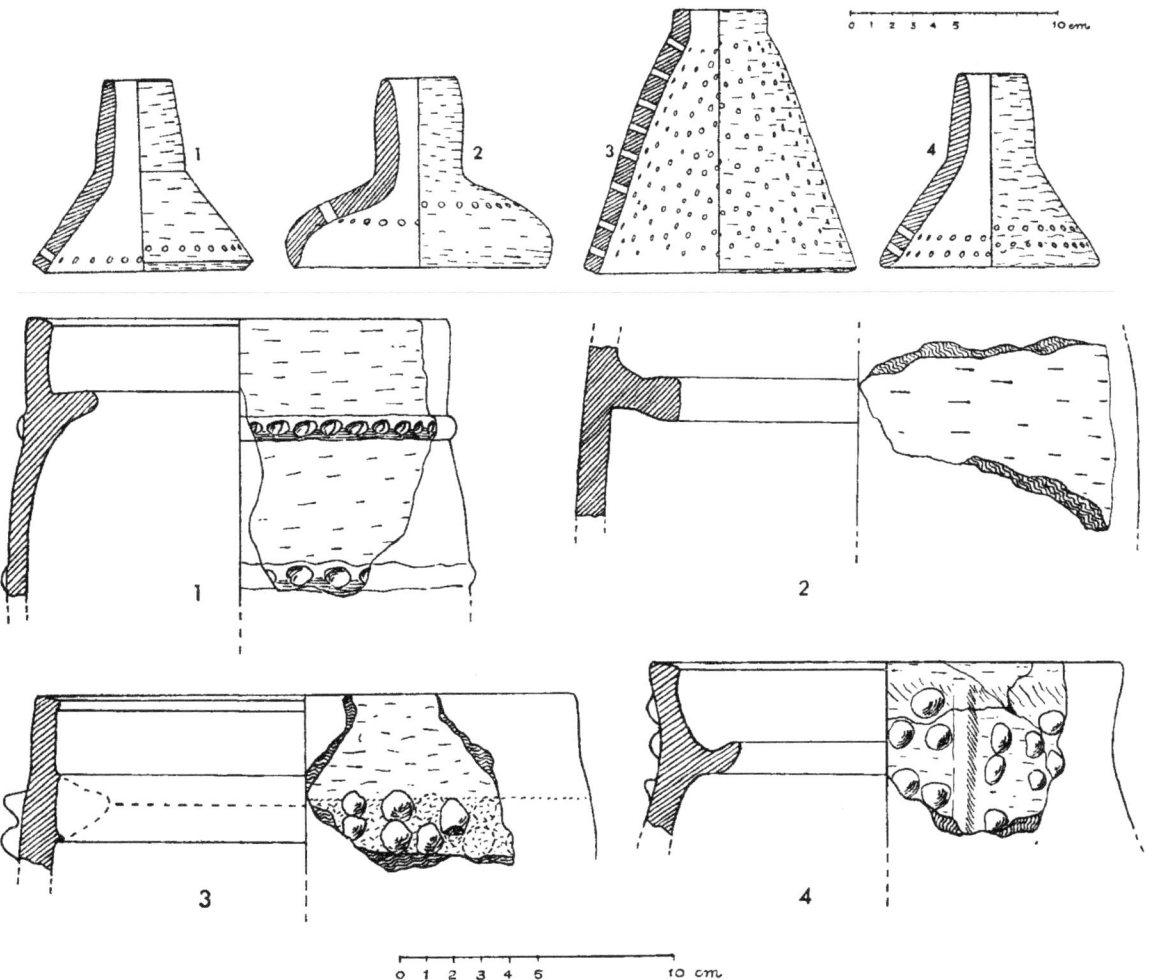

Figure 4.2 Apennine Bronze Age milk boilers. Pots with internal ledge: 1: Frasassi cave (Genga AN); 2: San Fortunato (Genga AN); 3: Fabriano (AN); 4: Conelle di Arcevia (AN). Lids: 1, 2 and 4: Belverde di Cetona (SI); 3: Casa Carletti (Cetona SI) (after Puglisi 1959, figs 4 and 7).

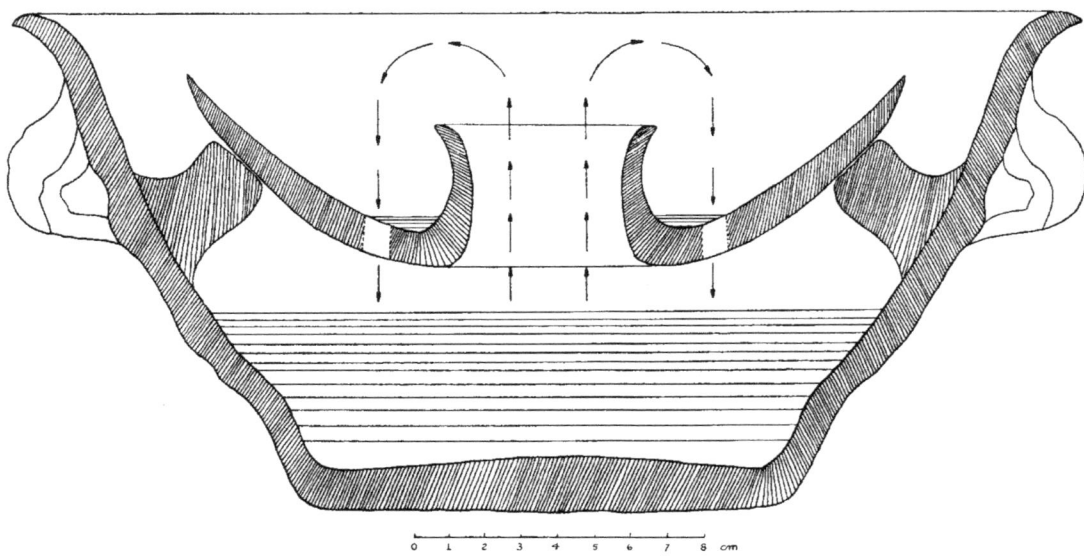

Figure 4.3 Puglisi's reconstruction of how an Apennine milk boiler may have functioned (after Puglisi 1959, fig. 11).

ist economy, such as those in the Sentino Gorge, and some sites would have been under snow for much of the winter, making them most likely to be transhumant pastoralists' summer camps, like Campo Pericoli (Pietracamela TE; *c.* 2000 m.a.s.l.) and the Grotta a Male (L'Aquila AQ; *c.* 950 m.a.s.l.) in the Gran Sasso massif (Trump 1966:110; Barker 1981:156).

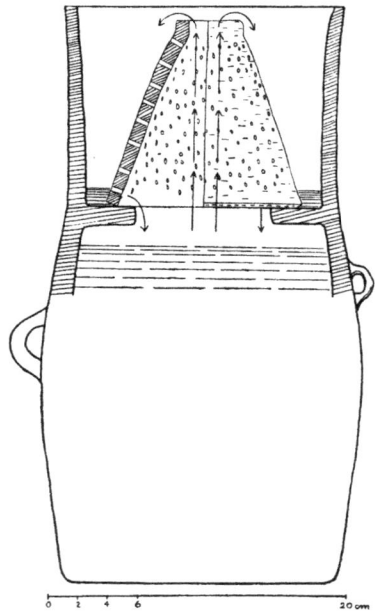

Figure 4.4 Puglisi's reconstruction of how an Apennine milk boiler may have functioned so as to avoid losing milk that boiled over (after Puglisi 1959, fig. 8).

Although it is now clear that the Apennine Bronze Age encompassed a wide range of economic behaviour, including sedentary lowland agriculture (Barker 1981:153–158), it is highly likely that cheese was produced in the Middle and Late Bronze Age Apennines. Now this does not of course mean that this cheese was necessarily matured and hardened, though in the absence of refrigeration hard cheese would be much easier to preserve for long periods and to transport. On the other hand, fresh, soft cheese would have to be regularly transported to consumers, necessitating regular journeys to the hot lowlands.

Secondly, let us consider the literary evidence. Here our sources seem to be very explicit. Homer describes the use of hard cheese in Book XI of the *Iliad*: the Thessalian hero Machaon is wounded, and Nestor takes him to his hut, where they are served a restorative cocktail of wine in Nestor's famous cup; goat's cheese was grated over this wine (like a dusting of parmesan on a pasta dish) and then barley was sprinkled (*Iliad* XI, 638–640). Although this drink seems strange to modern palates, it is also attested in the *Odyssey*, where the sorceress Circe serves a similar cocktail, sweetened with honey, to Odysseus' sailors (*Odyssey* X, 229–243).

So, if cheese was grated in the prehistoric Mediterranean, what is the evidence for cheese graters? Cheese graters are known from 9th century BC warrior's tombs at Lefkandi on Euboea and from rich 7th century BC tombs in western peninsular Italy (Ridgway 1997), and their forms are strikingly similar to that of a modern cheese-grater (Ridgway 1997: figs 2, 3, 5). It has been convincingly argued that the presence of these artefacts in warrior and princely graves indicates the practice of 'drinking from Nestor's Cup' in imitation

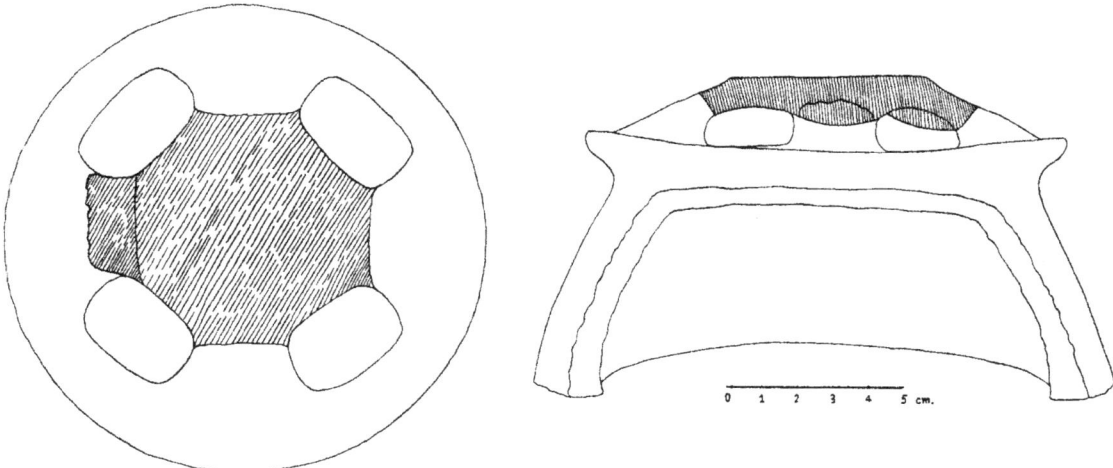

Figure 4.5   Pot stand, Grotta di Pertosa (SA) (after Puglisi 1959, fig. 15).

Figure 4.6   Whisk, Grotta di Pertosa (SA) (after Puglisi 1959, fig. 13).

of the Homeric heroes (Ridgway 1997). Cheese graters are also attested in later domestic contexts and classical literature. It is perhaps superfluous for me to point out that cheese must be relatively hard in order to be suitable for grating.

The *Iliad* and *Odyssey* were most probably written down in the 8th century BC, in Ionic Greek with 'some Euboean veneer' (Willcock 2012), which makes the cheese-graters found at Lefkandi on Euboea all the more significant. The Homeric epics may reflect practices of the later Bronze Age: certainly West (1998:190) argues that "… the grated cheese … belongs to a traditional account that is many generations older than our *Iliad*". Cheese is certainly mentioned in two Linear B tablets from Pylos, Un718 and Un1185, though it is not clear whether it was hard or soft (Ventris and Chadwick 1956:282–283; Bendall 2007:221).

If hard cheese was available in the Early Iron Age and perhaps the later Bronze Age Mediterranean, then why not in the Bronze Age Alps?

Archaeological evidence for high upland pastoral settlements

Let us now consider the evidence from the Bronze Age southern Alps. Firstly, there seems to be no doubt that high altitude Alpine grazing was used from at least the Early Bronze Age. There is evidence for seasonal pastoralism both in the upland areas of the Veneto (Migliavacca 1985:40–48; cf. De Guio 1994:166–168) and in the Trentino to the north. Here the Middle Bronze Age site of Dosso Rotondo (Storo TN) was established in a newly-deforested area at 1876 m.a.s.l. (Bassetti *et al.* 2008; Mottes and Nicolis 2004). The excavators have suggested that the large quantities of charcoal found at the site may be the result of boiling milk for cheese-making (Bassetti *et al.* 2008:123). Nearby, at 1810m above sea level, there was a contemporary insubstantial site at Malga Vacil (Storo TN), which is unlikely to have been occupied all year round (Marzatico 2001:379, note 61 on p. 411; 2007:169–173, figs 2–10). A flint sickle blade found at Malga Vacil has been interpreted as documenting the cutting of grass fodder (Marzatico 2007:169, 173, fig.7). It is likely that seasonal pastoralism was also practised at around 1700m in the Early and Middle Bronze Age at Mandrom de Camp (Brentonico TN; Mottes *et al.* 1999:89), and a similar interpretation has been suggested for the Castelir di Bellamonte (Predazzo TN), which is situated at 1548m above sea level in the Val Travignolo (Leonardi and Leonardi 1991:100).

A number of high altitude sites in the Alto Adige / Südtirol, deeper into the Alpine chain, are also best explained as being for seasonal grazing; these include a number of sites around the Sella pass at 2000–2250m above sea level (Selva di Val Gardena / Wolkenstein in Gröden BZ – Bagolini and Tecchiati 1993:50–51) and Lech Sant (Santa Cristina Valgardena / Sankt Christina in Gröden BZ) at 2096m above sea level; the material from this site includes a bronze sickle fragment, possibly for cutting fodder (*ibidem*). Finally, Umberto Tecchiati has argued that there is indirect evidence for cheese production at the Middle Bronze Age fortified settlement of Sotćiastel (Badia / Abtei BZ), a permanent site situated at *c.* 1400m above sea level. This includes evidence for the slaughter of high numbers of new-born or foetal cattle (17 of 33 individual cattle for which age at death could be determined, some 51%), fitting well Payne's (1973) classic kill-off pattern for milk production. This culling of calves was perhaps aimed at obtaining rennet for cheese-making (Tecchiati 1998:384–5; Riedel and Tecchiati 1998a:293–294), though it may also suggest slaughter motivated by a lack of winter fodder.

Sheep/goat dominate the Bronze Age faunal assemblages of both the Trentino and the Alto Adige / Südtirol: for example at Sotćiastel they are 45.8% MNI (=minimum number of individuals), compared to cattle 40.3% and pigs 5.5% (Riedel and Tecchiati 1998a:288, tab.1), while at Albanbühel (Bressanone / Brixen BZ) the MNI ratio is sheep/goat 60.71%, cattle 24.55% and pig 7.14% (Riedel and Tecchiati 1998b:325). There are a few exceptions, such as the relatively low site of Dos Grum di Cadine (Trento TN), at 647 m.a.s.l., where cattle are 56% of bone remains, sheep 30% and pigs 13% (Riedel and Tecchiati 2001:107–108, 112). At Fiavé sheep/goat were 51.9% of the bone remains, cattle 28.5% and pig 6.5% (Jarman 1976:543, table 20). More than 40% of cattle were slaughtered at Fiavé in the first two years, probably to promote milk production and almost all cattle were slaughtered before they reached 5 years old: it may be assumed that this was when they ceased to be productive milkers (Jarman 1976:543, table 10; Gamble and Clark 1987:427). There was a similar slaughter pattern for sheep/goat, with more than 50% culled in their first two years (Jarman 1976:544, table 11).

Simple calculations of the relative numbers of livestock present do not of course tell us their relative contribution to the economy of a site. For example, Barker (1983, table 5) suggests that prehistoric sheep/goat are likely to have each provided just 60lbs of meat each (some 27kg), while cattle provided some 500lbs (around 227kg) of meat each, which is some 8.3 times more. Clearly, animals slaughtered at a young age, such as the calves at Sotćiastel (as we have seen, around 51% of cattle) or Albanbühel (where almost half the cattle were killed under 4 months of age – Riedel and Tecchiati 1998b:325), will have made a much less significant meat contribution to diet.

Assessing milk yields of prehistoric cattle, sheep and goats is very difficult as these have changed dramatically through time as a result of selective breeding, and they vary according to animal diet, stage of lactation and also from individual to individual. Cattle milk yields are higher than those of goats, which are higher than those of sheep, not least because goats have a longer lactation (up to 300 days of milking, compared with sheep, which have up to 250 days of milking; Boyazoglu and Morand-Fehr 2001:6). Moreover the different species' milk has different properties, with sheep's milk the most suitable for cheese-making and goat's milk being the most easily digestible (Boyazoglu and Morand-Fehr 2001:6; Park *et al.* 2007:89, 92). This means that although sheep/goat were numerically more common in the Bronze Age Trentino and Alto Adige / Südtirol, they may not necessarily have made the most significant contribution to diet.

Salt

Salt is required for cheese-making as a preservative and a flavour-enhancer, and although Mediterranean sea salt would probably have been relatively easily available to pastoralists in the southern Alps it is useful to reflect that rock salt was mined at Hallstatt in the Upper Austrian Salzkammergut from the 15th century BC onwards (Reschreiter and Kowarik 2009). It is worth noting that in the 1st century BC, Varro (*Res Rusticae* 2, 11, 5) writes that rock salt is preferable to sea salt for cheese-making: "Qui aspargi solent sales, melior fossilis quam marinus" ("Those who sprinkle salt prefer mineral salt to sea salt"; translation: Hooper 1967:415). Moreover, salt is not just an important part of the cheese-making process: I have observed rock-salt licks used as a method of controlling stock left to graze unattended in the Apennines around U Puzzu di Ertola (Rezzoaglio GE; 23 May 2002); since the animals will return to the salt at regular intervals they can be easily recaptured if need be. Of course, such mechanisms for stock control mean that fencing is unnecessary and will be almost impossible to detect archaeologically. Finally, salt solution is used in the preparation of rennet extract (O'Connor 1993:6).

The origins of hard cheese production

None of these classes of evidence securely date the production of hard cheese in the southern Alps to the Bronze Age, but I have established 1) that there *was* high altitude pastoralism and 2) that hard cheese was known in the Italian peninsula in prehistory. It seems perverse to argue that despite the availability of the Mediterranean technology of hard cheese production, it was not adopted in the southern Alps to transform the milk produced on the Alpine pastures into a form that was suitable for storage and transportation. The simplest, the most parsimonious, explanation for the presence of high altitude sites in the Bronze Age Trentino and Alto Adige / Südtirol is that they were used in an Alpine economy based on dairy production, the production of hard cheese.

The pastoralism–metallurgy nexus

Let us now explore the importance of pastoralism in the Bronze Age southern Alps.

Over one hundred copper smelting sites have been identified in the Val Sugana area, in the southern Trentino (Preuschen 1973; Šebesta 1992; Perini 1992; Marzatico 1997; Pearce 2007:77–81). The smelting sites are located at more than 1000m above sea level, and are close to water, either ponds or streams. They seem to document metal production on an enormous scale, and Bayesian modelling of radiocarbon dates from the Acqua Fredda battery of furnaces at the Redebus pass (Bedollo TN) indicates that they were in use between the *second half of the 13th to the 9th century cal BC* (Marzatico *et al.* 2010:131–135).

Most of the smelting sites documented by slag heaps are in proximity to mineral outcrops, but there is an important concentration of smelting sites on the Mesozoic limestone Lavarone–Vézzena–Luserna plateau and, to a much lesser extent, on the northwestern margins of the Altipiano dei Sette Comuni nearby. These areas are situated at some distance from major copper resources (Preuschen 1973:144). In some sectors of the Lavarone–Vézzena–Luserna plateau there seems to have been much smelting activity, and there are many slag heaps and concentrations of batteries of furnaces, so much so that Preuschen (1973:134) calls the smelting evidence at Malga Principi in the Val Morta (Luserna TN) "la più importante installazione fusoria mai vista" ("the most important smelting facility ever seen"). It is not clear how long the sites remained in use or how many were in use at the same time, but we must imagine the presence of a substantial labour force on the plateau.

A number of hypotheses have been suggested to explain this concentration of smelting sites, located at some distance from copper resources. The floor of the Val Sugana, where the nearest major source of copper is situated (Preuschen 1973:126), is at about 500m above sea level while the Lavarone–Vézzena–Luserna plateau towers over it at 1200m above sea level; the plateau is reached by steep paths up the south side of the Val Sugana. It is likely that the ore was dressed at the mines to eliminate as much gangue and country rock as possible, as at the prehistoric copper mine of Monte Loreto (Castiglione Chiavarese GE – Maggi and Pearce 2005), but the transportation of the ore would be very difficult, especially in the absence of modern metalled roads.

Ernst Preuschen (1973:144) argued that the ore was transported to the Lavarone–Vézzena–Luserna plateau because of the need to seek alternative supplies of timber, arguing that fire-setting, the fabrication of props and scaffolding, ore roasting and smelting had caused a shortage of wood in the areas around the mines. We now know much more about traditional methods of woodland management (Moreno 1990) and it is no longer thought that prehistoric metallurgy necessarily led to deforestation (e.g. Marshall *et al.* 1999). Certainly roasting and smelting are complex operations which required large amounts of wood or charcoal and a substantial labour force (Marzatico 1997:575–576; Šebesta 1992:9–10; cf. Zschocke and Preuschen 1932:66–67).

Armando De Guio and I proposed an alternative, multi-factorial model to explain this transportation of copper for smelting to the Lavarone–Vézzena–Luserna plateau, high above the copper mines (Pearce and De Guio 1999). We agreed with Preuschen (1973:144) that the area provides a good supply of timber for fuel, but our argument is premised on the fact that the Lavarone–Vézzena–Luserna plateau and the nearby Altipiano dei Sette Comuni are ideal areas for summer grazing. Indeed De Guio (1994:166–168) argued that the Altipiano dei Sette Comuni (situated to the southwest of the Lavarone–Vézzena–Luserna plateau) was used for summer grazing by transhumant herders during the Middle and Recent Bronze Age just as it was in the Middle Ages and in fact its Middle and Recent Bronze Age material culture shows strong links to the Po plain below. We suggested that cheese produced on the upland plateaux during the summer months would have provided a major source of protein for the workforce engaged in smelting the copper ore – and Šebesta (1992:9–10) suggests that up to 200 people (woodcutters, charcoal-burners, smelters etc.) may have been involved in the whole process at a typical smelting facility.

Our model (Pearce and De Guio 1999), however, has not been accepted by all workers (Cierny and Marzatico 2002:264–265; Cierny *et al.* 2004:147–148; Marzatico 2007:174). In part this is because it takes a very modernist 'formalist' (Dilley 1996; Durrenberger 1996) approach, positing that the Bronze Age Po plain constituted a 'market' for Alpine copper production and that cheese was exchanged as part of these transactions (Cierny and Marzatico 2002:284; Marzatico 2007:174–175). However, surprisingly, our argument that hard cheese was produced in the later Bronze Age Alps and that this production mirrored the traditional Alpine economy of the historical period has also been contested. It is therefore appropriate to explore this further.

Cheese and smelting

Giuseppe Šebesta (1992:8) noted that areas of prehistoric copper smelting very frequently correspond to modern day Alpine pasture. He argued that tree-cutting for fuel and the poisoning of vegetation by the by-products of smelting created forest clearings that were then exploited by pastoralists for summer grazing, which led to the opening up of Alpine pasture. On the other hand, De Guio and I stressed that metallurgy and pastoralism seem to have a strong association in the archaeological record (Pearce and De Guio 1999; Maggi and Pearce 2010; cf. Orme 1981:269). Furthermore both areas, the Lavarone–Vézzena–Luserna plateau and the Altipiano dei Sette Comuni, are Mesozoic limestone: this means that the location of suitable grazing areas (*malghe*) is conditioned by the availability of water. Water is also extensively used in the smelting and as we have seen copper smelting sites were located near to water resources (Šebesta 1992:171). We argued that the continuing importance of these same areas for Alpine grazing and dairy production until the present day confirms their suitability for cheese production (Pearce and De Guio 1999).

We suggested that most of the milk products produced on the Lavarone–Vézzena–Luserna plateau, where the smelting sites were located, would have been consumed locally by the metallurgical and support workforce. The transhumant pastoralists from the Po plain who exploited the summer grazing on the Altipiano dei Sette Comuni, on the other hand, may have traded the cheese which they produced during the summer with the metallurgical workers to the north, as one commodity in the Bronze Age 'market economy' which we posited (they will clearly have also traded other commodities, such as the livestock they brought with them – Cierny and Marzatico 2002:265; Cierny *et al.* 2004:148).

The traditional Alpine economy: A 'storage culture'

The classic ethnography describing the traditional Alpine economy is *Balancing on an Alp* (Netting 1981). Netting (1981:34) describes the economy as based on storage and notes that "Without the techniques of preserving the products of summer for winter consumption and building up a stock of subsistence insurance against bad years and temporary climatic fluctuations, continued peasant life in the Alps would have been impossible". It is perhaps not superfluous to stress that hard cheese is an essential part of the system.

Conclusions

Much of what I have discussed is invisible to archaeologists and my argument is based on logic and surmise. However if we are to understand mountain landscapes we have to understand the economies which made their exploitation possible.

Sheep, goats and cattle all produce milk and for this resource to be useful for human communities it has to be made available for their consumption. Milk produced on high upland pasture can only be consumed in the lowlands if it is transformed into a form that can be easily stored and transported. Hard cheese is ideal for storage and transportation, and the availability of the technology (and salt) for its production suggests that it is highly likely to have been produced on the high summer pastures of the Bronze Age southern Alps.

Bibliography

Bagolini, B. and Tecchiati, U. 1993. Osservazioni sul popolamento delle valli ladine tra Neolitico ed età del Bronzo nel quadro della preistoria del bacino atestino. In *Archeologia nelle Dolomiti. Ricerche e ritrovamenti nelle valli del Sella dall'età della Pietra alla Romanità. Catalogo della Mostra*, pp. 47–55. Vigo di Fassa: Istitut Cultural Ladin 'Majon di Fashegn' / San Martin de Tor: Istitut Cultural Ladin 'Micurà de Rü'.

Barker, G. 1981. *Landscape and Society: prehistoric central Italy*. London: Academic Press.

Barker, G. 1983. Neolithic subsistence in the central Po plain. In P. Biagi, G.W.W. Barker and M. Cremaschi (eds.) *La stazione di Casatico di Marcaria (Mantova) nel quadro paleoambientale ed archeologico dell'Olocene antico della Val Padana centrale*, pp. 45–68, 116–119. Studi Archeologici 2. Bergamo: Istituto Universitario di Bergamo.

Bassetti, M., Dalmeri, G., Mottes, E. and Nicolis, F. 2008. La frequentazione delle alte quote nell'età del Bronzo. Il sito di Storo – Dosso Rotondo. In E. Mottes, F. Nicolis and G. Zontini (eds.) *Archeologia lungo il Chiese. Nuove indagini e prospettive della ricerca preistorica e protostorica in un territorio condiviso fra Trentino e Lombardia. Atti del 1 convegno interregionale, Storo, Teatro dell'Oratorio, 24–25 ottobre 2003*, pp. 107–127. Judicaria Summa Laganensis 18. Tione: Centro Studi Judicaria.

Bendall, L.M. 2007. *Economics of Religion in the Mycenaean World: resources dedicated to religion in the Mycenaean Palace economy*. Monograph 67. Oxford: Oxford University School of Archaeology.

Boyazoglu, J. and Morand-Fehr, P. 2001. Mediterranean dairy sheep and goat products and their quality: a critical review. *Small Ruminant Research* 40/1:1–11.

Cierny, J., and Marzatico, F. 2002. Note sulla cronologia relativa dei siti fusori e sulla circolazione del metallo. In A. Giumlia-Mair (ed.) *I bronzi antichi: produzione e tecnologia. Atti del XV Congresso Internazionale sui Bronzi Antichi organizzato dall'Università di Udine, sede di Gorizia, Grado-Aquileia, 22–26 maggio 2001*, pp. 258–268. Monographies instrumentum 21. Montagnac: Éditions Monique Mergoil.

Cierny, J., Marzatico, F., Perini, R., and Weisgerber, G. 2004. La riduzione del rame in località Acqua Fredda al Passo del Redebus (Trentino) nell'età del Bronzo Recente e Finale. In G. Weisgerber and G. Goldenberg (eds.) *Alpenkupfer – Rame delle Alpi*, pp. 125–154. Der Anschnitt, Beiheft 17. Bochum: Deutsches Bergbau-Museum.

De Guio, A. 1994. L'Altipiano di Asiago dal Bronzo Medio all'inizio dell'Età del Ferro. In *Storia dell'Altipiano dei Sette Comuni. I: Territorio e istituzioni*, pp. 157–178. Vicenza: Neri Pozza.

Dilley, R. 1996. Market. In D. Levinson and M. Ember (eds.) *Encyclopedia of Cultural Anthropology* 3:728–732. New York: Henry Holt and Co.

Durrenberger, E.P. 1996. Economic anthropology. In D. Levinson and M. Ember (eds.) *Encyclopedia of Cultural Anthropology* 2:365–370. New York: Henry Holt and Co.

Forster, E.S. and Heffner, E.H. 1954. *Lucius Junius Moderatus Columella: On Agriculture, vol. II, Res rustica V–IX*. London: Heineman and Cambridge, Massachusetts: Harvard University Press.

Gamble, C., and Clark, R. 1987. The faunal remains from Fiavé; pastoralism, nutrition and butchery. In R. Perini *Scavi archeologici nella zona palafitticola di Fiavé-Carera. Parte II: Campagne 1969–76: Resti della cultura materiale: metallo – osso – litica – legno*, pp. 423–445. Patrimonio storico e artistico del Trentino 9. Trento: Provincia Autonoma di Trento, Servizio Beni Culturali.

Hooper, W.D. 1967. *Marcus Porcius Cato On Agriculture; Marcus Terentius Varro On Agriculture*. Revised by Harrison Boyd Ash. London: Heineman and Cambridge, Massachusetts: Harvard University Press.

Jarman, M. 1976. Prehistoric economic development in sub-Alpine Italy. In G. de G. Sieveking, I.H. Longworth and K.E. Wilson (eds.) *Problems in Economic and Social Archaeology*, pp. 523–548. London: Duckworth.

Karg, S. 1998. Winter and spring-foddering of sheep/goat in the Bronze Age site of Fiavè-Carera, northern Italy. *Environmental Archaeology* 1:87–94.

Leonardi, P. and Leonardi, G. 1991. Il 'Castelir' di Bellamonte in Val Travignolo. In P. Leonardi (ed.) *La Val di Fiemme nel Trentino dalla Preistoria all'alto Medioevo*, pp. 68–100. Calliano: Manfrini.

Maggi, R. and Pearce, M. 2005. Mid fourth-millennium copper mining in Liguria, north-west Italy: the earliest known copper mines in Western Europe. *Antiquity* 79 (303):66–77.

Maggi, R. and Pearce M. 2010. Changing subsistence structures and the origins of mining in the Ligurian Apennine mountains. In P. Anreiter, G. Goldenberg, K. Hanke, R. Krause, W. Leitner, F. Mathis, K. Nicolussi, K. Oeggl, E. Pernicka, M. Prast, J. Schibler, I. Schneider, H. Stadler, T. Stöllner, G. Tomedi and P. Tropper, (eds.) *Mining in European History and its Impact on Environment and Human Societies; Proceedings for the 1st Mining in European History-Conference of the SFB-HIMAT, 12–15 November 2009, Innsbruck*, pp. 283–287. Innsbruck: Innsbruck University Press.

Marshall, P.D., O'Hara, S.L. and Ottaway, B.S. 1999. Early copper metallurgy in Austria and methods of assessing its impact on the environment. In A. Hauptmann (ed.) *The Beginnings of Metallurgy*, pp. 255–264. Der Anschnitt, Beiheft 9. Bochum: Deutsches Bergbau-Museum.

Marzatico, F. 1997. L'industria metallurgica nel Trentino durante l'età del bronzo. In M. Bernabò Brea, A. Cardarelli and M. Cremaschi (eds.) *Le Terramare; la più antica civiltà padana*, pp. 570–576. Milan: Electa.

Marzatico, F. 2001. L'età del Bronzo Recente e Finale. In M. Lanzinger, F. Marzatico and A. Pedrotti (eds.) *Storia del Trentino, I; la preistoria e la protostoria*, pp. 367–416. Bologna: Il Mulino.

Marzatico, F. 2007. La frequentazione dell'ambiente montano nel territorio atesino fra l'età del Bronzo e del Ferro: alcune considerazioni sulla pastorizia transumante e 'l'economia di malga'. *Preistoria Alpina* 42:163–182.

Marzatico, F., Valzolgher, E. and Oberrauch, H. 2010. Dating the later Bronze Age metal production in the south-central Alps. Some remarks on the relative and absolute chronology of the Luco / Laugen Culture. In P. Anreiter, G. Goldenberg, K. Hanke, R. Krause, W. Leitner, F. Mathis, K. Nicolussi, K. Oeggl, E. Pernicka, M. Prast, J. Schibler, I. Schneider, H. Stadler, T. Stöllner, G. Tomedi and P. Tropper (eds.) *Mining in European History and its Impact on Environment and Human Societies; Proceedings for the 1st Mining in European History-Conference of the SFB-HIMAT, 12–15 November 2009, Innsbruck*, pp. 129–143. Innsbruck: Innsbruck University Press.

Migliavacca, M. 1985. Pastorizia e uso del territorio nel vicentino e nel veronese nelle età del Bronzo e del Ferro. *Archeologia Veneta* 8:27–61.

Moreno, D. 1990. *Dal documento al terreno; storia e archeologia dei sistemi agro-silvo-pastorali*. Bologna: Il Mulino.

Mottes, E. and Nicolis, F. 2004. Storo – Dosso Rotondo (Trento): un sito di alta quota dell'Età del Bronzo in Valle del Chiese. *Annali del Museo, Gavardo*, 19 (2001–2):80–88.

Mottes, E., Nicolis, F. and Tecchiati, U. 1999. Aspetti dell'insediamento e dell'uso del territorio nel III e nel II millennio a.C. in Trentino–Alto Adige. In P. Della Casa (ed.) *Prehistoric Alpine Environment, Society and Economy; papers of the international colloquium PAESE 1997 in Zurich*, pp. 81–97. Universitätsforschungen zur prähistorischen Archäologie 55. Bonn: Rudolf Habelt.

Netting, R. McC. 1981. *Balancing on an Alp: ecological change and continuity in a Swiss mountain community*. Cambridge: Cambridge University Press.

O'Connor, C.B. 1993. *Traditional Cheesemaking Manual*. Addis Ababa: International Livestock Centre for Africa.

Orme, B. 1981. *Anthropology for Archaeologists; an introduction*. London: Duckworth.

Park, Y.W., Juárez, M., Ramos, M. and Haenlein, G.F.W. 2007. Physico-chemical characteristics of goat and sheep milk. *Small Ruminant Research* 68/1–2:88–113.

Payne, S. 1973. Kill-off patterns in sheep and goats; the mandibles from Aşvan Kale. *Anatolian Studies* 23:281–303.

Pearce, M. 2007. *Bright Blades and Red Metal: essays on north Italian prehistoric metalwork*. Specialist Studies on Italy 14. London: Accordia Research Institute.

Pearce, M. and De Guio, A. 1999. Between the mountains and the plain; an integrated metals production and circulation system in later Bronze Age northeastern Italy. In P. Della Casa (ed.) *Prehistoric Alpine Environment, Society and Economy, papers of the international colloquium PAESE 1997 in Zurich*, pp. 289–293. Universitätsforschungen zur prähistorischen Archäologie 55. Bonn: Rudolf Habelt.

Perini, R. 1992. Evidence of metallurgical activity in Trentino from Chalcolithic times to the end of the Bronze Age. In E. Antonacci Sanpaolo (ed.) *Archeometallurgia. Ricerche e prospettive; Atti del colloquio internazionale di Archeometallurgia, Bologna-Dozza Imolese, 18–21 ottobre 1988*, pp. 53–80. Bologna: Cooperativa Libraria Universitaria Editrice [reprinted in *Scritti di archeologia, II parte*, pp. 1121–1146. Trento: Provincia autonoma di Trento, Soprintendenza per i beni archeologici, 2004].

Preuschen, E. 1973. Estrazione mineraria dell'età del bronzo nel Trentino. *Preistoria Alpina* 9:113–150.

Primas, M. 1999. From fiction to facts; current research on prehistoric human activity in the Alps. In P. Della Casa (ed.) *Prehistoric Alpine Environment, Society and Economy; papers of the international colloquium PAESE 1997 in Zurich*, pp. 1–10. Universitätsforschungen zur prähistorischen Archäologie 55. Bonn: Rudolf Habelt.

Puglisi, S.M. 1959. *La Civiltà appenninica; origine delle comunità pastorali in Italia*. Origines. Florence: Sansoni.

Reschreiter, H., and Kowarik, K. 2009. The start of salt mining. In A. Kern, K. Kowarik, A.W. Rausch and H. Reschreiter (eds.) *Kingdom of Salt: 7000 years of Hallstatt*, pp. 50–51. Veröffentlichungen der Prähistorichen Abteilung 3. Vienna: Natural History Museum.

Ridgway, D. 1997. Nestor's cup and the Etruscans. *Oxford Journal of Archaeology* 16/3:325–344.

Riedel, A. and Tecchiati, U. 1998a. I resti faunistici dell'abitato della media e recente età del bronzo di Sotćiastel in Val Badia. In U. Tecchiati (ed.) *Sotćiastel; un abitato fortificato dell'età del bronzo in Val Badia*, pp. 285–319. Bolzano: Istitut Cultural Ladin 'Micuá de Rü' / Soprintendenza Provinciale ai Beni Culturali di Bolzano.

Riedel, A. and Tecchiati, U. 1998b. Gli insediamenti gemelli di Albanbühel (Bressanone) e Sotćiastel. Una comparazione delle faune. In U. Tecchiati (ed.) *Sotćiastel; un abitato fortificato dell'età del bronzo in Val Badia*, pp. 323–331. Bolzano: Istitut Cultural Ladin 'Micuá de Rü' / Soprintendenza Provinciale ai Beni Culturali di Bolzano.

Riedel, A. and Tecchiati, U. 2001. Settlements and economy in the Bronze and Iron Age in Trentino–South Tyrol. Notes for an archaeozoological model. *Preistoria Alpina* 35 (1999):105–113.

Šebesta, G. 1992. *La Via del Rame*. Supplement of Economia Trentina 3 (1992), Trento [reprinted: San Michele all'Adige: Museo degli Usi e Costumi della Gente Trentina, 2000].

Tecchiati, U. 1998. Osservazioni conclusive; problemi e prospettive della ricerca. In U. Tecchiati (ed.) *Sotćiastel; un abitato fortificato dell'età del bronzo in Val Badia*, pp. 379–386. Bolzano: Istitut Cultural Ladin 'Micuá de Rü' / Soprintendenza Provinciale ai Beni Culturali di Bolzano.

Trump, D.H. 1966. *Central and southern Italy before Rome*. Ancient Peoples and Places 47. London: Thames and Hudson.

Ventris, M. and Chadwick, J. 1956. *Documents in Mycenaean Greek*. Cambridge: Cambridge University Press.

Viviani, M. 1993. *Cultura alpina tra Passato e Futuro; Bagòs: una storia della montagna lombarda*. Brescia: Grafo.

West, M.L. 1998. Grated cheese fit for heroes. *Journal of Hellenic Studies* 118:190–191 [reprinted: *Hellenica; selected papers on Greek literature and thought: Volume I: Epic*. Oxford: Oxford University Press 2011:123–127].

Willcock, M.M. 2012. Homer. In S. Hornblower, A. Spawforth and E. Eidinow (eds.) *The Oxford Classical Dictionary*. 4th edition. Oxford: Oxford University Press. Available at http://www.oxfordreference.com/view/10.1093/acref/9780199545568.001.0001/acref-9780199545568-e-3138; date accessed 25 April 2014.

Zschocke, K., and Preuschen, E. 1932. *Das urzeitliche Bergbaugebiet von Mühlbach-Bischofshofen*. Materialen zur Urgeschichte Österreichs 6. Vienna: Anthropologische Gesellschaft in Wien.

Mark Pearce, Department of Archaeology, University of Nottingham, Nottingham NG7 2RD, GB
Email:mark.pearce@nottingham.ac.uk

5. Shepherds and miners through time in the Veneto Highlands: Ethnoarchaeology and archaeology

Mara Migliavacca

The eastern Italian Pre-Alps in the area between Lake Garda and the Brenta River have been exploited from historical times to the present day for many purposes that are typical of a mountain zone. Among the most important activities were mining and stock-raising. Ethno-archaeological and archaeological projects have been carried out in the study area in order to detect and document the traces of human activities, especially shepherds and sheep farming. To date, it has been possible to locate hundreds of sheep folds, shepherds' shelters and breeders' houses in the uplands, to discover that the most ancient traces of organised human exploitation in the uplands go back to the Bronze Age, while during the Iron Age a change in upland economy is evident, possibly connected with the organisation of larger territorial polities and their boundaries.

Le Prealpi dell'Italia nord-orientale comprese tra Lago di Garda e fiume Brenta sono state sfruttate nel corso del tempo per molti scopi tipici di un'area montana. Tra le varie attività attestate si segnalano la pastorizia e lo sfruttamento minerario. Da anni indagini etnoarcheologiche ed archeologiche vanno individuando e documentando le tracce delle attività umane nel passato, focalizzandosi particolarmente sugli alti pascoli dei Lessini a quote montane e sulla dorsale Agno-Leogra a quote collinari-montane. Negli alti Lessini, dove si sono individuati centinaia di ripari e costruzioni pastorali e ovili, le tracce più antiche di sfruttamento organizzato risalgono all'età del Bronzo, mentre un cambiamento interviene nell'età del Ferro probabilmente in connessione con il formarsi di organizzazioni territoriali protostatali più definite. Sulla dorsale Agno-Leogra si stanno individuando le tracce di uno sfruttamento minerario che risale alla protostoria. In entrambe le aree vanno emergendo anche gli elementi di un paesaggio cultuale, con caratteristiche diverse nelle età del Bronzo e del Ferro.

The Veneto Pre-Alps: Geographical and cultural landscape

The Italian Pre-Alps between Lake Garda and the river Brenta are a natural passage between the alpine world and the Po plain. Therefore they have been exploited from historical times to the present day for many purposes that are typical of a mountain zone: charcoal was made from wood; a poor agriculture was developed; mining activities were performed; stone quarry workers and stone dressers left remains of open quarries. One of the most important activities in the uplands was stock raising.

In this paper, the focus is on the Lessini Highlands, north of the city of Verona, where hundreds of sheepfolds and shepherds' shelters have been located (area 1); and on the Recoaro-Schio mining district (area 2).

Lessini Mountain (Fig. 5.1, area 1; Sauro *et al.* 2013) is the group of Venetian Pre-Alps which extend into the Po plain, at the foot of which the town of Verona is located: it was founded on a bend in the river where a meander of the river Adige touches the slopes. This explains the close relationships between the town and this mountain group.

The upper plateau area has always been accessible from the plains in one or two days walking, along the natural pathways following the main valleys, called '*vaj*'; a '*vajo*' is a deep canyon, like a cut, running south–north, starting from the plain and firstly wide and then more and more narrow. The grazing area corresponds to the upper part of the plateau, between 1300 and 1800m of altitude, with a total area of about 80km^2. Here the ridges are wide and rounded, interspersed by smooth basins and easily crossed in an east–west direction.

From the geological point of view, the plateau consists of a series of tectonic blocks made up of limestone formations of Jurassic and Cretaceous age, which mainly are: the group of 'Calcari Grigi'; the 'Calcari Oolitici di San Vigilio'; the 'Rosso Ammonitico' (that can be quarried in slabs used for the construction of the buildings); and the 'Maiolica'.

The Holocene climax vegetation of the high Lessini was characterised by beech forest, fading, in some summit areas, to shrubs. From proto-history, man

Keywords: Veneto Pre-Alps, Bronze Age, Iron Age, highland pastures, mining, ethnoarchaeology.

Parole chiave: Prealpi venete, archeologia ed etno-archeologia, età del Bronzo e del Ferro, pascoli d'alta quota, paesaggi minerari, paesaggi rituali.

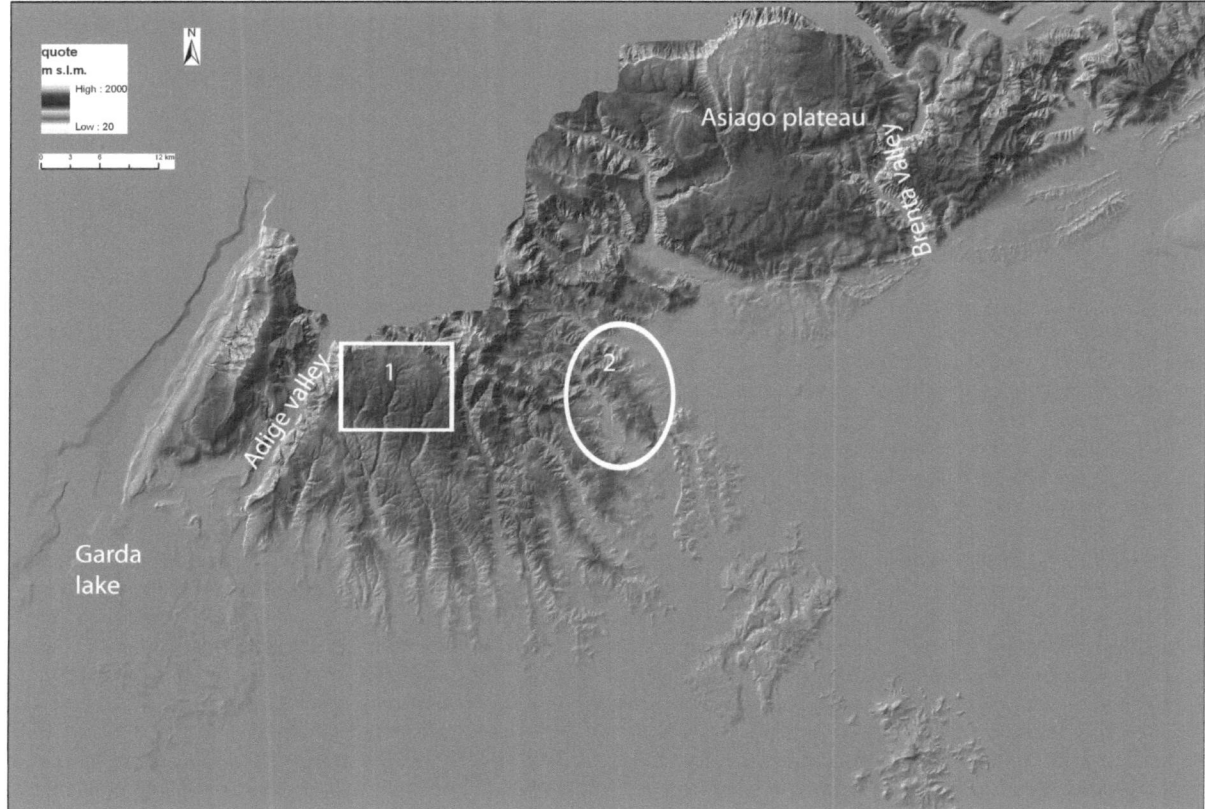

Figure 5.1 The area under study, with indication of the Lessini highlands (area 1) and of the Schio-Recoaro district (area 2).

has gradually cleared the forest to expand the grazing areas. 112 units of the summer grazing can be identified, today called '*malghe*', which can support a total of over 7000 adult cattle. The '*malghe*' territories are characterised by a predominance of pasture areas, but there are small islands of large beech trees, relicts of the original forest, called '*ripari*' in which livestock can shelter from sun and rain; and there are also small plots of forest delineated by walls, known as '*riserve*', providing in the past the firewood for the processing of milk and other necessities in the huts.

Very little is known about the exploitation of the high grazing area in Roman times and before, although a series of significant fortified hill top settlements (such as Monte Loffa, Monte San Giovanni, Le Guaite, Sottosengia, Monte Purga, San Vitale) were located on the southern fringe of the high plateau during the Iron Age . Some of them (Monte Purga for example) were occupied by Roman soldiers, and it is difficult to think that the Romans living in Verona were not interested in exploiting and controlling this mountain area. Since the tenth century, charcoal was made out of wood, which was also used as building material, but this exploitation was improved as German groups came from Bavaria and the Tyrol during the thirteenth century, occupying especially the zone between the Illasi and the Falconi valleys; this activity took place between 600 and 1300 m.a.s.l., leaving traces of abandoned charcoal pits and the huts of charcoal burners. A poor agriculture (devoted to cereals as well as fruits and vegetables) was developed by these German groups up to 1400m or more above sea level; traces of this activity are garden enclosures made of stone. Since the fifteenth and sixteenth centuries, stone quarry workers and stone dressers have been cutting the *Scaglia rossa*, which is very easy to work as building material, and also *Rosso ammonitico* in the central and eastern area of the plateau, leaving remains of open quarries and of the buildings in which they lived.

During the eighteenth and nineteenth centuries, the production of ice became important to satisfy the needs of the plain and of the town of Verona, so we can recognise the ice storage buildings. However, the most important activity on the plateau was stock raising; since the tenth century shepherds have been using the high pastures, controlled by important monasteries and influential families in Verona. The sheep went up to the high pastures from the plain, crossing the deep woodland belt; the shepherds used temporary shelters up in the high pastures, while the first permanent settlements spread between 900 and 1300 m.a.s.l. with the arrival of the German groups. During the eighteenth and nineteenth centuries, cattle husbandry prevailed over sheep rearing.

The Schio – Recoaro mining district (Fig. 5.1, area 2) corresponds to the eastern part of the Lessini mountains, embracing the upper Agno valley, the Leogra Valley, the Posina River Basin, the Tretto area and the Sinello valley in the Trentino region. In this zone there are exposures of the Crystalline Basement of Pre-Permian age as well as the sedimentary and volcanic strata lying above it (Zamperetti 2000; Mietto 2003). Three are the main types of mineral resources. The most important is between the volcanites and the earlier carbonate rocks such as the Calcare di M. Spitz, the Calcare di Recoaro and the *Gracilis* Formation. It contains zinc, lead, iron, copper and silver sulphides, lying from northeast to southwest from the Tretto area, through the Varolo mountain and the Mercanti Valley, to the Spitz mountain. The highest concentration of minerals is between Torrebelvicino and the Spitz (Frizzo 2001).

The second, less common type of mineral layer is composed of seams of mixed sulphides set within the volcanites. Typical elements are chalcopyrite and galena that are found especially in the Mercanti valley (Trisa, Varolo Mountains and Manfron Pass) and on Guizza-Faedo Mountain.

Along the River Leogra, between Pievebelvicino and Fonte Margherita, there is the third type of layer consisting of seams within the Crystalline Basement. They are quite rich in copper consisting of nodules of chalcopyrite, together with pyrites, ferriferous sphalerite, galena, pyrrhotine and occasionally haematite in a siderite, calcite and quartz gangue.

Another general remark can be applied to both the Lessini mountains and the Schio – Recoaro mining district. The territory lying under 800m above sea level is in the sub-alpine ecozone, where gentle ridges continue towards the Po plain, running down from the mountain tops to the north., It is very rich in water; in addition to the rivers Adige and Brenta there are very many streams, called '*progni*', flowing mostly into the Adige and the Bacchiglione rivers. In the sub-alpine ecozone the climate is mild, the dark soils are suitable for cultivation or as hay meadows: it is an environment very suitable for human settlement, very rich in resources, enhanced by the close links with the nearby Po plain.

But at about 800/900 metres above sea level the landscape becomes harsher: the fields are constructed on terraces, enclosed by slabs or stone walls; the economy is more self-sufficient and poorer in its relationships. "*Un nodo di relazioni umane e di ambiente preciso*"

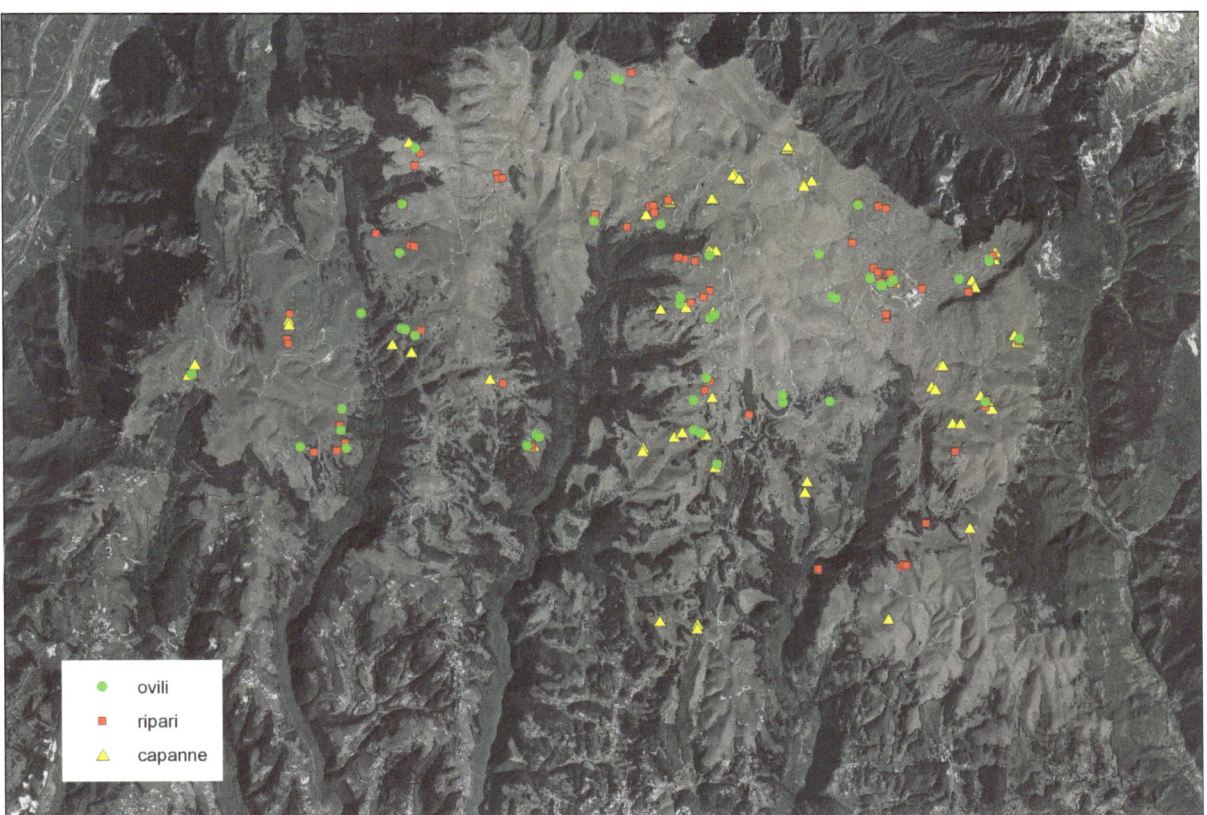

Figure 5.2 The breeders' houses (capanne); the shelters (*ripari*); the sheepfolds (*ovili*) found in the Lessini Highlands in 2010 (GIS elaboration by Francesco Ferrarese, Mara Migliavacca).

Figure 5.3 The remains of a wooden *casone* at Bagorno Nord (Lessini highlands) / at Magaello (*neve*) in the Lessini highlands (photo Ugo Sauro).

(Turri 1969:32–38) is connected with the transition from the hill to the mountain zone, a transition that is not only physical but also cultural. This transition was surely more sharply defined in ancient times, when the asphalted roads were not available and the mountain world seemed far away, even exotic in its farming economy consisting of sheep husbandry, forest exploitation and poor agriculture.

The Lessini highlands: breeders' houses, shelters, sheepfolds and archaeological remains

In the Lessini Highlands (Fig. 5.1, area 1, more than 1200 m.a.s.l.) a project has been underway since 2005, in order to gain more information about the most ancient exploitation of the zone, and also to detect the traces left on the ground by activities from historical times. Initially it focused on shepherds' activities. The aims of the project were to locate and document the traces of shepherds and sheep farming in the area, distinguishing them from the traces left by the other activities performed in the territory, such as cattle farming; to understand the patterns of these traces within the framework of the different elements making up the evolving landscape of the high pastures; to understand the changes that pastoral structures underwent through time as well as the interaction between sheep farming in the mountain ecosystem and the environmental and landscape features of other ecosystems; to single out the specific characteristics of pastoral sites, selecting the shared and recurring features of shepherds' occupation through time.

A systematic field survey covered the whole of the Lessini highlands. In this way about 600 pastoral structures were discovered, recognised and recorded in databases collecting their geomorphological location and architectural features (Sauro *et al.* 2013).

The buildings were also mapped with GPS technology. A GIS was achieved using ArcView GIS 9.0 software (Fig. 5.2).

Breeders' houses; shelters; sheepfolds

Three types of buildings have been recognised as connected with sheep husbandry: the breeders' houses; the shelters; and the sheepfolds (Fig. 5.2).

The breeders' houses are the most difficult to define, because of the great variability of the traces, that are also not easy to detect, consisting of simple – if any – stone alignments remaining on the ground.

A first distinction can be made looking at the dimensions, as the length of these structures is bi-modal, with a group averaging around 3 metres and another around 12 metres.

Figure 5.4   The remains of a small building, connected with sheep farming, in the Lessini Highlands (photo Ugo Sauro).

The second group (Fig. 5.3) consists of large buildings, from 9 to 17m. long and 5–6m wide. The geology associated with these buildings is varied, as is their height above sea level, which does not perfectly respect the lower limit of the high pastures, and which is more or less the upper limit of permanent dwellings (1350 m.a.s.l.). They were always built in dominant locations from where it was possible to control the surrounding pastures. They are probably the remains of the wooden *casoni*, large buildings connected with cattle husbandry where many activities (certainly involving milk processing in one room, and cheese storage in another) were performed. They are mentioned in historical sources from the fifteenth century, and were replaced by the *malghe* built in stone during the nineteenth century, but some of them were still in use in the 1950s; stone footings supported the wooden *Blockbau* structure. The expertise in wood working is probably due to the German skilled workers who arrived in eastern Lessinia during the thirteenth century. The highlands were controlled by the town organisation of Nobile Compagnia through *massari*, *casari* and *vaccari*. 40% of the *casoni* have one curved short side, possibly to resist the wind or to support the radial beams of the roof: it is possible that the structures with an apse are the most ancient.

A second significant group consisting of square buildings of very small area (from 4 to 11 square metres), suggesting a function of simple shelter; they were connected with sheep farming (Fig. 5.4). It is intere-

Figure 5.5   A shelter at Campo Rotondo in the Lessini Highlands (photo Mara Migliavacca).

sting to note that they are located on 'calcari grigi', 'calcari oolitici di San Vigilio', and 'maiolica', and only in few cases on the limestone formation of Rosso Ammonitico, on which natural shelters would have been utilised. In most cases, the traces of these small buildings are shallow hollows in the soil, sometimes with remains of stone walls that supported a wooden superstructure. In some cases, these structures are not far from charcoal pits. In fact, the *casotto/kasun*, a very rough wooden or leather made shelter observed by Baragiola at the end of the nineteenth century in the Asiago plateau (Baragiola 1908, fig. 15, 67, 93, 94), was used equally by shepherds, woodcutters or charcoal-burners. In other cases these small buildings were completely and skilfully constructed in stone, probably by the stone-cutters who appeared in the high pastures in the sixteenth century when there was a change from the wooden-built structures to the stone-built structures. They carried out the construction of the most impressive buildings in the pastoral economy, the cheese store-houses (*casello / cassina / casera*), the first stone-built buildings in the highlands. These buildings have an area ranging from 10 to 40 square metres and are located in most cases in situations exposed to the wind in order to keep the cheese fresh. The same concern explains the high (up to 1.70m–1.90m), thick (up to 70/80cm), often cavity stone walls and sometimes semi-subterranean.

As far as the shelters are concerned (Fig. 5.5), the most simple natural shelters firstly used by the shepherds were the few caves and the many natural rocky overhangs, but it was not possible to carry out archaeological excavations, so there are no data available to date their first exploitation. More recent shelters are those constructed with stone slabs, probably starting from the fifteenth and sixteenth centuries, when the stonecutters began to open quarries in the summer grazing areas and to build in limestone. Most of the natural shelters are situated within the outcrops of the limestone formation of Rosso Ammonitico. Normally, they are very low and only a few square metres in area, good for one or two persons to lie down in. Probably, the first users improved their protective characteristics by canopies covering the outside. Only with the arrival of the stonecutters would many shelters have been partly enclosed with stone slabs or drystone walls. Most of the shelters used were within sight of a sheepfold, to allow the control the flock during the night. Dairy and general farm-production using the milk of both sheep and cattle was carried out elsewhere, and not by the shepherds. In the past, the dairy structures utilised for this were large wood huts called '*cason*', but now replaced by stone buildings called a '*baito*'. Another type of building, the '*casara*', was the storage place for the butter and the cheese.

A sheepfold (Fig. 5.6) is a closed structure bounded by a natural barrier or by a fence, normally situated in a natural niche. To improve or complete the enclosure of a fold, beside the natural obstacles, such as the small cliffs of the Rosso Ammonitico blocks, the shepherds were able to construct wooden fences; starting from the fifteenth century the stonecutters, interacting with the shepherds, built many new boundaries with drystone and slab walls. It was important to keep the flock concentrated both to prevent loss of sheep, and to defend them against attacks by bandits or wild animals; but a fold was also useful to keep the flock together for the daily milking, and for the wool shearing. Some folds are subdivided into multiple compartments in which the flocks were kept separated by sex and age. In the Lessini Mountain, over the centuries the sheep grazing was gradually 'marginalised' from the finest and permanent pastures, such as those on the chalk type limestone, to improve the grazing of the more productive dairy cattle. So the sheep were confined to the slopes with very rocky outcrops, such as those of Rosso Ammonitico.

Figure 5.6   A sheepfold at Campo Rotondo in the Lessini Highlands (photo Mara Migliavacca).

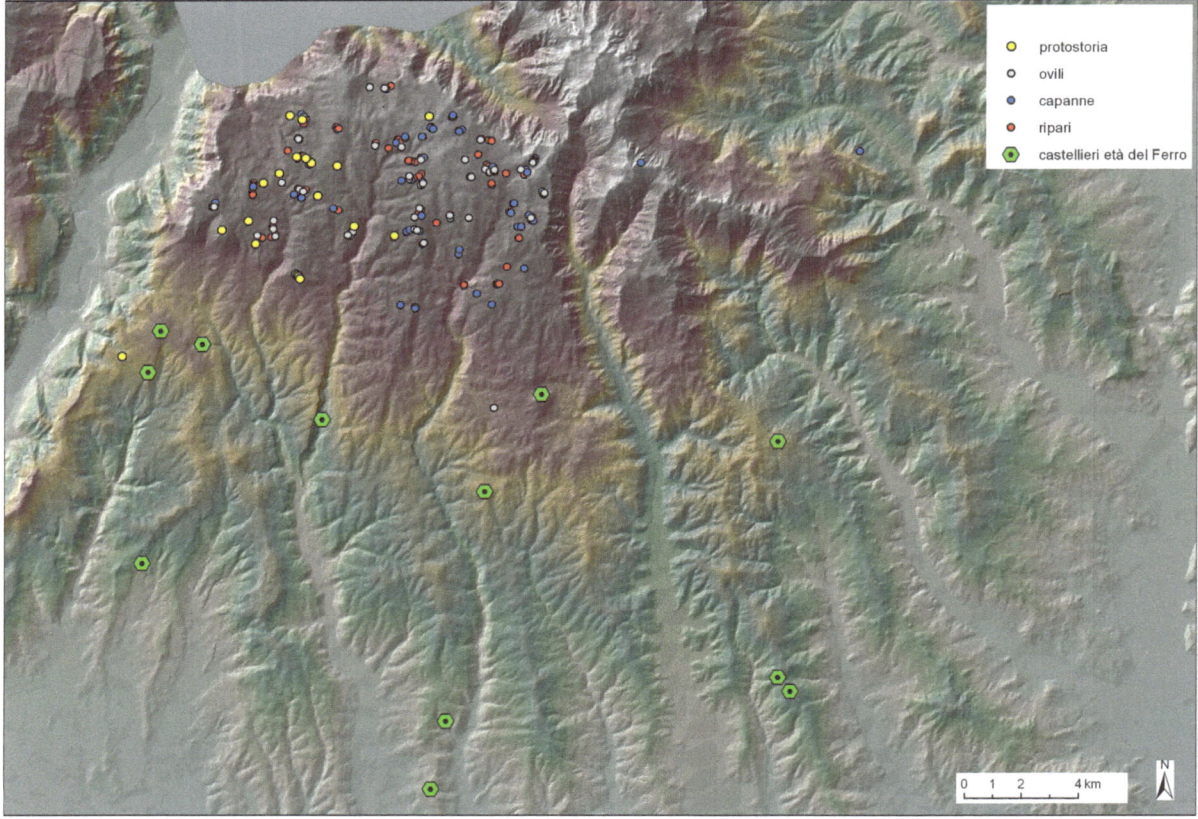

Figure 5.7 Distribution of breeders' houses, shelters, sheep folds and Bronze Age artefacts in the Lessini highlands; also Bronze and Iron Age fortified settlements are indicated (GIS elaboration by Francesco Ferrarese, Mara Migliavacca).

The folds identified are very different in size; the average is around 1500m^2, but the variability is very high also because of the restrictions of some natural niches.

Some folds are located in karstic and periglacial hollows, others in karst dolinas, while others are located in karst corridors in Rosso Ammonitico, locally called '*vallina*'; there are also folds constructed inside quarry depressions. A large number of the folds are closely linked to one or more shelters from which it was possible to control the flocks. To check some folds lacking in visible shelters, the shepherds probably used small transportable wooden huts (called '*baitei*').

Archaeological remains

No significant traces of the presence of humans were detected in these highlands before our research, which both directly discovered new archaeological remains through some digging and relocating many previous discoveries. Tools for wood cutting, cattle bells, clothing components, weapons, vessels, horse fittings from historical times were found, together with coins dating from Roman times to the nineteenth century (Sauro *et al.* 2013). As far as protohistory is concerned, a number of bronze axeheads, daggers and spearheads was discovered, all dating to the Recent-Final Bronze Age (thirteenth to eleventh centuries BC) (Fig. 5.7). Before our research, there were only two bronze axes, found in 1865 (Martinati 1876) on the top of Monte Purga di Velo, a conical mountain where traces of a village dating to the Late Iron Age were also detected (Zorzi 1950). One of the two bronze axeheads is lost; the other dates to thirteenth – twelfth centuries BC.

In 2008 an important chance discovery was made on the Busimo mountain (1375 m.a.s.l.; Fig. 5.8). This is a little grazing plateau with large rock outcrops, 2km northwest of Erbezzo in a central position in the Lessini highlands; from the mountain top you have a good view down to the Po plain to the south, up to Lake Garda to the west, of the central and eastern Lessini to the east, and of the highlands to the north. Here three daggers and two razors were found around a rock, 20cm under the surface; another smaller dagger was found not far from the other objects; one iron spearhead and one iron axehead, both dating to the Late Iron Age, were also found nearby. The bronze objects date to the Middle – Recent Bronze Age: the razors, now dispersed, are both of the Peschiera type according to Salzani (Salzani 2012); two daggers are of the Oria type, the third is of the Tredossi type. The association of daggers and

Figure 5.8 Artefacts found on the top of Busimo mountain: three bronze daggers, one iron spearhead and one iron axe (from Sauro *et al.* 2013:203).

Figure 5.9 A bronze axehead found in the Fittanze mountain (Erbezzo; from Sauro *et al.* 2013:205,4).

razors is significant and suggests a funerary or ritual deposition (Peroni 1994:73) so also the other new bronze find in the Lessini highlands appears in a new light.

Some of them were found around the Fittanze Pass, one of the main ways from the highlands to the Adige valley, an alternative to the narrow and dangerous Ceraino weir through which the Adige flows to the plain. An isolated axehead (Fig. 5.9) dating to the Final Bronze Age was found along the road going from the Pass to Sega di Ala, not far from the border with the Trentino – Alto Adige region (Ponte san Giovanni type, Peroni 1996, fig.71.5). Three bronze axeheads, dating to the Recent – Final Bronze Age, were found in the Roccopiano hill, on the way towards the Fittanze Pass from Bivio del Pidocchio (Fig. 5.10, 11, 12). In the same area Roman coins were also found.

Other important finds were made in the Erbezzo commune. Close to the Castilverio summer farm, on the way towards Monte Cornetto, another bronze axehead was found; it dates to the same period. Bronze daggers of the Torre Castelluccia type, dating to the Recent Bronze Age, were found both close to the Coe Veronesi summer farm (Fig. 5.13) and close to the Modetto summer farm (Fig. 5.14); close to the Modo summer farm another bronze dagger was found (Fig. 5.15); it is of the Montegiorgio or Baierdorf type, also dating to the Recent Bronze Age.

In the Boscochiesanuova commune an isolated spearhead, dating to the Final Bronze Age, was found close to the Gasparine di Mezzo summer farm (Fig. 8.16), that is right on the way to the Lessini Highlands from the Trentino alto Adige region through Val Bona.

The mining district: a work in progress report

The Recoaro-Schio mining district (Fig. 5.1, area 2) is a mountain ridge between the Agno and Leogra Valleys. At the top of it, fieldwork carried out from 2005 to 2010 in the high altitude ridge of Montefalcone-Cima Marana (more than 1500m above sea level) identified traces of human exploitation from 100,000 B.P. up the Second World War (De Guio and Migliavacca 2010). During the Bronze Age the route from the Veneto Pre-Alps to the very rich copper deposits in Trentino –

Figure 5.10 A bronze axehead found in the Roccopiano mountain (Erbezzo: from Sauro *et al.* 2013:204,1).

Figure 5.11 A bronze axehead found in the Roccopiano mountain (Erbezzo: from Sauro *et al.* 2013:204, 2).

Alto Adige is marked by a series of metal artifacts (including a rough copper ingot) indicating a traffic in half-worked metals and a route still used by shepherds and smugglers in historical times. No traces of human exploitation date to Iron Age, although there are important Iron Age sites not far from the ridge, along the Chiampo and Agno valley (Migliavacca 2009–2012).

The Iron Age people seem to have preferred other routes. This is especially evident in the Agno Valley, where finds dating to the First Iron Age indicate the importance of the connection between the Agno Valley and the Leogra one, where there is evidence of iron working at the site of Santorso (Fig. 5.17).

Therefore, the general scarcity of finds in the highlands during the Iron Age could be explained with the loss of importance of the copper and bronze trade with Trentino – Alto Adige through the pre-alpine area, as the focus was perhaps now in the Recoaro-Schio mining district, where not only copper but also iron sources were available.

In this area archaeological fieldwork, both surface survey and excavation, has been carried out in 2011 and 2012 (De Guio and Migliavacca 2012; Migliavacca *et al.* 2013). The surface survey was especially devoted to the identification and recording of the traces left by the exploitation of the numerous mines of the area (Fig. 5.18), mainly worked during the domination of Venice (fifteenth – seventeenth centuries) and afterwards until the beginning of the twentieth century. It is very probable though that the exploitation of the mineral resources had begun in antiquity, as many hints suggest. Roman coins dating to the time of Augustus were found amongst the slag of an old blast furnace in the Astico Valley and near ancient mines in the Tretto area (Fabiani 1930); bronze axes dating to the Final Bronze Age and the beginning of Iron Age were found in the eastern (Pievebelvicino) and in the western (Novale, San Quirico) piedmont area. The axes were used to clear trees in order to gain more space for agriculture and husbandry, but wood was important also for the initial processing of the extracted minerals. The

Mara Migliavacca: Shepherds and miners through time in the Veneto Highlands

Figure 5.12 A bronze axehead found in the Roccopiano mountain (Erbezzo: from Sauro *et al.* 2013:205, 3).

Figure 5.13. A bronze dagger found at Montagna Coe Veronesi (Erbezzo: from Sauro *et al.* 2013:205, 7).

Figure 5.14 A bronze dagger found at Montagna Modetto (Erbezzo: from Sauro *et al.* 2013:205, 5).

Figure 5.15 A bronze dagger found at Montagna Modo (Erbezzo: from Sauro *et al.* 2013:205, 6).

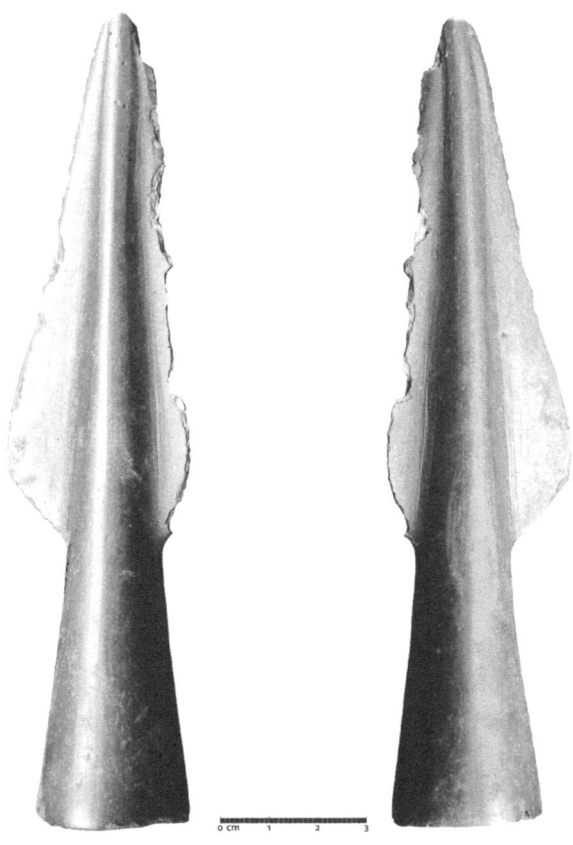

Figure 5.16 A bronze spearhead found at Montagna Gasparine di Mezzo (Boscochiesanuova: from Sauro *et al.* 2013:206).

important settlement of Magrè was founded in the Middle – Final Bronze Age in the eastern piedmont area of the mining district, and became a sanctuary in the Final Iron Age; not far from the district, at Bocca Lorenza, three axes were found together with ceramic fragments of the Square Mouth Jar Culture which are at the moment the most ancient metal artefacts of the area; during the Final Iron Age Santorso became a centre for iron processing (AA.VV. 1988; Balista, Ruta Serafini 1988; Panozzo 2004). The 2011 survey identified some twenty mines spread along Val Mercanti, the longest and most passable route on the eastern side, the narrow and steep valley where contrada Manfron and Monte Cengio are located, the larger but higher Val Riolo and the northeastern side of the Civillina Mountain (Fig. 5.19). A hidden, multistratified minescape showed the twentieth century systems of galleries and shafts continuing and effacing the fifteenth – seventeenth century exploitation; the twentieth century mineral village of Valbella completely engulfed by the beech forest; near the entrance to the galleries twentieth century washing tanks together with not easily datable terraces for the deposition of the excavated minerals and heaps of rubble; remains of cableways that replaced in the last century the tiring hand-pulled cart and the horse or mule wagon (Migliavacca *et al.* 2013).

Some of these mines could be ancient: in the highest part of Val Riolo there is a large flat area just before the passage towards Val Mercanti, where a large, not very deep trial pit for chalcopyrite suggests an ancient form of prospecting. In the northeastern side of Civillina mountain, in a place called Sassi Neri, the Montecatini was excavating a gallery in 1922 when traces of an ancient '*punta e mazza*' exploitation of a lead galena vein were discovered, possibly of Venetian, or even Roman times according to local sources. In the same area, large and deep trial pits, similar to the Val Riolo one, can be found (Casolin 2000:35–36).

Therefore it is quite interesting that on the top of Civillina mountain (Fig. 5.20), not far from these traces of possibly ancient mining trial pits, many pieces of broken pottery vessels were found during the 2011 archaeological excavation. The clay, of a reddish brown / dark reddish brown colour, containing small crystals of quartz, lamellae of mica and many pieces of chamotte, was not glazed or treated in other ways and was obtained from local raw materials (De Antoni 2011–2012). The vessels were big bowls probably for domestic use and can be dated to the beginning

Figure 5.17 The Schio-Recoaro minerary district with the main archaeological sites quoted in the text.

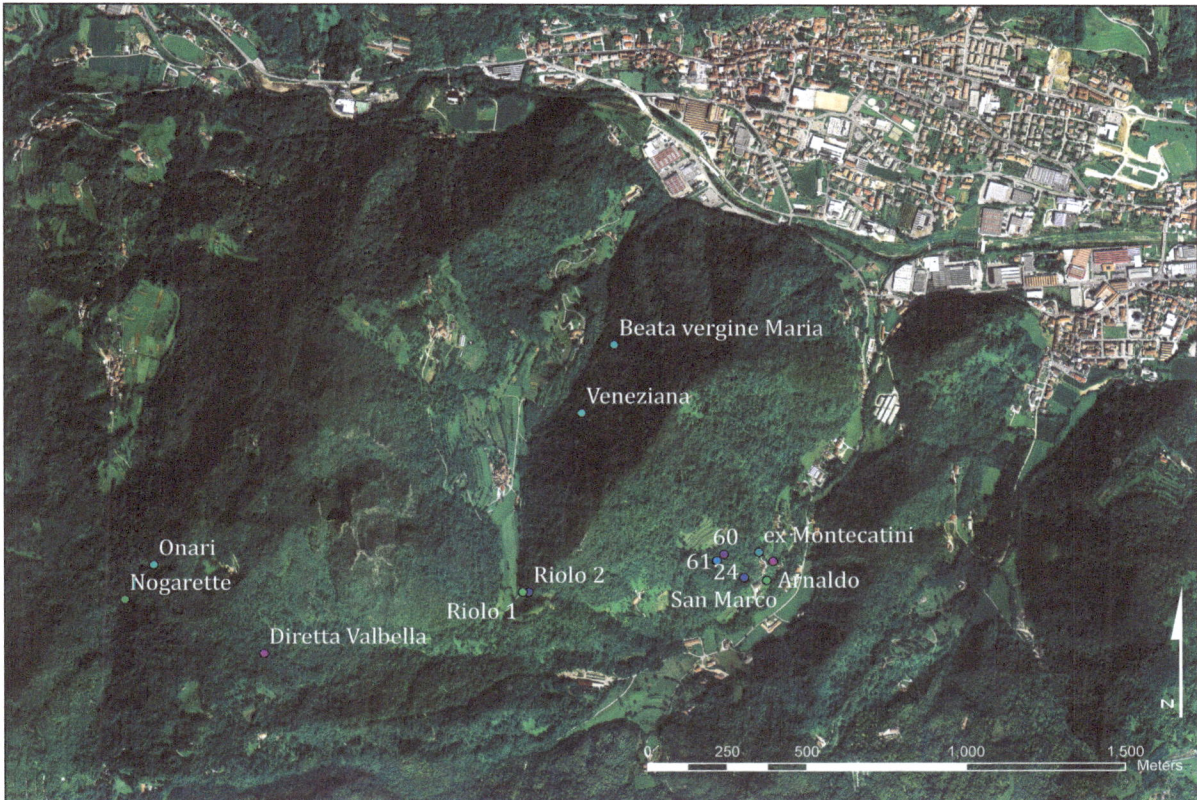

Figure 5.18 The mines explored in the Schio-Recoaro mining district during the 2011–2012 field survey (GIS elaboration by Filippo Carraro).

Figure 5.19 The entrance of Beata Maria Vergine mine (Torrebelvicino, Vicenza).

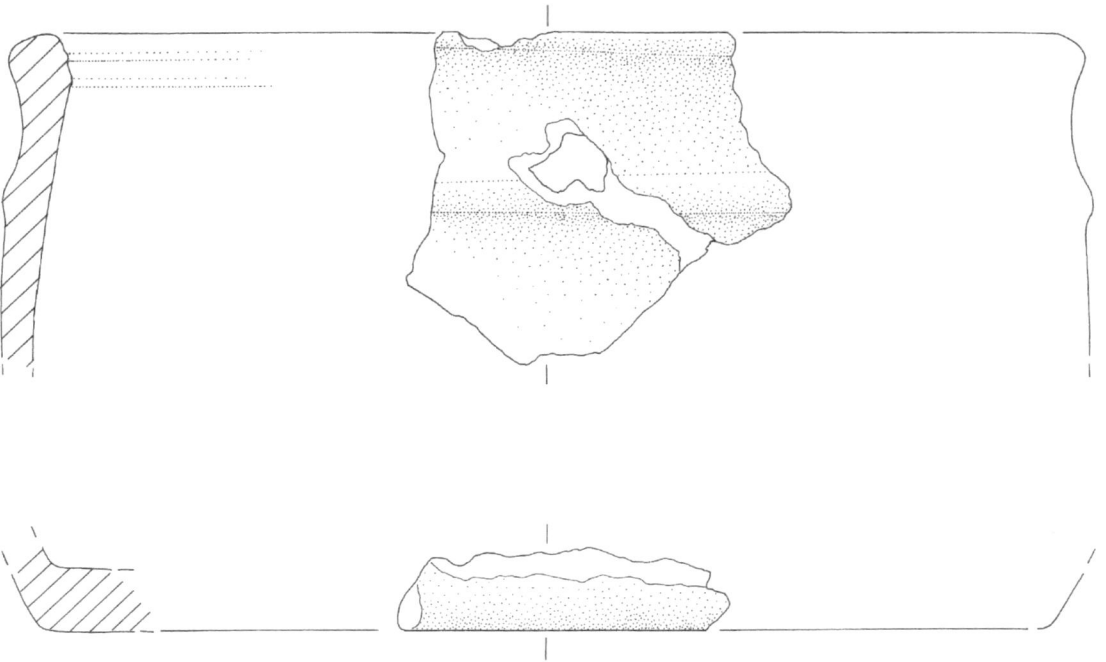

Figure 5.20 Fragments of First Iron Age pottery found at Monte Civillina (from Migliavacca 2009, fig.5).

of the Iron Age, as comparisons with pottery vessels found at Montebello Vicentino, at the piedmont of the mountain ridge where Civillina is placed, Castion di Erbè, Oppeano and Padua, in the Po plain, seem to suggest. The archaeological strata in the Civillina were disturbed by later human interference, the major of which were the military defence works dating to the First World War, so that it was not possible to link the First Iron Age pottery to other remains of occupation.

Not far from Civillina, in 2012 the archaeological fieldwork discovered another interesting site at Passo Mucchione, where a small, possibly fortified protohistoric settlement was identified, with hundreds of broken pieces of First Iron Age pottery and dry-stone structures still *in situ*. The site is still under study, but the hill is a volcanic peak not rich in water sources, so that use for permanent settlement does not seem probable. Some hints, such as the presence of carbonised animal bones and of petroglyphs of different ages, suggest that it could have been used for strategic or ritual purposes. It is worth remembering that also the Agno-Leogra ridge, which is the core of the Schio-Recoaro mining district, was a frontier zone: at the end of the Bronze Age, between the Luco Culture from Trentino and the Protovillanovan Culture in the plain; in the Iron Age, between the peoples of Raeti from the mountains and the Veneti in the plain.

General remarks

Some general remarks can be made about the discoveries of these two different areas:

1. The highlands of the Lessini Plateau, between 1300 and 1800m of altitude were seasonally exploited in ancient times (Recent – Final Bronze Age, thirteenth to eleventh centuries BC). It is worth remembering that some settlements existed in the same zone at lower levels: Monte Loffa (Middle Bronze Age); Guaite (Middle Bronze Age), Folesani (Middle Bronze Age); M. Croce (Recent Bronze Age). In fact the Middle and Recent Bronze Ages are considered the phases of the greatest expansion of the settlement in the Veneto pre-alpine area (Leonardi 2010); some of the sites were surely involved in seasonal sheep movements.

2. The type of artifacts found in the upper lands is significant: the weapons could be connected with the defence of a border, across which were carried both metals coming from the Trentino area, rich with minerals, and animals moving seasonally between different grazing areas. The axes are also connected with the key moment in the transformation of the woodland in a landscape of open summer pastures, and also the metal processing requires charcoal obtained from wood. The Recent – Final Bronze Age daggers found in the highlands are typical of the Garda and Trentino regions; in the ethnographic record a dagger / knife is common in the shepherd's equipment, used for many purposes, such as personal defence in the isolated highlands; the cleaning of animals' hooves; shearing; the slaughter of animals; and the processing of meat and leather.

3. The zone of Bronze Age discoveries coincides with the area of sheep husbandry traces found in the

fieldwork in the Lessini Highlands (Fig. 5.7). In the Modo, Modetto and Coe Veronesi Mountains many natural shelters and some sheepfolds situated within the outcrops of the limestone formation of Rosso Ammonitico were found; these could be the most ancient types of sheepfolds and shelters in the area.

4. All the discoveries in the Lessini Highlands come from the highest locations on isolated mountains, or in key positions on ridges from which a large view is possible; in some of these places (Busimo, Roccopiano, Monte Purga) clusters of finds were found, dating to the same Recent – Final Bronze Age period, and artifacts of more recent times show the continuity of use of these places. The idea of ritual places, where the cult of the mountain tops was performed, is suggested by the presence of razors together with daggers from the Busimo Mountain and fits well with the exploitation of highlands for metals and seasonal grazing.

5. The scarcity of Iron Age finds in the highlands is also remarkable: it is not accidental, as the fieldwork was very thorough, and could be explained in different ways. We could think that the exploitation of the highlands for animal husbandry became more important, and this kind of exploitation leaves very poor, transitory traces as the analysis of breeders' houses, shelters and sheepfolds has demonstrated. The increase in high level animal husbandry could be one of the reasons for the development of Iron Age *castellieri* at the upper limit of permanent settlements, close to highland pastures. On the other hand it is possible to hypothesise the loss of importance of the copper and bronze trade through the Lessini pre-alpine area, as the focus was now on iron working, carried out in the piedmont area of the Schio-Recoaro mining district.

6. In fact, the fieldwork in progress in the Schio-Recoaro mining district is bringing to light the traces of an exploitation dating to the very end of Bronze Age and the beginning of Iron Age. It was connected with the presence of mines, as the Civillina finds seem to suggest; but it could also have other causes, maybe ritual, maybe connected with the process of definition of a borderland, not always obvious to our way of thinking.

Bibliography

AA.VV. 1988. *Carta archeologica del Veneto*, Vol. I. Modena, Panini editore.

Balista, C. and Ruta Serafini, M.A. 1988. Percorsi di indagine analitica di una struttura polifunzionale della II eta' del Ferro a Santorso (VI). In *L'interpretazione funzionale dei dati in paletnologia.* Roma, Atti del Convegno.

Baragiola, A. 1980. *La casa villereccia delle Colonie Tedesche Veneto-Tridentine.* Vicenza, Comunità Montana dei Sette Comuni (reprint of the 1908 book).

Casolin, G. 2000. *Anfiteatro Dolomitico; le miniere, le cave, le fonti.* Schio, Tipografia Menin.

De Antoni, M. 2011–2012. *Indagini archeometriche di produzioni ceramiche locali della Valle dell'Agno e zone limitrofe come strumento nell'autenticazione della 'testina di Valdagno'.* University of Padua, M.A. Dissertation.

De Guio, A. and Migliavacca, M. 2010. Basto al Campetto (Recoaro Terme, Vicenza): Risultati della campagna 2009. *Quaderni di Archeologia del Veneto* 26:108–114.

De Guio, A., Migliavacca, M. 2012. Progetto Agno-Leogra: Le indagini 2010-2011. *Quaderni di Archeologia del Veneto* 28:132–136.

Fabiani, R. 1930. *Le risorse del sottosuolo della provincia di Vicenza.* Vicenza, Ed. G. Peronato.

Frizzo, P. 2001. Giacimenti minerari e attività estrattive della Valle dell'Agno. In G.A. Cisotto (ed.) *Storia della Valle dell'Agno: l'ambiente, gli uomini e l'economia*: 79–82. Valdagno: Ed. Comune di Valdagno.

Leonardi, G., 2010. Le problematiche connesse ai siti d'altura nel Veneto tra antica età del Bronzo e romanizzazione. In L. Dal Rì, P. Gamper and H. Steiner (eds.) *Höhensiedlungen der Bronze- und Eisenzeit: Kontrolle der Verbindungswege über die Alpen. Abitati dell'età del Bronzo e del Ferro. Controllo delle vie di comunicazione attraverso le Alpi* Trento, Temi Editrice. Pp. 251–276.

Martinati, P. 1876. Storia della Paleoetnologia Veronese. *Atti Memorie Accademia Verona* 53:50–128.

Mietto, P. 2003. Aspetti geologici del Recoarese (Prealpi Vicentine) con particolare riguardo all'area del Tretto (Schio). In P. Frizzo (ed.) *Atti della giornata di Studio: l'argento e le 'terre bianche' del Tretto e della Val Leogra, Schio, 15 aprile 2000.* Schio: Edizioni Menin. Pp. 11–38.

Migliavacca, M. 2009. Frequentazione antica nella Lessinia vicentina. In *La Lessinia – Ieri Oggi Domani. Quaderno culturale* 32:105–112.

Migliavacca, M. 2009–2012. On the edge of urbanism: northern Italian mountain communities in an age of change. *Accordia Research Papers* 12:57–70.

Migliavacca, M., Carraro, F. and Ferrarese, A. 2013. Nelle viscere della montagna: paesaggi pre-industriali sulla dorsale Agno-Leogra. *European Journal of Postclassical Archaeologies* 3:247–280.

Panozzo, N. 2004. *Grotta Bocca Lorenza*. Santorso, Comune.

Peroni, R. 1994. *Introduzione alla protostoria italiana*, Bari, Laterza.

Peroni, R. 1996. *L'Italia alle soglie della storia*, Roma-Bari. Laterza.

Salzani, L. 2012. Monti Lessini: rinvenimenti di manufatti di bronzo; depositi votive o oggetti dispersi? *Quaderni di Archeologia del Veneto* 28:151–159.

Sauro, U., Migliavacca, M., Saggioro, F., Pavan, V., and Azzetti, D. (eds.) 2013. *Tracce di antichi pastori negli alti Lessini.* Vago di Lavagno, La Grafica Editrice.

Turri, E. 1969. *La Lessinia*. Sommacampagna, Verona, Cierre Edizioni (republished 2005).

Zamperetti, G. 2000. *Sentiero Geologico Mineralogico.* Schio, Comunità Montana Leogra-Timonchio.

Mara Migliavacca, University of Padua. via Fazio 31/a, 36078 Valdagno (VI), Italy.
Email: mara.migliavacca@unipd.it

6. Seasonal settlements and husbandry resources in the Ligurian Apennines (17th–20th centuries)

Anna Maria Stagno

This paper compares the results of different investigations devoted to the historical reconstruction of seasonal settlements (locally called casoni) that were widespread in the western Ligurian Apennines during the post-medieval period, and in doing so evaluate their potential as archaeological sources for the study of the management systems of husbandry resources. Between the 17th and 20th centuries, casoni became important components of pasture organization in the eastern Ligurian Apennines. As buildings where cattle and shepherds found shelter, it is possible to locate various casoni through place names, the analysis of historical and current cartography or by field-survey. The research summarized in this paper was carried out in four areas (Casoni Lagorara, Casoni di Bargone, Perlezzi sites and Casone del Giazzo) using archaeological (particularly architectural archaeology and survey) and documentary sources (historical cartography, archival documentation about jurisdictional conflicts). In each area, the archaeological analysis of the buildings revealed important transformations between the 18th and early-19th centuries. The comparison between the case studies suggests that these changes were the result of more general transformations in husbandry during this period, particularly those associated with transformations in agricultural systems.

Il contributo mette a confronto i risultati di differenti indagini dedicate alla storia di particolari insediamenti stagionali (localmente definiti "casoni"), diffusi in tutto l'Appennino ligure durante il periodo postmedievale e, nel farlo, tenta di valutare il loro potenziale come fonti archeologiche per lo studio dei sistemi di gestione delle risorse dell'allevamento. Nell'Appennino ligure orientale, tra XVII e XX secolo, i casoni diventano importanti elementi dell'organizzazione dei pascoli. Edifici dove il bestiame e i pastori trovavano ricovero, data la loro diffusione, è possibile localizzarli sia su base toponomastica, con l'analisi della cartografia storica e attuale, sia con indagini di terreno. La ricerca presentata è stata condotta in quattro aree (Casoni Lagorara, casoni di Bargone, casoni di Perlezzi e Casone del Giazzo) attraverso l'uso di fonti archeologiche (archeologia dell'architettura e di superficie) e documentarie (cartografia storica, fondi archivistici su conflitti giurisdizionali). In tutte le aree, l'analisi archeologia degli edifici rivela importanti trasformazioni tra l'inizio del XVIII e il XIX secolo. Il confronto tra i casi di studio suggerisce che questi cambiamenti siano stati il risultato di più generali trasformazioni nei sistemi dell'allevamento associabili a modifiche nei sistemi agricoli.

"Ogni innovazione è sempre una trasformazione di qualcosa che già esisteva"

"Every innovation is always a transformation of something that already existed"

(Tiziano Mannoni, 1997)

1. Introduction

This paper examines the contributions of the archaeological study of seasonal settlement in the reconstruction of the management systems of the environmental resources in particular the relationship between animal husbandry and agriculture. It will focus on the chronology of seasonal settlements, known locally as *casoni* and which are widespread in the eastern Ligurian Apennines. Archival (and iconographical/cartographical) sources testify to their presence since at least the 16th century, whilst the oldest archaeological evidence discussed here dates to the early-18th century.

The *casoni* still conserved today are buildings comprising of two (or three) floors, located along terraced slopes, often at the upper edge of the terraces, a short distance (2–4 km) from permanently inhabited settlements. They are often gathered in small groups and are only rarely isolated buildings. Interviews with present owners of the *casoni* studied stated that during the most recent phase of their usage, often between the 1950s and 1960s, their use was not only connected with animal husbandry, but also with the storage of agricultural products. They were used as stables on the ground floor, and as a hay-barn and storage on the first floor. This utilization corresponds both with that described in the 1960s geographical monograph on the subject, and in the historical written documentation, both with

Keywords: rural archaeology, husbandry systems, historical cartography, terraces, Post-Medieval period
Parole chiave: *archeologia rurale, sistemi di allevamento, cartografia storica, terrazzamenti, postmedioevo*

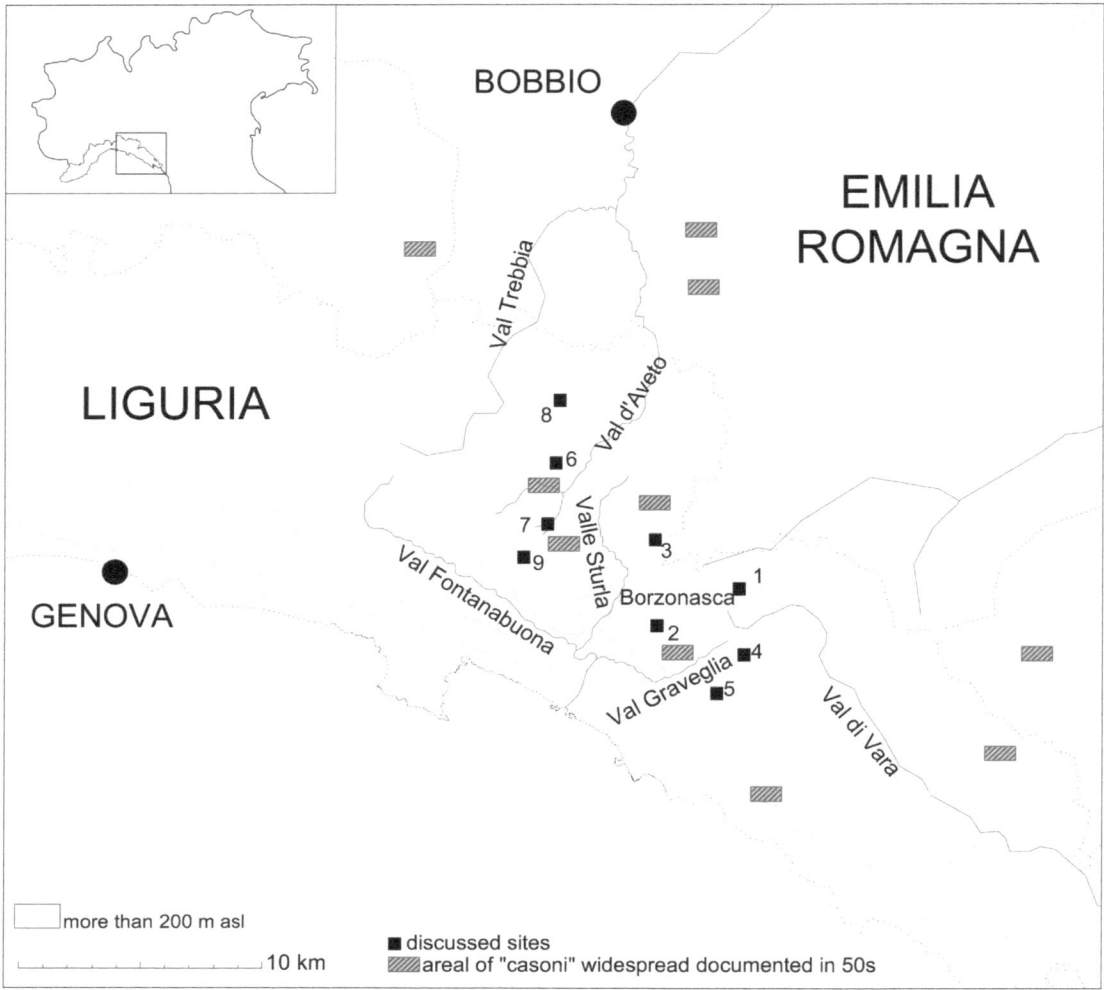

Figure 6.1 Location map of the sites cited. The area of the *casoni* distribution documented in the 1950s is indicated with grey colour (source Cevasco 2014).

that documented by the archaeological investigations conducted in the Ligurian Apennines by ISCUM, and at *Casoni* Lagorara (see below).

The multipurpose use of these structures highlights the close relationship between farming and pastoral activities. In Italian historical studies the connection between husbandry and agricultural systems has been extensively discussed, particularly in relation to customary regulations regarding the access of animals to cultivated areas and their role in fertilising crops (Ambrosoli 2011; Lorenzini 2011; for Liguria, references are Raggio 1995 and Moreno 1990). Recently, for the mountain areas, the historical study of the relationship between agricultural and pastoral practices in the transfer of fertility and its role in the processes of bio-diversification has been developed through the methods of historical ecology (Cevasco 2007).

Through the study of rock shelters (Angelucci *et al.* 2009 for an overview), archaeological research has examined seasonal settlement for the pre- and protohistoric periods, whilst research focused on the medieval and post-medieval periods have been less frequent (De Maestri and Moreno 1980; Andrews and Cima 1984; Milanese and Biagini 1999; Giovannetti 2004). Recently, research focusing on the study of seasonal settlements and pastoral spaces have become more frequent (De Guio and Migliavacca 2012; Gaio 2014), and have included the investigation of rock engravings, writings and inscriptions produced by shepherds during the medieval (Troletti 2014) and post-medieval periods (Rossi and Gattiglia 2007; Bazzanella *et al.* 2014).

This intensification of study can be interpreted in various ways, on one hand to the increase in the number and scale of rural heritage projects (Brogiolo *et al.* 2012; Carrer *et al.* 2013; Redi and Di Blasio 2010) and on the other to the growth of 'environmental themes' in Italian archaeology, as has occurred at a European level in the social sciences in recent years (Ingold 2011; Torre 2008). From a methodological point of view, this growth has led to a new discussion of the problems encountered in the study of mountain areas,

Figure 6.2 *Casoni* of Lavaggi di Chiappozzo (820 m.a.s.l.), upper Graveglia valley (from Ferrando Cabona and Mannoni 1989:157, fig. 167).

both in terms of spatial and temporal high resolution, and of the dialogue between archaeology, palaeo-ecology, historical ecology and, more recently, history.[1]

The research presented in this paper seeks to engage with these themes and is focused on the study and role of rural buildings as suitable historical sources in the reconstruction of the agro-silvo-pastoral systems and their transformations over time (Stagno 2012). Investigations took place within a wider framework of historical ecology and rural archaeology projects which were devoted to the reconstruction of the history of environmental practices and processes connected to the husbandry systems in the Ligurian Mountains and in northwest Italy more generally, conducted by the Laboratory of Environmental Archaeology and History (LASA) at the University of Genoa (Moreno *et al.* 2010; Cevasco 2013).

2. Literature review and research hypothesis: *casoni* from pastures to terraces

[1]. See the First International Workshop on Landscape Archaeology of European Mountain Areas, ICAC – Institut Català d'Arqueologia Clàssica, Tarragona 4th–6th June 2008; *2éme Workshop International d'Archéologie du Paysage des Montagnes Européennes*, GEODE – Université de Toulouse II – Le Mirail (8–11 October 2009). *Table ronde internationale de Gap Archéologie de la montagne européenne*, 29.09–1.10 2008 (Tzortis and Delestre 2010); International Workshop on Archaeology of European Mountain Landscape, 'Carved mountains; engraved stones; contribution to the environmental resources archaeology of the Mediterranean Mountains, Borzonasca (20–22 October 2011), organized by LASA, University of Genoa (Stagno 2014).

The distribution and broad spread of *casoni* between the Trebbia and Vara valleys (Fig. 6.1) is well known, and was identified by typology and place-names during the 1950s, during human geography enquiries in the Ligurian Apennines (Cevasco 2014).

Pasture organization in the Vara valley was studied by Diego Moreno through archival documentation concerning pastorie, soccide and fidarezze, leading to the hypothesis that *casoni* spread during the 16th-century as a sort of summer farm in common lands, as well as documenting the development of local cattle breeding, next to (and in competition with) the large number of transhumant sheep, goats and cattle. A map depicting the Vara Valley, produced by Cristoforo De Grassi (Abbozzo dimostrativo di Varese, ca. 1597–1603), supported this hypothesis due to the fact that it features and specifically names a number of *casoni* in the pastures between the Zatta and Verruga Mountains (Moreno 1990:242–243). In this work, Moreno showed that during the Ancien régime, *casoni* were one of the two points of the 'monticazione' system, in which cattle were brought to the mountain pastures during the summer, whilst being stabled in permanent settlements during the winter months.

The organization of this system fits with the interpretation of the vertical nature of the exploitation of mountain areas (Moreno and Raggio 1991), connected to the extensive spread of common lands and to the scattered settlement patterns ('ville' in Ancien Régime archival documentation) that characterized this part of the Apennines since at least the Middle Ages. Here, 'ville' did not correspond to formal administrative centres and indicate permanent settlements with lim-

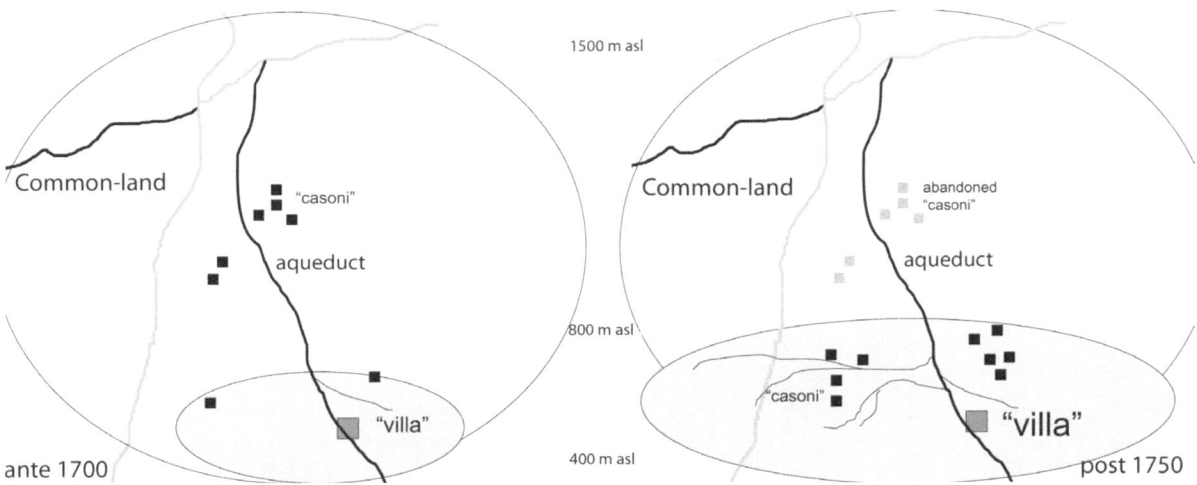

Figure 6.3　Sketch of the location models of '*casoni*' before and after 1700–1750, derived from the Perlezzi case study (upper Sturla Valley).

ited demographical consistencies, tied to parishes and oratories, and that, during the Ancien Régime, were more often connected to single kinships (Raggio 1990; 1995).

Twenty years of archaeological research by the Istituto di Storia della Cultura Materiale (Institute of the History of Material Culture) of Genoa conducted in the Vara and neighbouring valleys (Graveglia, Aveto, Sturla, Fontanabuona, etc.) have shown that seasonal settlements – *casoni* – were widespread during the 18th century (Ferrando Cabona and Mannoni 1989; Mannoni *et al.* 1988:42–48), two centuries after the first references located in archival documentation. The oldest study site was at Lavaggi di Chiappozzo in the upper Graveglia valley (820 m.a.s.l.) and is dated to the early-18th century (Fig. 6.2).

The hypothesis of this paper is that the observed differences in the chronology between archival and archaeological sources is connected to a shift in the locations of the *casoni*, from common land pastures to terraced slopes on private properties (Fig. 6.3).

Through the examination of three examples, it is my intention to demonstrate how these changes occurred from the late 17th and throughout the 18th and the 19th centuries, and that they were caused by a contemporary process of slope terracing and the consequent changes in the organization of pasture and agricultural activities. In line with this process, *casoni* were built at lower altitudes (between 600 and 800 m.a.s.l., instead of at more than 1000), inside or on the upper ridge of terraces. This hypothesis was formulated during a study devoted to the reconstruction of a series of 18th century controversies concerning the water access rights between four hamlets in the Sturla Valley, and confirmed by the archaeological data of previous investigations in the neighbouring Petronio (*Casoni* di Bargone) and Lagorara (*Casoni* della Pietra) valleys.

A fourth case study, Casone del Giazzo, is also discussed in order to illustrate the difficulty in making generalizations. The history and characteristics of the Casone del Giazzo (Aveto Valley) differ from the other investigated '*casoni*', consisting of a building, dated to the late-17th century, located around 1100 m.a.s.l., within the common lands used by the inhabitants of the hamlet of Salto, an area that was to a limited extent terraced only during the 19th century.

Excluding the investigations at the *Casoni* della Pietra (a discussion based on the archaeological literature of a rescue archaeology project) the case studies were all examples of original research into landscape archaeology and architectural archaeology carried out within the framework of multidisciplinary projects of the Laboratory of Environmental Archaeology and History. The examples discussed have been investigated to different extents, with some work still in progress. The objective is to suggest how the study of the buildings as sources for the history of resources management, could be realized moving from a geographical and typological study to a strictly analytical and contextual perspective. In doing so, such an approach permits a closer reading of archaeological and archival documentation, aiding the reconstruction of the transformations and changes in post-medieval husbandry systems otherwise hidden by terminological and typological similarities associated with the same practice of 'monticazione'.

3. Methods of investigation

As previously mentioned, during all the investigations, particular interest was paid to transformations in the management systems of the environmental resources. For each case study Table 6.1 summarizes the investigation methods, the choice of which was dependant on the specific research context.

As noted above, in the original research the analyses of the buildings were linked with archaeological field surveys of the surrounding slopes (as well as historical ecology surveys in the *Casoni* di Bargone case

	Surveys			Site analysis		Lab analysis		Documentary sources analysis	
	Archaeology	Historical ecology	Oral sources interviews	Archaeology of architecture	Archaeological excavation	Palynological analysis	Geomorphology	Archival sources	Cartographical sources
Casoni di Perlezzi	x		x	x		x		x	x
Casoni della Pietra			x	x	x		x		x
Casoni di Bargone	x	x	x	x					x
Casoni del Giazzo	x			x					x

Table 6.1 Methods of investigation employed in the case studies. The grey colour highlights investigation based only on the literature.

study). Investigations were carried out with non-systematic methods, with an intensive survey.[2] Architectural archaeology methods were employed in all case studies. Overall, twenty *casoni* were analysed, through an expeditious cataloguing of each building, in order to identify the phase of construction and the sequence of the subsequent most important modifications (additions of new parts to building, elevations, reconstructions, insertion or substitution of openings). Configuration analyses (Mannoni 1984; 1998) were conducted in the Perlezzi and Bargone case studies, and associated with the stratigraphic analysis of the standing walls in the other two cases.

The analysis of the openings was particularly important for their potential use as dating elements. It is well known that door lintel typologies have been changed in the course of time, and this fact allowed the construction of specific chrono-typologies for each area (Ferrando Cabona *et al.* 1989; Gobbato 2004). The chrono-typology defined in the previous investigation of the Ventarola hamlet in Aveto valley (Stagno 2009a; 2009b) was used during the other research in order to date building phases.

Chronological clues were also derived from the analyses of other documentary sources. This comparison made it possible to verify the chronology established from archaeological investigations and, particularly for the Perlezzi case study, to indirectly date buildings and other artefacts, thanks to the comparison with the analysis of historical cartography.

In the *Casoni* della Pietra, Perlezzi and Bargone case studies, the present owners of the buildings (*Casoni* della Pietra, Bargone) and of the lands investigated (Perlezzi) were interviewed. Interviews focused on the final stages of the use of buildings and terraces and on the environmental resource management practices still used during the 1950s and 1960s. The results of the interviews have been crucial in constructing the interpretation grid of the use of the *casoni* spaces. Based on the interpretation and the characteristics of the last recognizable function, this grid aided hypotheses concerning the function and changes of internal spaces (as identified through the analysis of the walls and the openings). There is a short discussion about the archaeological markers of the functions of *casoni* in the last paragraph.

4. Material characteristics of *casoni* buildings

The analysis of *casoni* buildings showed that they share some physical characteristics. The standing buildings are usually on two floors (maximum height 4.5 metres), but phases of super-elevation are also documented. The roof can be a single sloped (45°) sloping down to the valley, or two asymmetrical slopes. When *casoni* are located on a drop, the access to the ground floor and to the first floor is made up of two different entrances. The openings are all with architraves. The windows on the ground floor are of small size and often square

---
2. Archaeological research in mountain areas has now reached its maturity at an European level, both from a theoretical point of view, and a methodological one, above all thanks to increasingly strict linkage between environmental archaeology and the 'classical' practice of archaeological survey (for an application see Rendu 2003; Le Couédic 2010). Concerning the necessity to define a precise methodology for surveys in mountain areas, see Van Leusen *et al.* 2011; Tzortzis and Delestre 2010. For landscape archaeology in so-called low visibility areas, and the necessity also to investigate the elements traditionally considered as obstacles to the investigation like vegetation and not to separate historical ecology surveys and archaeological ones, see Giovannetti 2004; Stagno 2015, in press. This approach, not so developed in Italy, is today much more common, see Moscatelli 2011 for an overview.

Figure 6.4 Domenico Carbonara map "*Tipo geometrico delli Condotti, o Corse d'acqua fra Perleggi, Careggi e Caroso*", 1752 (from Stagno and Tigrino 2012).

shaped. The small size of the ground floor windows indicates that the housing of livestock was not continuous. The stables continuously used during winter have larger windows, as confirmed in the investigations of the hamlets of Ventarola and Casanova in the Ligurian Apennines (Stagno 2009a:161; Tigrino *et al.* 2013). The first (and second) floors have openings of considerable size, compatible with their use as hay-barns and storage houses. In all investigated *casoni*, no evidence suggesting changes in their functions were observed.

The walls are double row with a rubble core, with irregular work and parameters, consisting of broken, irregularly shaped and medium sized stones always bonded with lime mortar mixed with earth, and arranged in sub-horizontal courses with abundant subwedges. The cantons are indicated by larger sculpted stones. As is frequently the case in late and post-medieval rural buildings, the wall surfaces have no significant transformations divulging chronological information. More information, as noted above, was derived from the analysis of the openings when preserved (in many cases the original lintel is removed).

The analysis of the architectural complexes consisting of multiple bodies has shown that often these are built at the same point in time which may provide clues to construction agreements between different owners so as to reduce costs linked to the presence of one or two shared walls and, certainly in the Perlezzi case study, can be linked to the strategies of terracing slopes.

5. The case studies

Below, I will discuss the four case studies of the *casoni* at Perlezzi, Lagorara, Bargone and Giazzo. During the discussion, investigations are briefly outlined and some archaeological information is summarized in a specific table.

Casoni di Perlezzi (Sturla Valley)

The study of *Casoni* di Perlezzi originated from research developed from the study of archival documentation and later through fieldwork (Stagno and

Tigrino 2012).[3] The study started with the reconstruction of an 18th-century controversy concerning the common lands shared by different hamlets (ville) located in the upper Sturla Valley, under the jurisdiction of the Capitanato di Chiavari. During the Ancien Régime this area belonged to the Repubblica di Genova, and constituted a boundary area with seigniorial estates of the Doria family. Conflicts over common lands were caused by problems of access rights to the same stream used for feeding the different irrigation aqueducts of the hamlets Perlezzi, Prato, Caroso and Careggi. In 1752, during the controversy, a map ("Tipo geometrico concernente le prese d'acqua..." drawn by 'Domenico Carbonara, Ingegnere', Fig. 6.4) was produced in order to collect elements necessary for defining access rights on common lands (and so to water). For this reason the map indicated *casoni* and pieces of lands of private people ('particolari') in a very accurate manner. Obviously, this precision does not correspond with our criteria: it is impossible geo-referencing the map, but the precision is connected with the extensive index of 93 details concerning place-names, hydrography, meadows, '*casoni*', etc. as described by the inhabitants of Perlezzi, Caroso and Careggi.

However, this precision has permitted the identification of the features represented on current cartography (Fig. 6.5) and to date phases in the construction of the aqueduct identified during archaeological surveys. It is important to highlight that only the reconstruction of the controversy permitted the understanding of the jurisdictional and certifying value of this non-geo-referenced map, and that only for this reason was it possible to use archival data as a tool in establishing chronological information through a direct comparison with the field survey results.

The analysis of the map clearly showed the abandonment of all the eight *casoni* located inside common lands, and the presence of 18 *casoni* at lower altitude, inside the 'particolari' lands of Perlezzi inhabitants, eight of which were near to the water-work branches (Fig. 6.6; Table 6.3). The co-existence of abandoned and used *casoni* suggests a shift in their location, which was explained through field surveys.

Archaeological surveys carried out with intensive and systematic methods did not identify any traces of buildings in the Perlezzi common lands (Beltrametti *et al.* 2014),[4] while on terraced slopes they made it possible to:

1. establish that the standing *casoni* were all located on terraced slopes, where the aqueduct branches brought water;
2. identify three different phases of terrace construction;
3. observe the physical relationship between the aqueduct channels and the terraces, which showed that construction of the water-works was realized at the same time as the terraces.[5]

Thanks to this contemporaneity, it was possible to use the chronology of the Perlezzi aqueduct (which was dated through archival documentation and historical cartography to the first half of the 18th century) for dating the corresponding phase of the terracing process, that was the second of the three identified phases (Table 6.2).

Perlezzi terrace phases (Fig. 6.7) were built with the following chronology:

Phase 1, end of 16th century and 17th century. Terraces adjacent to the permanent settlement built during the hamlet development.

Phase 2, end of 17th century to the first half of 18th century. North and west Perlezzi slopes terraces developed contemporary to the aqueduct network.

Phase 3, second half of the 18th century and the 19th century. Terracing of the south and west slopes.

This reconstruction confirms that the 18th-century conflicts over water resources depended on and happened at the same time as the second phase of terracing, which is the more significant. Water, obviously, was used not in common lands, but where aqueducts bring water, on private lands that we know were just terraced. So the conflict over water resources depended on the new demand for water in consequence to the construction of new agricultural terraced spaces. The surveyed *casoni* were constructed precisely on the terraces belonging to this second phase, an observation that provides an important clue in establishing the chronology of the *casoni* themselves.

---

3. The *casoni* of the inhabitants of Perlezzi were studied during a research devoted to 'water perimeters', the technical and juridical devices which allow access to water by social groups who need to exploit it (University of Genova- PRA2008, director O. Raggio, supervisor A.M. Stagno, see Stagno 2009a).

4. It should be noted, that it was possible to suggest a continuity of use of the common-lands from the 18th century. In fact, surveys allowed the identification of all the areas represented in the map of 1752 as private pieces of land inside common-lands, thanks to the presence of particular features: hawthorn trees, alignments of stones or of bushes (Beltrametti *et al.* 2014). For this reason, the impossibility of finding evidence of the abandoned '*casoni*' that the same map shows in the same areas suggests that they could be made of wood or of some other non-durable material. Obviously, in the sloping areas, it is also necessary take in account the question of erosion and of formation processes for the interpretation (this problem was addressed at an early stage for the study of Ligurian mountain areas – see the ever fundamental Mannoni 1970).

5. Both for the identification of the relationships between water-works and terraces, and the use of documentary and cartographical sources for dating archaeological remains, the study of the irrigation system of Perlezzi was inspired by the procedures of Spanish hydraulic archaeology, employed in the study of *Al-Andalus* society and the spatial and social organization of *campesinos* (Kirchner and Navarro 1993:135–146; Asins-Velis 2006:29–31). The archaeology of terraces is rapidly maturing, above all thanks to the development of Spanish and French agrarian archaeology; see Guilane 1991; Harfouche 2005; 2007; Kirchner 2010 and, in connection to water archaeology, Barceló *et al.* 1996.

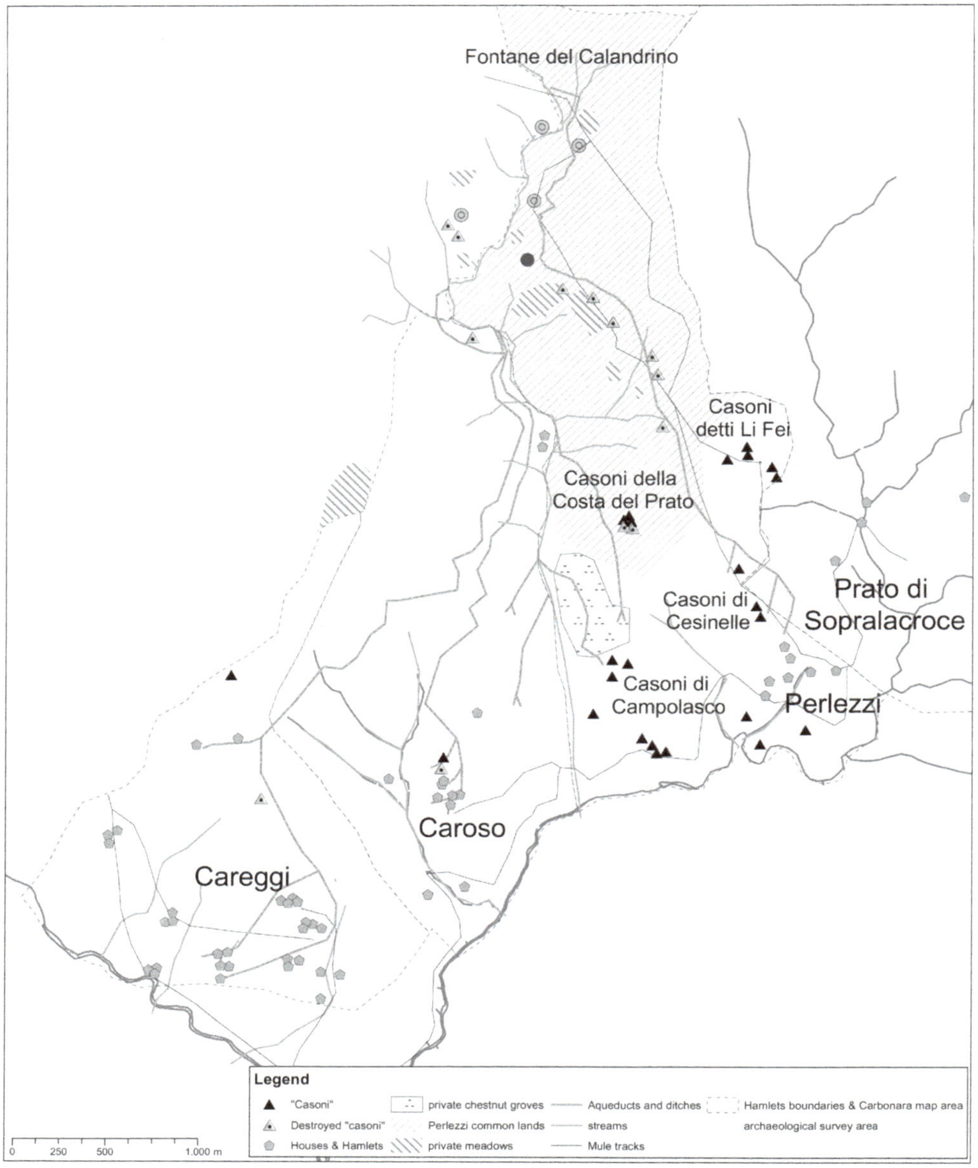

Figure 6.5 Localization on the present cartography of the features represented in the map of D. Carbonara (after Stagno and Tigrino 2010).

Figure 6.6 Detail of the '*casoni*' represented in the '*Tipo geometrico*' of Domenico Carbonara (1752). See Table 6.3 for the list.

Aqueduct phases	chronology sources	Terrace phases	chronology sources	*casoni*	chronology sources
16th – beginning of the 17th	(Chronology of terraces; aqueducts bring water to hamlet terraces)	Phase 1: 16th – beginning of the 17th century	Dated door lintels in Perlezzi hamlet		
1700 – 1752	Archival documentation of controversy; Historical maps	Phase 2: end of 17th – first half of 18th	Physical relationship of contemporaneity with the aqueduct	First half of 18th century	Relationship with the terraces; door lintels; historical cartography
1820	Historical maps	Phase 3: second half of 18th – beginning of the 19th century	Relationship with the aqueduct, stratigraphic relationship between terraces	Late 18th century	Relationship with the terraces; historical cartography
				second half of 19th	Relationship with other *casoni*; door lintels
1939	Historical maps				

Table 6.2 Comparison between the chronology of the aqueduct, terraces and *casoni* (with the indication of the sources for the chronology).

1641	1752	1789
– Terra seminativa detta il Chiapone con casone di Antonio Grillo e di Filippo Grillo (mezzo casone a testa) confina con la valle – Terra seminativa detta Nonalaga con casone di Benedetto Massa qu Michele – Terra hortiva e arborata di noci detta il Cozale e l'horto con un casone rotto di Pietro Costa qu Alessandro – Terra seminativa detta Li Fei con casone di Batta.ce Massa qu Lucca (confina con il Commune)	– Abandoned and partially destroyed *casoni* in Perlezzi common lands near 'Moglia del Fango' (6) – Casoni della Costa del Prato at the boundary between common and private lands (5, of which 3 destroyed) – Casoni della Costa del Coscione (2) – Casoni delle Cesinelle: 4 *casoni* – Casoni de Particolari di Perlezzi detti Le Fei (10)	– Casoni di quei di Perelezzi, nel loro teriotorio detti di Campo-lasco ove sono de' vigneti (11) – Casoni del Fei e Casoni del Lerino nel Territorio di Perlezzi (17)

Table 6.3 List of the '*casoni*' of Perlezzi and of their owners located in private and permanent cultivated lands as described in the '*Caratata*' of 1641, in the *Tipo geometrico* of Domenico Carbonara (1752), and in the *Tipo geometrico* of Giuseppe Ferrretto (1789).

Eighteenth-century *casoni*

Archival sources provide some clues about the process with *casoni* widespread since the 17th century. The analysis of a fiscal cadastre (Caratata) of 1641,[6] shows that in that period there were only four *casoni* privately owned by individuals at Perlezzi, located inside the private and permanently cultivated lands, one of which was shared by two owners (Table 6.3). There is no mention of *casoni* in areas of common land, because common lands were not included in this cadastre.

As described above, in 1752 there were 18 *casoni* in use, located on private and cultivated areas, and eight abandoned *casoni* in common lands. As shown by another map of 1789, a few decades later there were no more *casoni* documented in common lands, and the 29 *casoni* at Perlezzi were all located in cultivated and (thanks to archaeological surveys we can also add) terraced land.[7] This second map, connected to a new

---

6. ASG, *Magistrato Comunità*, 718, *Caratata della Cappella di Valle Sturla, Ordinaria di Perlezzi*.

7. *Tipo geometrico dei rispettivi territorî delle due Comunità di Gazzolo e di Perlezzi elevato l'anno 1789 in settembre per ordine degli Ecc.mi Camerali Com.ti per le differenze vertenti fra dette Comunità a motivo delle acque denominate Calandrine* drawn by Giuseppe Ferretto (ASG, *Mappe e Tipi Cartografici*, b. 9bis, n. 302, I Gazzolo).

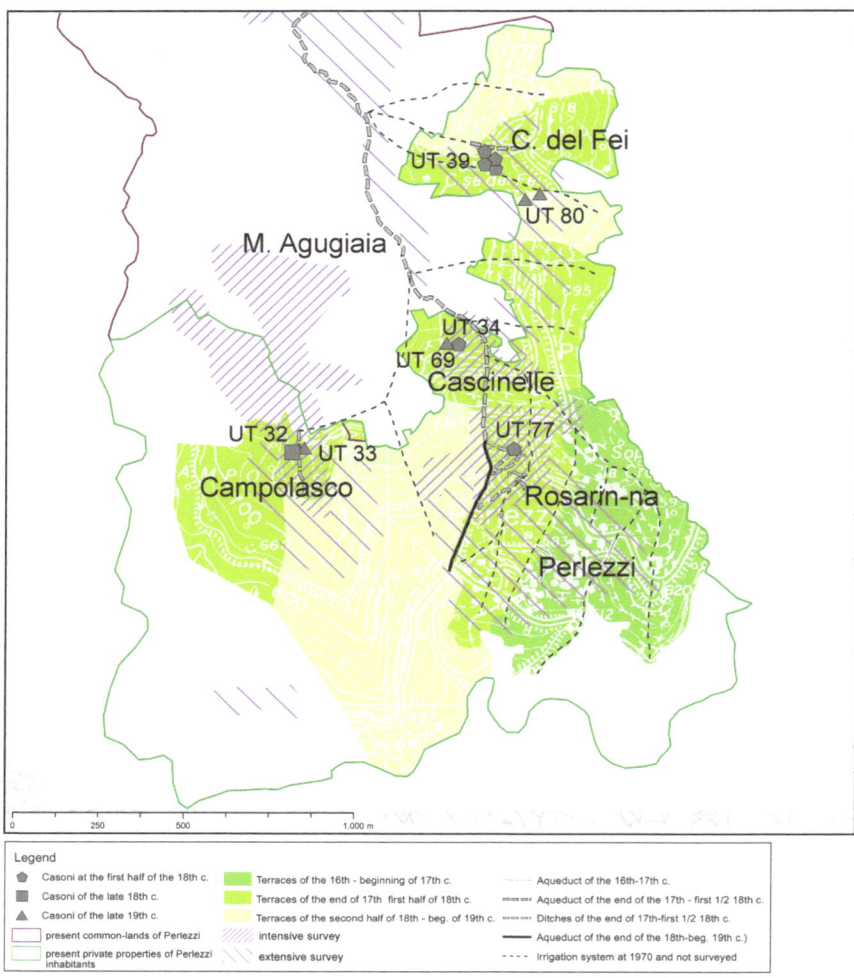

Figure 6.7 Hypothesis of terrace and water-works phases, and the chronology of *casoni*, derived from the comparison between historical cartography analysis and fieldwork results. The extension of terraces belonging to the different phases is hypothesized thanks to this comparison and to the distribution of the branches, as documented in a 1970s map (Stagno and Tigrino 2012, modified).

controversy on water access rights involving Perlezzi (against another hamlet 'Gazzolo') was drawn with the same perspective as the first one and shows that the water network had been expanded.

From the first half of the 18th century, the increase in the number of *casoni* located in cultivated areas and their contemporary disappearance in common lands was evident during the cartographical comparisons. These maps represent *casoni* distributed in small groups of two, three or more buildings (Table 6.3). In particular, for the group of Casoni detti li Fei, the map of 1752 indicates the owner of each casone.[8] It is an interesting element that could also indicate how terraces were built. In fact, the localization of *casoni* belonging to different owners in the same area at the top of terrace systems could be a clue of a co-participation of different owners in the construction of the same sets of terraces. This type of agreement is certainly documented for the building of the aqueduct at the Caroso hamlet (the only one for which we have found this type of information). This water work was built by a 'master mason' (maestro muratore), whose remuneration was paid in accordance with quotas established by a specific written agreement ('instrumento di Convegno') between different parties and the future users of the water. The documented contemporaneity between terraces and water-works suggests that agreements could have also been signed for the construction of terraces.

During archaeological fieldwork along the Perlezzi slopes, a total of eleven *casoni* were surveyed (Table 6.4). Seven of them were built on terraces that belong to the second phase of terracing (late 17th to 18th century, see Table 6.2, Fig. 6.7) and so it is possible to confirm

---

8. '*Particolari*' dei Casoni detti li Fei: Casoni del Reverendo Cesare Mazza; Casone di Agostino Mazza; Casone di Gio Grillo; Casone di Girolamo Mereto; Casone di Santino Grillo; Casoni degli eredi di Gio Grillo.

		Site 9 Casinelle	*Site 15 'Cas del Fei'*	*Site 8 Campolasco*
Characteristics	Altitude	825	820	820
	Number of buildings (architectural complex or building element)	UT 34 (one building of 2 elements); UT 69 (one building)	UT 39 (4 buildings, of which one completely destroyed) (UT 80 (2 buildings)	UT 32 (one building of 4 elements); UT 33 (one building)
	Building dimensions	UT 34: 15, 46×15, 46 46×6, 54m; UT 69: 4, 68×4 64m		UT 32: 23×6.60m UT 33: 7×6.6m
	Evidence of feeders	Yes, of the basement	Yes, preserved	Not surveyed
	State of preservation	Standing	Restored and standing	Standing except a collapsed stucture
Methods	Investigation methods	Survey and configurational analysis		
	Chronology methods	Indirect: archaeological survey crossed with documentary sources analysis		
Chronology	Chronology elements	UT 69 external wooden door lintel (half 19th century)	UT 39 restoration with new door with wooden door lintel UT 80 external wooden door lintel	UT 33 external wooden door lintel
	Chronology of construction	End of 17th century – beginning of 18th century	Late 17th century	Second half of 18th century (Ferretto map)
	Transformation	New *casoni*	Enlargements and new *casoni*	Enlargements and new *casoni*
	Continuity of use	Used	Used	Used
	Evidence of previous buildings	Not demonstrated	Not demonstrated	Not demonstrated
Functional relationships	Distance from the hamlet	2 km from Perlezzi	3 km from Perlezzi	3.5 km from Perlezzi
	Enclosure for manure collection	Yes	No	No
	Presence of terraces	UT 62, 1800 sq m (terraces width 4–7m)	North UT 78 (terraces), south UT 66 / (embankments)	UT 68, 2800 sq m (terraces width 7m)
	Presence of water-works	Yes (UT)	Yes (UT 66)	Yes (UT ...)

Table 6.4 Synthesis of the archaeological investigation on *casoni* of Perlezzi.

that these *casoni* are definitely contemporary or later than the terraces. It was possible to directly date only one building (UT 34), thanks to the presence of an 18th century door lintel. In the other cases, the comparison with the historical cartography offered an important chronological element because the maps indicate the presence of *casoni* in the same areas where they were identified during fieldwork. Therefore, even if we cannot be sure that the present buildings of *casoni* are those of the 18th century, we can be sure that *casoni* were in these areas, an adequate conclusion for our discussion of the change in the location of *casoni*. The analysis of historical maps compared with the results of archaeological survey showed its connections with the progressive process of terracing, and also provides information about the organization of *casoni* located on the boundary between terraces and upper common lands.

Surveys showed that the branches of the aqueduct serve not only cultivated terraces but also terraces for chestnut groves. An archival document of the controversy confirms also that during the 18th century the irrigation of chestnut groves was practiced (Stagno and

Tigrino 2012:276). Palynological analysis conducted in a sample of the soil of these terraces demonstrated that the fertilisation of the parcel was assured by grazing and that the parcel was also cultivated with cereal sowing in a cycle that provided wood, charcoal, fruit, hay and grass (Molinari 2010:159–173).

Nineteenth century *casoni*

Archaeological surveys demonstrated that during the second half of the 19th century new scattered *casoni* were built along the private and cultivated terraces (UT 33, UT 69 and UT 80), not far from the most ancient *casoni* (UT 32 Campolasco, UT 34 Casinelle, UT 39 Case del Fei, see Fig. 6.7). Thanks to the presence of the external wooden door lintel of the stable, it was possible to date these buildings (Ferrando Cabona *et al.* 1978:108–118; Stagno 2009a:158–159). A 19th century phase of *casoni* construction was also documented in the other case studies and have been connected to a new phase of the expansion of cattle breeding, and of an intensification of the production of fodder resources (which caused the need for more space for hay-barns), already documented in the Ligurian Apennines (Moreno 1990), as well as in Italy and Europe generally (Vecchio *et al.* 2002) and which can be connected with more general transformations that will be discussed in the conclusion.

Seasonal settlements at the middle of the 20th century

Since the 1950s, new seasonal settlements have been built on the common lands in Perlezzi: 'Malga Perlezzi' owned by the Comitato di gestione dei Beni Frazionali (Management Commitee of the Common Lands). This committee holds the access rights to the common land of Perlezzi. The malga is used as a summer farm, whilst its stables are used for nightly summer stabling of cattle owned by different families of Perlezzi and of the neighbouring hamlets. The 'malga' is managed through a rota system by the members of the Consorzio di Miglioramento Zootecnico, that brings together the Perlezzi owners of cattle and sheep, and to which the Comitato di gestione dei Beni Frazionali grants, through a free loan, common lands for grazing. It is interesting to note that this organization was funded in a period of the progressive reduction of the importance of agricultural activities, and in which the exploitation of terraces for permanent cultivation was abandoned, and so it was therefore no longer necessary to reconcile agricultural and pastoral activities. Therefore, since the 1950s, the terraces on the Perlezzi slopes were only used for vegetable gardens, and for hay mowing, and the studied *casoni* along terraces have been progressively abandoned and only used for the deposition of farm tools and as hay-barns, and no longer as stables.

This 'new' seasonal settlement and its connection with a collective organization of pastoral activities have a positive environmental impact. The common

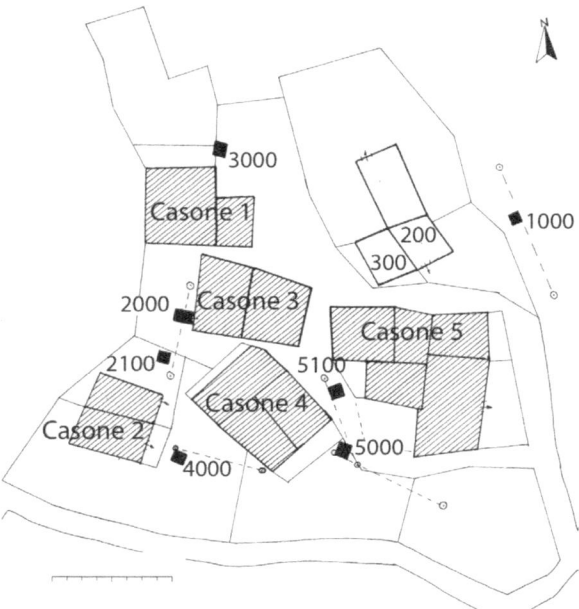

Figure 6.8 Casoni della Pietra. Sketch of the site (after Milanese and Biagini 1999, figs. 9 and 43, modified).

lands of Perlezzi are today occupied by beech forests and prairies of mountain pastures which have not been affected by the advance of secondary forest growth as has happened in the ancient common land belonging to the 'ville' of the valley and more generally in mountain areas of Liguria where the grazing was already abandoned (Stagno and Tigrino 2012).

Casoni della Pietra, Val Lagorara

The site of Casoni della Pietra (Maissana, SP) occupies an area of 1500 square metres (Fig. 6.8), at 790 m.a.s.l., on the western side of the Lagorara valley (a valley in the sub-basin of the Torza stream in the upper Vara Valley). It is comprised of six building complexes, five of which are abandoned and partially collapsed, and one of which is completely destroyed, located at the upper boundary of a large terraced area used at the time of the investigations for hay mowing.

In order to collect extensive and stratigraphical data for the reconstruction of the diachronic development of the settlement, investigations were predominately carried out through non-destructive methods: core drillings, shovel pits, stratigraphical analysis of standing walls, micro-morphological analysis.[9] Archaeological excavations have been carried out only in two of the three structures constituting the ruined

---

9. Archaeological researches on Casoni della Pietra were carried out by the Chair of Methodology of Archaeological Research of the University of Genova, in the framework of the project *Laboratori di Archeologia Montana*, realized by *Polo etnobotanica e storia* (now *Laboratorio di archeologia e storia ambientale*) at the same University in collaboration with the *Soprintendenza per i Beni Archeologici della Liguria* (Milanese and Biagini 1999).

Figure 6.9 Casoni della Pietra. Sketch of areas 100 and 200 (after Biagini and Milanese 1999, figs. 24 and 29, modified).

casoni (area 100 and area 200, Fig. 6.9), with the aim of reconstructing its functions and to date the construction. Fieldwork investigations were associated with a preliminary analysis of the historical cartography and with several interviews with present owners of *casoni*. Micro-morphological analyses were carried out on two undisturbed samples of the straw bedding for the cows sampled from a collapsing casone (casone 4, area 4000) and from a stratigraphic layer sampled in area 100.

These investigations permitted the reconstruction of the phases when the site was in use. The more ancient phase is connected with structures preceding the *casoni*: a part of a wall (identified in area 100), built directly on the artificially flattened bedrock used as a foundation. The presence of this wall suggested the existence of a settlement before the present *casoni*. A collapsed wall, documented during shovel tests (area 5000, Fig. 6.8), confirmed the presence of the same bedrock and hints of previous buildings (Ottomano 1999:34–35).

After a period of abandonment documented by colluvium which had hidden the preceding remains, between the end of the 17th century and the beginning of the 18th century, the more ancient part of the excavated casone was built (area 100). This chronology was extended to the whole settlement, because the same palaeo-surface of colluvium was identified in all the shovel-pits, and the footings of the buildings have been laid on this palaeo-surface.

The excavated casone (Fig. 6.9) comprised of three structures (area 100, area 200, and area 300 not investigated), with independent accesses. The largest one had a rectangular shape, and is on two floors, with an external staircase for access up to the first floor. The ground floor was paved with pebbles and stones (US 110), partially obliterated by a layer (US 109) with predominantly clay matrix, interpreted as the result of the deposition of the straw bedding, thanks to the micro-morphological analysis, these analyses indicated that the stable housed cattle, and that they were not fed with dry fodder, but rather grazed/pastured and stabled in the 'casone' only seasonally and a few times, possibly during the night (Ottomano 1999:36).[10] For this reason this room was interpreted as a stable. The adjacent area 200 was smaller, and belonged to another cadastral parcel, possible evidence of a different property, as already documented in the *casoni* of Perlezzi. It is possible to interpret this room as also a stable. The presence of almost exclusively transport pottery, and the absence of permanent fireplaces inside confirm a seasonal/temporary use and not permanent habitation in this building.

Unlike the other buildings of the settlement, the excavated casone was abandoned at the end of the 19th century. This fact documents a diversification in the use and in the way of abandonment of *casoni* connected to a specific family history. The current owners, in fact, stated that the majority of these structures were abandoned during the 1950s. The observation of the still standing *casoni* shows that at least one of them was built in the second half of the 19th century, as documented by the presence of an external door lintel with wooden architraves surmounted by segmental arches and jambs made with off-cuts, which are comparable to the ones at Ventarola, Perlezzi and Zignago.[11]

The surveys carried out in area 4000 (outside *casone* 4, Fig. 6.8) documented the presence of litter developed within an open enclosure, evidence, for the 1950s, of the practice of collecting manure for the fertilization of the surrounding terraced and cultivated areas. The micro-morphological analysis conducted on this litter shows that it has the same composition as US 109, allowing the same interpretation.

---

10. About the problem of archaeological study of the stables and possible indicators to characterize their uses see Charles *et al.* 1998.

11. This is a new observation that I made on the basis of the published images (Milanese and Biagini 1999:16, fig. 7), because the results of the stratigraphic analysis of the standing walls have not been yet published. Those investigations have been focused on the micro-stratigraphy of the mortars, which did not allow concrete results (Milanese and Biagini 1999:50–51). During the study conducted in the hamlet of Ventarola, I could compare the results of previous micro-morphological analysis on the samples of mortar, with the results of the stratigraphic analysis of the standing walls. The comparison showed the impossibility of obtaining chronological and stratigraphical information from the differences of the mortar, such as the different colours and compositions of the mortars which do not necessarily correspond to different wall phases (Stagno 2009:175–176).

		Survey on 'Casone della Pietra'	Excavation (area 100 and 200)
Characteristics	Altitude	790	790
	Number of buildings (architectural complex or building element)	6 architectural complexes	1 architectural complex, consisting of two structures
	State of preservation	Collapsing	Collapsed
Methods	Investigation methods	Standing wall archaeological analysis, shovel pits	Archaeological excavations
	Chronology methods	Direct: pottery from the shovel pits, stratigraphical relationships with the layers of the excavation, architectural chronological elements (door-lintels), historical cartography	Direct: pottery in the archaeological excavation
Chronology	Chronology of construction	End of the 17th – beginning of the 18th century	
	Transformation	Enlargement of old *casoni* (with new structures), construction of new *casoni*	Construction of a new structure
	Continuity of use	Until 1950s	Until second half of the 19th century
	Evidence of previous buildings	Yes, a collapsed wall in the shovel-test area 5000	Old wall in area 100
Functional relationships	Distance from the hamlet	3 km from Maissanae hamlet	
	Evidence for cheese production	No	
	Enclosure for manure collection	Yes, in '*Casone* 4'	
	Presence of terraces	Yes	
	Presence of water-works	Not surveyed	

Table 6.5 Synthesis of the investigation at 'Casoni della Pietra'

The life-stories collected from present owners about the use of the *casoni* date from the beginning of the 20th century to the 1950s, when, as noted above, most of the buildings were gradually abandoned. In that period, Casoni della Pietra were used by some families of Maissana as summer settlements involved in cultivation and grazing. The memoirs tell of intense summer activities, in which the *casoni* were their 'pivot'. Men went up daily to *casoni*. Cattle were daily brought by young shepherds from *casoni* to the upper pastures of Mount Porcile (and sometimes also on to the terraces). In the meantime, the terraces below the *casoni* were in a cultivation cycle with crops, potatoes, corn, hay, etc. The milk was brought down daily from *casoni* to the Maissana hamlet, where cheese was made at the homestead. At lunchtime, 'the lucky ones' were joined by their wives with lunch, whilst the others ate what they had brought from home. The animals (cattle) were left up throughout the summer, while the men came down to the village every day. In winter, the animals were kept in the stables in the hamlet, but the hay was left in *casoni*, from where it was periodically removed. In conclusion, the data obtained from surveys, excavations, micro-morphological analysis, and oral interviews converge with the seasonal (spring and summer) function of these structures, and confirm that for the 20th century, as for previous periods from their construction, they were used within a system in which the animals (in this case cattle) were left on a nightly basis (the *casoni*), while shepherds/farmers went back to sleep in the valley (hamlet). In this system, already called 'monticazione', herding also had a role in the terraced agriculture (fertilization evidenced by the collection of manure).

The new casone built in the second half of the 19th century could be evidence of the intensification of agro-pastoral activities towards monoculture, which can be compared with the transformations of Monte Santa Maria, located on the other side of the same valley. The historical ecology investigations documented that, after the first half of the 19th century, on these slopes, terraces with chestnut were inter-planted with crops. Previously chestnut trees were scattered in a system of wooded meadow pastures (Cevasco *et al.* 1997–1999; Cevasco 2007; Molinari 2010).

Casoni di Bargone (Val Petronio)

The study of the *casoni* of the Bargone hamlet was carried out in the framework of research on the reconstruction of the management systems of the environ-

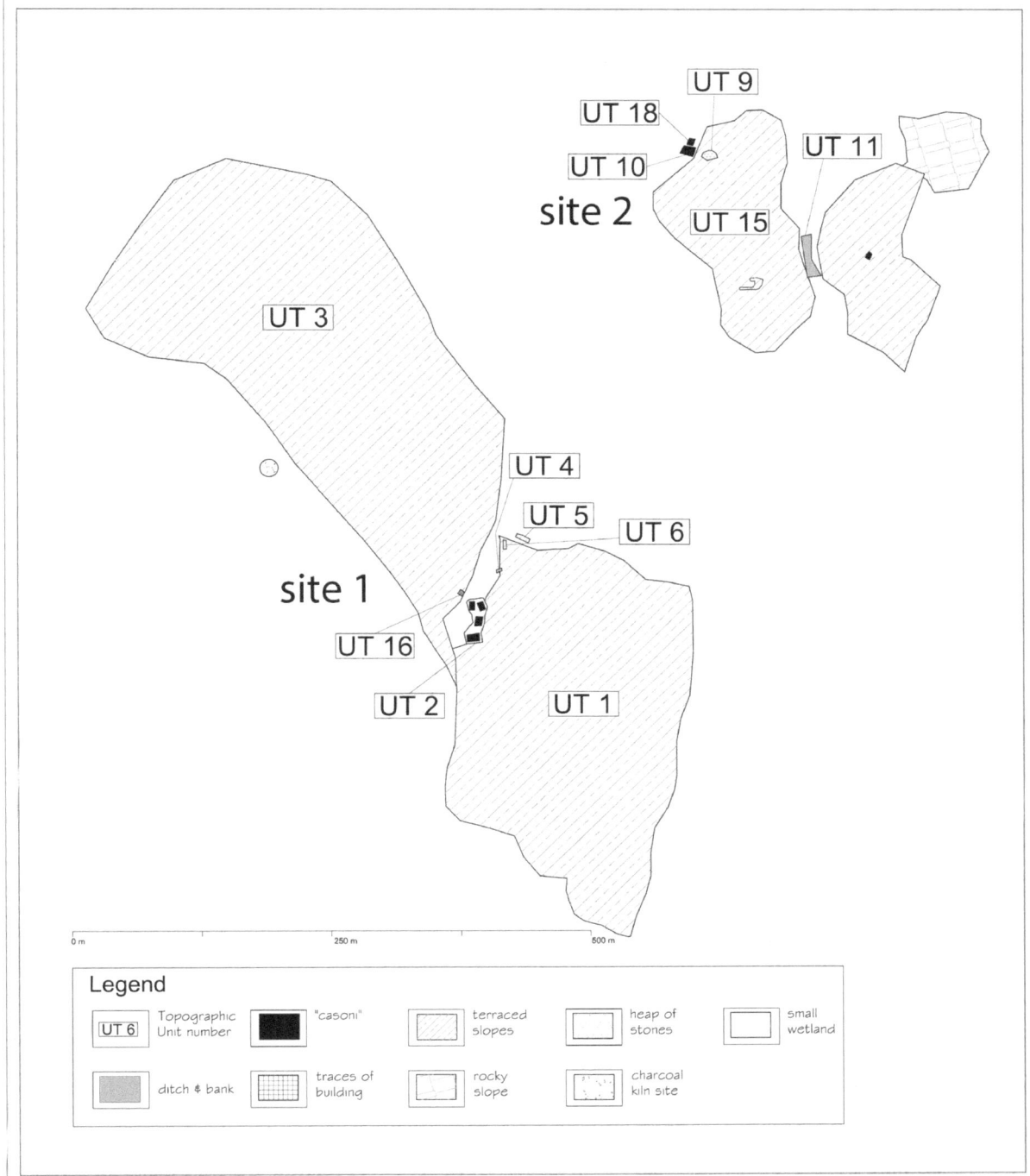

Figure 6.10 Historical cartography analysis. Transect Case delle Barche – Pian del Lago.

mental resources in the high Petronio valley.[12] Particular attention was devoted to the transformations of land use and land cover. Preliminary analysis of historical maps suggested the possible presence of a wooded pasture that was replaced by a chestnut-grove between 1818 and 1853, at the head of the valley of the Rio Bucato (Cevasco *et al.* 2005).

Historical ecology and rural archaeology surveys verified this transformation and identified two groups

---

12. These investigations were carried out in the framework of a research project applied to the management of the Site of Communitarian Interest 'Roccagrande – M. Pu'. See LASA 2005 for the complete archaeological (A.M Stagno) and historical ecology reports (R. Cevasco and D. Moreno).

		Site 1 Case delle Barche	**Site 2 Prato di Pozzo**
Characteristics	*Altitude*	582	660
	Number of buildings (architectural complex or building element)	4 architectural complexes, consisting of different structures (UT 2)	Two buildings: one partially standing (UT10 divided into three rooms) and one totally collapsed (UT 18)
	Building dimensions		UT 10: 5.5×10m UT 16: 4×4m
	State of preservation	All collapsing, excluding one still standing	1 almost erased and 1 almost collapsing
Methods	*Investigation methods*	Archaeological survey and configurational analysis	
	Chronology methods	Indirect: relationships with dated terraces and historical cartography analysis	
Chronology	*Chronology elements*		
	Chronology of construction	End of the 18th century	First half of the 19th century
	Transformation	Enlargement and construction of new *casoni*	Not identified
	Continuity of use	Until 1950s	Until 1930s
	Evidence of previous buildings	Yes, remains of a dry-stone building (UT 16), in an area now occupied by a domesticated chestnut tree dating from the first half of the 19th century	Not identified
Functional relationships	*Distance from the hamlet*	3 km from Bargone	3.2 km from Bargone
	Enclosure for manure collection	Not documented	Enclosure made of stones
	Presence of terraces	Yes (UT 1 and UT 3)	Yes (UT 15)
	Presence of water-works	Irrigation system constructed through branches from the stream (UT 4)	Irrigation system constructed through branches from the stream (UT 11)

Table 6.6. Synthesis of the investigation at Casoni di Bargone.

of buildings referred to as *casoni*, closely related to the surrounding terraced slopes: site 1 Case delle Barche; and site 2 Prato di Pozzo (Fig. 6.10; Table 6.6).

Site 1: Case delle Barche

The site 1, known locally as Cà delle Barche, Casoni delle Barche or Barche di Bargone, consisted of a group of four collapsing buildings (except for one structure that is still standing) and the surrounding terraced slopes (chestnut to the north, hay meadows still grazed to the south).[13]

The chestnut grove that extends north and west of Case delle Barche (and up to the top of Monte Tregin) is now abandoned. As noted above, the historical cartography analysis (Fig. 6.11) dated this terraced chestnut grove to between 1818 and 1854. Along the terraced slope, historical ecology surveys have documented the traces of 'lunettature', earlier crescent-shaped revetment walls made between rocky outcrops, constructed around trees of oak and chestnut to reduce erosion and conserve moisture; these are probably related to an earlier stage when the slope was still occupied by wooded meadow pastures.

The *casoni* of Case delle Barche were built on a flat area continuing southeast in a system of terraces and embankments with traces of water-works. The location of the buildings suggests that they were made together with the terraces. The system of terracing has been dated between the second half of the 18th and the beginning of the 19th century thanks to the pottery found in the heaps of stones (UT 5, UT 6, Fig. 6.10) located on the upper terrace and interpreted as resulting from stone clearance activities made during the first phases of cultivation of the terraces, and so shortly after their construction.[14]

---

13. The place-name '*barca*' indicates particular wooden-structures used as barns (associated with meadow-pasture systems) not existing today in the settlement of Case delle Barche, wholly constituted by *casoni*. It is possible to suppose that this place-name belongs to a previous phase of the settlement constituted by '*barche*', traces of which were identified during surveys. On the study of place-names as a source for environmental management system reconstruction, in particular referring to the upper Petronio valley, see Marullo 2012.

14. Two fragments of glazed bowl in the same fabric date between

The structural analysis of the building (made during archaeological surveys) has shown that these different structures were made at the same time (probably in consequence of agreement between the different owners) as already documented in the Perlezzi and Lagorara case studies. The analysis of the historical maps fits with this hypothesis (Table 6.7). The Napoleonic cadastre (c. 1805–1813) shows that the core of the buildings has already been registered at the beginning of 19th century as 'Barche Hameau'. The cadastre shows four buildings, one of which consisted of several structures, each one identified by a different cadastral parcel (and therefore presumably belonging to different owners).[15]

Two of the present owners of the Case delle Barche *casoni* stated that the terraces had been cultivated and that *casoni* have been used as seasonal and night stables for cattle until the 1950s–1960s. After this period, the abandonment of the buildings happens at the same time as the abandonment of agricultural activities in the surrounding areas (UT 1 and UT 3, Fig. 6.10). One of the owners also relates that one structure of one of the buildings, the only one still standing, was used during 1950s as a permanent house, and the other one was used as a drying shed for chestnuts (dialect '*a gré*'). It is possible to suppose that this drying shed was built after the planting of the chestnut grove, during the second half of the 19th century.

Site 2: *Prato di Pozzo*

In site 2 are located two buildings in an advanced state of decay (UT 10 and UT 16, Fig. 6.10) and the surrounding terraces (UT 15). Buildings appears at the contact point between an abandoned terraced chestnut grove, and a system of terraces with traces of agricultural irrigation, now used for grazing and hay mowing. One of the two buildings is in an advanced state of collapse, while the second is only recognizable by walls which seem to have been razed to the ground surface. Probably this structure was already destroyed in the 1930s, as documented by historical cartography (see Table 6.7). The best-preserved structure consisted of three small rooms and was originally of at least two storeys. Outside of the building, there is a fence, probably connected with the collection of manure for fertilization. This feature shows its use for cattle stabling.

Similar to site 1, the topographical relationship between terraces and buildings make it possible to extend the chronology of the terraces as well as of the buildings. Also, in this case, the dating of the terraces was suggested through the chronology of pottery fragments discovered in a big heap of stones (UT 9)

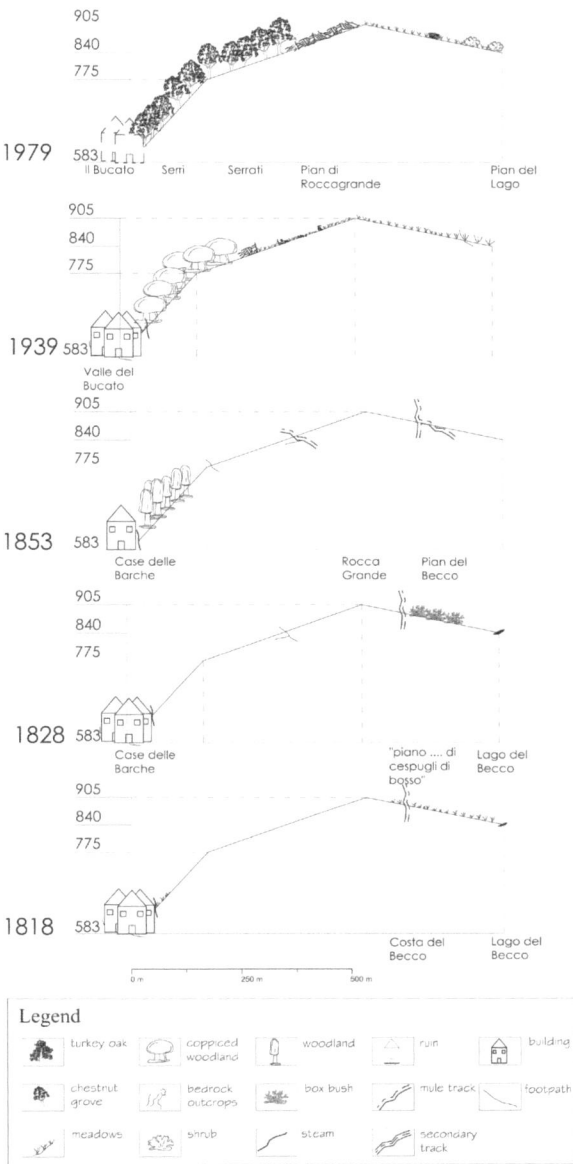

Figure 6.11 Maps of the surveys carried out in the upper Petronio Valley (after Stagno in LASA 2005, modified).

located on the upper terrace, interpreted as the results of the stone clearance just before the cultivation of the terraces, with sherds dated to the beginning of the first half of the 19th century, and therefore the terraces as well.[16] The analysis of historical cartography confirms this chronology (Table 6.7).

Pottery analysis offered some evidence for a reorganisation of the exploitation of the area. The signs of fracture visible on some ceramic fragments are attributable to the use of fork hoes for tilling. The hypothesis of the use of terraces for permanent crop cultivation is

---

the end of the 18th century and the 19th century, and a body fragment of a glazed closed vessel dated between the end of the 18th century and the early 19th century (Milanese and Biagini 1999:37).

15. Archivio di Stato di Genova, *Raccolta Cartografica*, 667 Casarza "1/3: "Section A de Bargone. En une feuille" " (c. 1805 – c. 1813), dim. Supporto: carta Altezza (mm): 66 Larghezza (mm): 982.

16. Two fragments of glazed and *ingobbiate* open table and transport wares (cf. Milanese and Biagini 1999).

Year	Cartography	Scale	Bargone Sito 1 *'Case delle Barche'*	Bargone Sito 2 *'Prato di Pozzo'*
1979	CTR 1979	1:1000	Il Bucato, two collapsed buildings and one standing	One building
1950s	Cadastral map (Bargone)	1:2000	Three buildings divided between several parcels	One standing, and one collapsing building, *Prato di Pozzo*
1938–1939	IGM sheet	1:25000	Five buildings	One building
1877	IGM sheet	1:25000	*C. delle Barche*	Two buildings *Pian della Zeppa*
1852–1854	Gran Carta degli Stati Sardiardi di Terraferma	1:50000	*Case delle Barche*	Not identified
1828	Tavoletta Manoscritta	1:20000	*Case delle Barche*, four buildings	No
1816–1827	Tavoletta Manoscritta	1:9450	Four buildings	No
1803–1812	Catasto Napoleonico	1 :...	*Barche hameau* (four buildings, one of them divided between different parcels)	Not identified

Table 6.7 Historical cartography evidence for the *casoni* of 'Case delle Barche'.

confirmed by a land register of 1935 (LASA 2005:99–107) and by information from oral sources about the abandonment of this area in the 1950s–1960s. Pottery fragments show different chronologies until the beginning of 20th century, which could be related to various stone clearance activities, connected to the cultivation of permanent crops. These fragments belong to cooking and transport wares and probably originate from the use of the nearby buildings, and were possibly employed for carrying ready-made meals from the Bargone homestead, a daily use of these structures, similar to that of the Casoni Lagorara.

Also these buildings have been used since the second half of 19th century inside the 'monticazione' system, where the relationships between cultivation (here including chestnut grove cultivation) and grazing were very strong (e.g. collection of manure used for fertilizing).

Casone del Giazzo (Val d'Aveto)

The research was part of investigations that sought to reconstruct the history of the cattle breeding system on the Aveto-Trebbia watershed.[17] Casone del Giazzo consists of a building located 300m southeast of the watershed, inside the common lands historically held by the hamlet ('villa') of Salto in the Parish of Priosa (Rezzoaglio) in the upper Aveto Valley. This valley borders the Sturla Valley, but till 1798 it was part of Feudi Imperiali of the Doria family, and not under the jurisdiction of the Repubblica di Genova.

The building existed in 1683, as we can see carved in the north door lintel (Fig. 6.12) that belongs to the oldest wall. Casone del Giazzo was continuously used from its construction (around 1683) to c. 1980, when it was abandoned. In the last phase, the first floor was used as a hay-barn (as probably was also the second floor), and the ground floor as a shed/stable.

Different phases of the building were defined through the stratigraphic analysis of standing walls. Three main transformations were identified, thanks to modifications in the wall characteristics.

First Phase: end of 17th century, not defined. During the construction, the building had two floors and was smaller; in this phase a large use of sandstone chips and slabs can be observed.

Second phase: not defined (?), late 19th century. The building was afterwards enlarged in height and width, with the prevalent use of larger stones. At this stage of the investigation it was not possible to be precise concerning the chronology of this enlargement that can only be placed between the first and the third phase.

Third phase: late 19th – today. The building reached the current height and appearance only in the late 19th century. The subsequent enlargement of Casone del Giazzo could have been due to the lack of space for feeding the cattle or as a result of a change in the function of the building. It is possible to assume a connection between this enlargement and the construction of small terraces near the building. These terraces could therefore correlate with the creation of permanent upland cultivation areas and to a definite utilisation of these areas by private individuals. Apparently in this case the use of the casone changed from a grazing purpose (and to store hay products) to being connected also to agricultural exploitation of the surrounding slopes.

---

17. LASA carried out investigations in the framework of the project "Parco dell'Aveto e Rete Natura 2000: boschi e biodiversità" (2006–2007). A first discussion of the results was presented in Cevasco *et al.* 2007.

Anna Maria Stagno: Seasonal settlements and husbandry resources in the Ligurian Apennines

\multicolumn{2}{|c|}{}	**'Casone del Giazzu'**	
Characteristics	Altitude	1110
	Number of buildings (architectural complex or building element)	One building
	Building dimensions	
	Evidence of feeders	Yes, preserved
	State of preservation	Standing
Methods	Investigation methods	
	Chronology methods	Direct: archaeology of architecture
Chronology	Chronology elements	Door lintel dated 1683
	Chronology of construction	Last 25 years of the 17th century
	Transformation	Building super elevation and enlargement (18th and 19th century)
	Continuity of use	Until 1950s
	Evidence of previous buildings	Not identified
Functional relationships	Distance from the hamlet	2.5 km from Salto
	Enclosure for manure collection	Not identified
	Presence of terraces	Nearby the building
	Presence of water-works	Not identified

Table 6.8   Synthesis of Casone del Giazzo archaeological investigations.

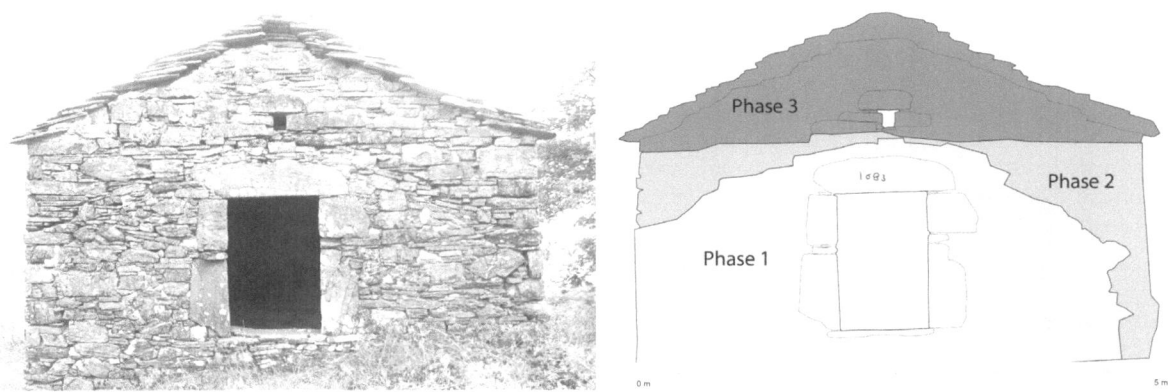

Figure 6.12   Sketch of the standing wall stratigraphic analysis at Casone del Giazzo.

6. Management systems of environmental resources

The history of Casone del Giazzo is in part different from the others; its location corresponds to *casoni* as documented by archival sources before the 18th century, when they were mainly built within common land pastures. This element is an indicator of the need to deepen the investigations in order to better understand these local management models and their transformations.

In fact, as discussed for the other settlements studied, location appears connected to wider changes in the local the management systems of environmental resources occurring between the 17th and 19th centuries. From the 18th century, *casoni* were no longer built in uplands and common lands, but rather at lower altitudes (between 600 and 850 m.a.s.l.), close to terraced slopes, and often at the boundary between private terraces and common land pastures.

All the case studies show substantial continuities in the use of the structures. With different chronologies, all the case studies offered many markers to support conclusions suggesting the seasonal, daily use of the *casoni* (Table 6.9), a use where shepherds occasionally stayed overnight, as well as their prolonged use as housing for cattle (only at night and not constantly). The latter of these uses relates to patterns where manure was collected to be used as fertilizer for the neighbouring terraces. First (and second) floors with big windows are related to their use as hay-barns for storing fodder and agricultural products (probably used also during winter, as testified by oral sources for Casoni Lagorara). During the summer, cattle were grazed in the upper common lands, while in winter they were stabled in stables in the hamlet. These clues confirm the presence of a system of 'monticazione', described by oral sources for the first half of the 20th century that had started to become organized in this way from the 18th century, with the shift of the construction of *casoni* to lower permanently cultivated and terraced slopes.

The modification in the location of *casoni*, due probably to conciliating agricultural and pastoral activities, highlights an intensification of agricultural activities (maybe related to the local demographic growth of the 18th century) and suggests a transformation in their use. *Casoni* from a use primarily connected with animal husbandry (temporary shelter for shepherds and animals and temporary storage for hay from meadows on the higher (common) lands, as it is clearly described by archival sources during the 16th to the first half of the 18th century), became more associated with permanent agricultural activities regularly realized in terraces: permanent crops and vegetable garden cultivation, fodder production, and chestnuts. This hypothesis could be applied also to other archaeologically investigated *casoni*, including some not discussed here, for instance Lavaggi di Chiappozzo (Ferrando Cabona and Mannoni 1989) and Casoni delle Piove (815 m.a.s.l.; Cevasco *et al.* 2006).

The shift in the location of *casoni* also had an effect on access to common land pastures (and probably in their management); cattle no longer stayed overnight in common land pasture, but rather in stables located on private lands, and from there moved daily to common land. Therefore, the shift in the location of *casoni* reveals the colonization of the slopes through the construction of terraces and the spread of irrigation systems, an intensification that also had a direct effect on the use of common lands and in their management.[18]

All the case studies can be compared with a well-documented phenomenon in Mediterranean mountain areas since the Middle Ages: the occupation of common lands. The Casone del Giazzo case study finds a precise parallel with the phenomenon of cabane and cortal studied in the French Pyrenees, within the framework of the transformation of estivage practices (Rendu 2003), and from a juridical point of view (Conesa 2012; Bille *et al.* 2007). Precisely from this point of view, the comparison is very close, even if in different periods (the appropriation of common land with terracing at Casone del Giazzo seems to have become permanent during the 19th century). However, there is a large difference: the case of Casone del Giazzo is a single family action, while the other cases document wider actions involving several families or the whole community of a hamlet ('villa'), as happened in Perlezzi.

During the 18th century, local husbandry systems underwent other modifications that have already been documented by the archaeological study of functional husbandry spaces carried out in two 'ville' located in Aveto and Trebbia Valleys: Ventarola (Stagno 2009a; 2009b); and Casanova di Rovegno (Tigrino *et al.* 2013). From the end of the 18th century, the primacy of local cattle breeding over sheep (and cattle) transhumant breeding is testified by the construction of new stables for bovines, with effluent discharge systems not documented in previous periods (when the differentiation between houses and spaces devoted to live-

---

18. Concerning the problem of the study of the organization of work in relation to seasonal settlement and access patterns studied from an ethno-archaeological point of view see Burri 2010.

Marker	Interpretation	Documented at	Chronology
Absence of permanent fireplace	Temporary use	Excavation at Casoni della Pietra	18th – second half of 19th the century
Presence of feeder	Cattle stabling	All the case studies (not in the excavation)	18th – 20th century
Exclusive presence of transport and table wares	Temporary use	Excavation at Casoni della Pietra; *casoni* of Bargone surveys	18th – second half of the 19th century
Small dimension of the ground-floor windows	Not prolonged cattle stabling	All the case studies	18th – 20th century
Wider dimension of the first-floor windows	Barns for hay and cultivation products, maybe not only seasonally	All the case studies	18th – 20th century
Limited caw litter	Not prolonged cattle stabling, temporary use	Micro-morphological analysis at Casoni della Pietra	19th – 20th century
Enclosures for manure collection	Collection of manure for fertilization	All the case studies	(surely) 19th – 20th century

Table 6.9   Summary of the markers of *casoni* functions.

stock was not so clear, and sheep breeding prevailed). During the 19th century at the Ventarola hamlet, new stables were built in the southeastern part of the hamlet near to the stream and to the vegetable gardens, in order to facilitate, during the fertilization period, the direct employment of the liquid manure, probably collected in slurry pits (from the discharge system of the stables – Plomteux 2000:66–67). In both cases during the 19th century, the growth in the size and number of stables and hay-barns testifies to a process of intensification in cattle breeding and the production of fodder resources (Cevasco 2014) that can be associated with the growth of seasonal settlements as documented for the *casoni* studied here.

In general the construction of new stables and barns can be correlated to the definitive change from transhumance breeding, that connected the coastal grazing land (winter quarters) to the Apennine pastures (summer quarters), to resident bovine breeding and the consequent need for more space for hay to feed cattle during the winter, and therefore of more stables (Moreno and Raggio 1991). These changes also can be linked to a change from a multiple to a mono-cultural management system of agro-sylvo-pastoral resources (Stagno 2009a; Moreno 1990). In the eastern Ligurian Apennines the LASA group in previous historical ecology and environmental archaeology research has documented the progressive diminishing of the local multiple land-use systems from the first half of the 19th century, where the same parcel could be used for temporary cultivation, wooded meadows and grazing. Very specialised agro-sylvo-pastoral cycles existed, one of which was analytically reconstructed for the Aveto Valley (Moreno *et al.* 1998; Bertolotto and Cevasco 2000; Beltrametti *et al.* 2014), on sites where alder trees prevailed. Since the first half of 19th century, it is possible to verify the spread of a monoculture system, characterized by spaces permanently and exclusively devoted to cultivation or meadows or pasture or forest (Cevasco 2007; Cevasco and Molinari 2009; Stagno 2011; Cevasco *et al.* 2005), or chestnut groves as well, as shown in the Lagorara and Bargone case studies.

This transformation cannot be linked directly to an increase in rural population. In these areas, the mountain population grew from the 18th century until the mid-19th century. The second half of the 19th century saw a decline, with a swift decrease at the beginning of the 20th century. In preference to seeing a close relationship between demography and the intensification of the exploitation in monoculture environmental resource systems, the transformation could be better linked to the actions promoted by new regulations adopted in the kingdom of Sardinia after 1815 that sought to homogenise, in a modern monoculture, the customary management of the forest and pastoral resources (Moreno 1990:54–56, 222). From the point of view of mountain economies it could also be the effect of the decline of the transhumance systems, but also of the commercial role of the Apennine passes that linked coastal porti to the inland (the pianura padana), in which all the areas studied were involved through different routes that local communities controlled at least until the mid-19th century.

Acknowledgements

This study has been carried out as part of the project "History and Archaeology of Rural Society" (D.R.1080, 16.11.2010, University of Genoa), in the frame of the researches of Laboratory of Environmental Archaeology and History (LASA, Genoa University, Italy, www.lasa.unige.it) focused on the historical reconstruction of environmental resources management systems by geographers, archaeologists, historians and archaeobotanists. I wish to thanks Dr. Robert Hearn for language assistance. However, any errors are my exclusive responsibility.

Bibliography

Ambrosoli, M. 2011. Pastorizia e agricoltura nel Friuli in età moderna. In A. Mattone and P.F. Simbula (eds.) *La pastorizia mediterranea: storia e diritto secoli XI–XX*. Pp. 667–690. Roma: Carocci.

Andrews, D. and Cima, M. (eds.) 1984. *Dal villaggio alla malga: primo contributo per un'archeologia degli insediamenti storici in valle Orco (scavi a Uvera e Pian Cravere)*. Courgné (TO): Edizioni Corsac.

Angelucci, D.E., Boschian, G., Fontanals, M., Pedrotti, A. and Vergès, J.M. 2009. Shepherds and karst: the use of caves and rock shelters in the Mediterranean region during the Neolithic. *World Archaeology* 41/2:191–214.

Asins Velis, S. 2006. Linking historical Mediterranean terraces with water catchment, harvesting and distribution structures. In J.-P. Morel, J. Tresserras Juan and J.C. Matamala (eds.) *The Archaeology of Crop Fields and Gardens*. Santo Spirito (BA): Edipuglia. Pp. 21–40.

Barceló, M., Kirchner, H. and C. Navarro, 1996. *El agua que no duerme: fundamentos de arqueología hidráulica andalusí*. Granada: Fundación El legado andalusì.

Bazzanella, M., Kezich, G., Pisoni, L., Toniutti, L. 2014. Montagne dipinte: le scritte dei pastori fiemmesi (TN) tra etnoarcheologia e studi di cultura materiale. In A.M. Stagno (ed.) 2014:357–367.

Beltrametti, G., Cevasco, R., Moreno, D. and Stagno, A.M. 2014. Cultures temporaires entre longue durée et chronologie fine: traces des pratiques dans les sols, la végétation et les textes. In Christine Rendu and Roland Viader (eds.) *Cultures temporaires et féodalité: les cycles culturaux et l'appropriation du sol dans l'Europe Médiévale et Moderne*, Toulouse. Pp. 235-258.

Bertolotto, S. and Cevasco, R. 2000. Fonti osservazionali e fonti testuali: le 'Consegne dei Boschi' e il sistema dell''Alnocoltura' nell'Appennino Ligure Orientale (1822). *Quaderni Storici* 103, a. 35/1:87–108.

Bille, É., Conesa, M. and Viader, R. 2007. L'appropriation des Espaces Communautaires dans l'Est des Pyrénées Médiévales et Modernes: enquête sur les Cortals. In P. Charbonnier, P. Couturier, A. Follain and P. Fournier (eds.) *Les espaces collectifs dans les Campagnes, XIe–XXIe siècle*. Clermont-Ferrand: Presses Universitaires Blaise-Pascal. Pp. 177–192.

Brogiolo, G.P., Angelucci, D.E., Colecchia, A. and Remondino, F. (eds.) 2012. *Apsat 1. Teoria e metodi della ricerca sui paesaggi d'altura*. Mantova: SAP Società Archeologica srl.

Burri, S. 2010. Le problème de la mobilité des artisans, pasteurs et usagers de l'incultum en Basse-Provence centrale (XIIe–XVIe siècle). In *Des sociétés en mouvement: migrations et mobilité au Moyen Âge, XLe Congrès de la SHMESP (Nice, 4–7 juin 2009)*. Paris: Publications de la Sorbonne, Histoire ancienne et médiévale 104. Pp. 131–136.

Carrer, F., Angelucci, D.E. and Pedrotti, A. 2013. Montagna e pastorizia: stato dell'arte e prospettive di ricerca. In D.E. Angelucci *et al.* (eds.) *Apsat 2. Paesaggi d'Altura del Trentino: evoluzione naturale e aspetti culturali. Progetti di Archeologia*. Mantova: SAP, Società Archeologica srl. Pp. 125–140.

Cevasco, R. 2007. *Memoria Verde: nuovi spazi per la geografia*. Reggio Emilia: Edizioni Diabasis.

Cevasco, R. (ed.) 2013. *La natura della montagna: scritti in ricordo di Giuseppina Poggi*. Sestri Levante: Oltre Edizioni.

Cevasco, R. 2014. Archeologia dei versanti montani: l'uso di fonti multiple nella ricerca geografica. In E. Dai Prà (ed.) *Approcci geo-storici e governo del territorio. Vol. 2. Scenari nazionali e internazionali*. Milano: Franco Angeli. Pp. 361–375.

Cevasco, A., Cevasco, R., Gemignani, C.A., Marrazzo, D., Spinetti, A. and Stagno, A.M. 2007. Archaeological and ecological evidence of rearing practices: fodder and water resources management in post-medieval Ligurian Apennines (NW Italy). In *Medieval Europe Paris 2007. 4ème Congrès International d'Archéologie Médiévale et Moderne*. http://medieval-europe-paris-2007.univ-paris1.fr.

Cevasco, R., Marrazzo, D., Marullo, E., Spinetti, A. and Stagno, A.M. 2007. *Archeologia e storia dei pascoli dei Siti di importanza comunitaria Monti Caucaso e Ramaceto*. Poster, Comunità Montana Fontanabuona.

Cevasco, R., Marullo, E. and Stagno, A.M. 2005. L'analisi della cartografia storica per lo studio delle variazioni della copertura vegetale nel SIC RoccaGrande-M.te Pu (Liguria orientale). In *Atti della 9a Conferenza Nazionale ASITA (Federazione delle Associazioni Scientifiche per le Informazioni Territoriali e Ambientali) Catania 15–18 novembre 2005*. Catania: ASITA. Pp. 683–688.

Cevasco, R. and Molinari, C. 2009. Microanalysis in woodland historical ecology: evidences of past leaf fodder production in NW Apennines (Italy). In E. Saratsi, M. Bürgi, E. Johann, K. Kirby, D. Moreno and C. Watkins (eds.) *Woodland Cultures in Time and Space: tales from the past, messages for the future*. Athens: Embryo Publications. Pp. 147–154.

Cevasco, R., Moreno, D., Poggi, G. and Rackham, O. 1997–1999. Archeologia e storia della copertura vegetale: esempi dell'Alta Val di Vara. *Memorie della Accademia Lunigianese di Scienze 'Giovanni Cappellini'* 67–68–69:241–261.

Charles, M., Halstead, P. and Jones, G. (eds.) 1998. Fodder: archaeological, historical and ethnographic studies. *Environmental Archaeology* 1:1–123.

Conesa, M. 2012. *D'herbe, de terre et de sang: La Cerdagne du XIVe au XIXe siècle*. Perpignan: Presses Universitaires de Perpignan.

De Guio, A. and Migliavacca, M. (eds.) 2012. Archeologia di montagna a Recoaro (VI): la frequentazione delle alte quote in Età postmedievale. Risultati delle campagne di ricognizione e scavo 2006–2010. *Archeologia Postmedievale* 14 (2010):153–168.

De Maestri, S. and Moreno, D. 1980. Contributo alla storia della costruzione a secco nella Liguria rurale. *Archeologia Medievale* 7:319–342.

Ferrando Cabona, I. and Mannoni, T. 1989 (1993). *Liguria: ritrato di una regione*. Genova: Sagep Editrice.

Ferrando Cabona, I., Gardini, A. and Mannoni, T. 1978. Zignago 1: gli insediamenti e il territorio. *Archeologia Medievale* 5:273–374.

Ferrando Cabona, I., Mannoni, T. and Pagella, R. 1989. Cronotipologia. *Archeologia Medievale* 16:647–661.

Gaio, S. 2014. Archeologia e storia di un fienile della valle di Primiero (TN): un approccio pluridisciplinare allo studio de un contesto insediativo rurale (sec. XV–XX). In A.M Stagno (ed.) Montagne incise: archeologia delle risorse nella Montagna Mediterranea. *Archeologia Postmedievale* 17 (2013):369–380.

Giovannetti, L. 2004. Archeologia e storia della Montagna della Garfagnana e delle sue risorse: il caso di Gorfigliano nel più ampio contesto apuano e appenninico. In J.A. Quirós Castillo (ed.) *Archeologia e storia di un castello apuano: Gorfigliano dal Medioevo all'età moderna*. Pp. 225–252. Firenze: All'Insegna del Giglio.

Gobbato, S. 2004. L'architettura di Gorfigliano in età postmedievale. In J.A. Quirós Castillo (ed.) *Archeologia e storia di un castello apuano: Gorfigliano dal Medioevo all'età moderna*. Firenze: All'Insegna del Giglio. Pp. 205–224.

Guilaine, J. (ed.) 1991. *Pour une Archéologie agraire. À la croisée des sciences de l'homme et de la nature*. Paris: A. Colin.

Harfouche, R. 2005. Retenir et cultiver le sol sur la longue durée: les terrasses de culture et la place du bétail dans la montagne méditerranéenne. *Anthropozoologica* 40/1:45–80.

Harfouche, R. 2007. *Histoire des paysages méditerranéens terrassés: aménagements et agriculture*. Oxford: Archaeo-press.

Ingold, A. 2011. Écrire la nature: de l' histoire sociale à la question environnementale? *Annales, Histoire, Sciences Sociales* 66:11–29.

Kirchner, H. (ed.) 2010. *Por una arqueología agraria: perspectivas de investigación sobre espacios de cultivo en las sociedades medievales hispánicas*. Oxford: British Archaeological Reports International Series, 2062.

Kirchner, H. and Navarro, C. 1993. Objetivos, metodos y practica de la arqueología hidráulica. *Archeologia Medievale* 20: 121–150.

LASA 2005. OB.2 Misura 2.6 b 'Realizzazione Rete Natura 2000' Progetto Roccagrande. La storia dell'uomo e della natura. Valorizzazione del sito 'Roccagrande – M. Pu'. Unpublished final report available at the LASA Lab., Università degli Studi di Genova.

Le Couédic, M. 2010. *Les pratiques pastorales d'altitude dans une perspective ethnoarchéologique: cabanes, troupeaux et territoires pastoraux pyrénéens de la préhistoire à nos jours*. Thèse de doctorat, Tours.

Lorenzini, C. 2011. Monte versus bosco, e vice-versa. Gestione delle risorse collettive e mobilità in area alpina: il caso della Carnia fra Sei e Settecento. In G. Alfani and

R. Rao (eds.) *La gestione delle risorse collettive: Italia settentrionale, secoli XII–XVIII*. Milano: Franco Angeli. Pp. 95–109.

Mannoni, T. 1970. Sui metodi dello scavo archeologico nella Liguria montana. *Bollettino Ligustico* 32:51–64.

Mannoni, T. 1984. Metodi di datazione dell'edilizia storica. *Archeologia Medievale* 11:396–403.

Mannoni, T. 1997. Archeologia globale e archeologia postmedievale. *Archeologia Postmedievale* 1: 21–28.

Mannoni, T. 1998. Analisi archeologiche degli edifici con strutture portanti non visibili. *Archeologia dell'Architettura* 3:81–85.

Mannoni, T., Cabona, D. and Ferrando, I. 1988. Archeologia globale del territorio: metodi e risultati di una nuova strategia della ricerca in Liguria. In G. Noyé (ed.) *Structures de l'habitat et occupation du sol dans les pays méditerranéens: le méthodes et l'apport de l'archéologie extensive. Actes du colloque de Paris (12–15 novembre 1984)*. Roma–Madrid: École française de Rome. Pp. 43–58.

Marullo, E. 2012. *Fonti topografiche e toponomastica per lo studio del patrimonio storico-ambientali delle valli Aveto e Petronio*. Tesi di dottorato in 'Geografia storica per la valorizzazione del patrimonio storico-ambientale', Università degli Studi di Genova.

Milanese, M. and Biagini, M. 1999. Archeologia e storia di un 'alpeggio' dell'Appennino ligure orientale: i Casoni della Pietra nella valle Lagorara (Maissana, SP) (XVII–XX sec.). *Archeologia postmedievale* 2 (1998):9–54.

Molinari, C. 2010. *Ricerche palinologiche per l'identificazione di sistemi agro-silvo-pastorali storici*. Tesi di Dottorato in Geografia Storica per la Valorizzazione del Patrimonio Storico-Ambientale, Università degli Studi di Genova.

Moreno, D. 1990. *Dal documento al terreno: storia e archeologia dei sistemi agro-silvo-pastorali*. Bologna: Il Mulino-Ricerche.

Moreno, D., Cevasco, R., Bertolotto, S. and Poggi, G. 1998. Historical ecology and post-medieval management practices in alder woods (Alnus incana (L.) Moench) in the northern Apennines, Italy. In K. Kirby and C. Watkins (eds.) *The Ecological History of European Forests (vol. 2)*. Oxford: CABI. Pp. 185–201.

Moreno, D., Montanari, C., Stagno, A.M. and Molinari C. 2010. A plea for a (New) Environmental Archaeology: the use of the geographical historical microanalytical approach in mountain areas of NW Italy. In S. Tzortzis and X. Delestre (eds.) *Archéologie de la montagne européenne. Actes de la table ronde international de Gap, 29 septembre–1er octobre 2008*. Condé Sur Noireau: Éditions Errance (Bibliothèque d'Archéologie Méditerranée et Africaine 4). Pp. 75–83.

Moreno, D. and Raggio, O. 1991. The making and fall of an intensive pastoral land-use-system in eastern Liguria, 16th–19th centuries. In R. Maggi, R. Nisbet and G. Barker (eds.) Archeologia della pastorizia nell'Europa meridionale, Chiavari 22–24 settembre 1989. *Rivista di Studi Liguri* 56/1–4 (1991–1992):193–217.

Moscatelli, U. 2011. Tra dibattito teorico e prassi operativa: lo studio del paesaggio medievale nel progetto R.I.M.E.M. In G. Capriotti and F. Pirani (eds.) *Incontri: storie di testi, immagini, oggetti*. Pp. 89–112. Macerata: EUM Edizioni.

Ottomano, C. 1999. Osservazioni geoarcheologiche sul sito. In M. Milanese and M. Biagini Archeologia e storia di un 'alpeggio' dell'Appennino ligure orientale: i Casoni della Pietra nella valle Lagorara (Maissana, SP) (XVII–XX sec.). *Archeologia postmedievale* 2 (1998):30–37.

Plomteux, H. 2000. *Cultura contadina in Liguria: La Val Gravegia.* Sagep, Genoa (1st edition 1980).

Quiròs Castillo, J.A. (ed.) 2004. *Archeologia e storia di un castello apuano: gorfigliano dal Medioevo all'età moderna*. Firenze.

Raggio, O. 1990. *Faide e parentel: lo stato genovese visto dalla Fontanabuona*. Torino: Einaudi.

Raggio, O. 1995. Norme e pratiche: gli statuti campestri come fonti per una storia locale. *Quaderni Storici* 88/1:155–194.

Redi, F. and Di Blasio, L. 2010. *Segni del paesaggio agropastorale: il territorio del Gran Sasso-Monti della Laga e dell'Altopiano di Navelli*. L'Aquila: L'Una (Archeologia e Territorio, 1).

Rendu, C. 2003. *La montagne d'Enveig: une estive pyrénéenne dans la longue durée*. Perpignan: Ed. Du Trabucaire.

Rossi, M. and Gattiglia, A. 2007. Pierre, écriture et figure dans le vallon du Longis (Molines-en-Queyras, Hautes-Alpes). In T. Mannoni, D. Moreno and M. Rossi (eds.) Pietra scrittura e figura in età postmedievale nelle Alpi e nelle regioni circostanti. *Archeologia Postmedievale* 10 (2006):17–40.

Stagno, A.M. 2009a. *Archeologia rurale: spazi e risorse. Approcci teorici e casi di studio*. Tesi di dottorato in 'Geografia storica per la valorizzazione del patrimonio storico-ambientale', Università degli Studi di Genova.

Stagno, A.M. 2009b. Geografia degli insediamenti e risorse ambientali: un percorso tra fonti archeologiche e documentarie (Ventarola, Rezzoaglio GE). In G. Macchi Janica (ed.) *Atti del convegno 'Geografie del Popolamento'. Casi di studio, metodi e teorie (Grosseto 2008)*. Siena: Edizioni dell'Università. Pp. 301–310.

Stagno, A.M. 2011. Mapas históricos y recursos ambientales: la filtración cartográfica de área y el caso de Riomaggiore (Cinque Terre, Italia). *Investigaciones Geográficas* 53 (2010):189–215.

Stagno, A.M. 2012. Casa rurale e storia degli insediamenti. Un approccio geografico per l'archeologia dell'edilizia storica. In F. Redi and R. Forgione (eds.) *Atti del VI Congresso nazionale di archeologia medievale (L'Aquila 2013)*. Firenze: All'Insegna del Giglio. Pp. 23–27.

Stagno, A.M. (ed.) 2014. Montagne incise, pietre incise: archeologia delle risorse nella montagna mediterranea. *Archeologia Postmedievale* 17 (2013):5–438.

Stagno, A.M. 2015. *Lo spazi locale dell'archeologia rurale: risorse ambientali e insediamenti nell'Appennino ligure tra XV e XX secolo*. Firenze.

Stagno, A.M. and Tigrino, V. 2012. Beni comuni, proprietà privata e istituzioni: un caso di studio dell'Appennino ligure (XVIII–XX secolo). In P. Nervi (ed.) *Annali del Centro studi e documentazione sui demani civici e le proprietà collettive*. Pp. 261–302. Milano: Giuffré editore (Archivio Scialoja-Bolla, 1).

Tigrino, V., Beltrametti, G., Stagno, A.M. and Rocca, M. 2013. Terre collettive e insediamenti in alta val Trebbia (Appennino Ligure): la definizione della località tra Sette e Novecento. In P. Nervi (ed.) *Annali del Centro studi e documentazione sui demani civici e le proprietà collettive*: Pp. 105–156. Milano: Giuffré editore (Archivio Scialoja-Bolla, 1).

Torre, A. 2008. Un tournant spatial en histoire? In Paysages, regards, ressources. *Annales HSS* 63:1127–1144.

Troletti, F. 2014. Rapporto tra incisioni di epoca storica e frequentazione umana in alcuni siti con incisioni rupestri della Valcamonica. In A.M. Stagno (ed.) 2014. Montagne incise, pietre incise: archeologia delle risorse nella montagna mediterranea. *Archeologia Postmedievale* 17 (2013):345–356.

Tzortzis, S. and Delestre, X. 2010. Avant-propos. In S. Tzortzis and X. Delestre (eds.) *Archéologie de la montagne européenne. Actes de la table ronde international de Gap, 29 septembre–1er octobre 2008*:9–10. Condé Sur Noireau: Éditions Errance (Bibliothèque d'Archéologie Méditerranée et Africaine 4).

Van Leusen, M., Pizziolo, G. and Sarti, L. (eds.) 2011. *Hidden Landscapes of Mediterranean Europe. Cultural and methodo-logical biases in the pre- and protohistoric landscape studies. Proceedings of the International Meeting (May 25–27, 2007)*. Oxford: British Archaeological Reports.

Vecchio, B., Piussi, P. and Armiero, M. 2002. L'uso del bosco e degli incolti. In R. Cianferoni, Z. Ciuffoletti and L. Rombai (eds.) *Storia dell'agricoltura italiana III*: L'età contemporanea*. Firenze: Edizioni Polistampa. Pp. 128–216.

Anna Maria Stagno, Laboratory of Environmental Archaeology and History, DAFIST, University of Genoa, Italy.
Research Group on Heritage and Cultural Landscapes (GIPyPAC), University of the Basque Country (UPV/EHU).
Email: anna.stagno@unige.it

7. The 'invisible' shepherd and the 'visible' dairyman: Ethnoarchaeology of alpine pastoral sites in the Val di Fiemme (eastern Italian Alps)

Francesco Carrer

Pastoral groups are often considered 'invisible' by archaeologists, as their mobility is supposed to affect the archaeological visibility of their sites. In order to tackle this invisibility issue, ethno-archaeological research was carried out in the eastern Italian Alps (Val di Fiemme, Trentino province). It enabled the identification of two husbandry strategies, one focused on dairying animals (reared for their milk) and the other on non-dairying animals (reared for their wool and meat). It was noticed that the seasonal sites related to the 'dairying' strategy are more complex and less ephemeral than those related to the 'non-dairying' strategy. This led to the conclusion that the 'non-dairying' pastoralists are less visible in the archaeological record than the 'dairying' pastoralists. This inference enhances the understanding of specific mountain archaeological sites, and also confirms that ethnoarchaeology has the potential to solve specific archaeological problems, such as those related to archaeological visibility.

Pastoralism: The invisible culture?

Pastoral groups have often been perceived as invisible cultures within the archaeological framework (Cribb 1991:67–80). This perception depends mainly on the limited archaeological visibility of their sites and facilities, related to their mobility. This invisibility biases the archaeological investigation of pastoralism, as the absence of remains does not enable in-depth evaluations of mobility trends and settlement patterns. Furthermore this issue is not limited to Asian (Chang 2006; Hole 1978) or African (Gifford 1978) contexts, but can be referred also to European mountain zones. Here a short range pastoral mobility (a few hundred kilometres) is still carried out nowadays, especially in the Iberian Peninsula (mainly in the Asturias and in the Pyrenees), in the Alps, in the Apennines, in the Balkans, in the Carpathians and in some parts of Anatolia. Khazanov (1984:22–23) called this strategy *yaylag pastoralism*, but it is more usually defined as 'transhumance' in the northern Mediterranean area (Boyazoglu and Flamant 1990). It is characterised by a seasonal shift between valley bottoms or plains (in winter) and mountain high plateaux (in summer) (Davies 1941). This strategy forces mobile groups to exploit summer and winter grazing areas for a few months every year, and for this reason transhumant shepherds are supposed to use ephemeral sites. Such ethnographic overview suggests that pastoral groups had ephemeral and so invisible sites also in the past.

However, in the last thirty years archaeological research, focused on pastoral landscapes and high European mountain plateaux, documented many dry-stone structures exploited during the prehistoric, protohistoric, Roman and medieval periods (Angelucci *et al.* 2013; Gassiot *et al.* 2010; Hebert and Mandl 2009; Walsh *et al.* 2007; Walsh *et al.* 2005; Rendu 2003; Horvat 1999; Barker and Grant 1991). This evidence demonstrates that the main problem of pastoralism in European archaeology is the lack of focused research, rather than the actual dearth of seasonal pastoral sites. Nevertheless, this suggestion does not solve the problem. As a matter of fact, in some areas and for some historical and prehistoric periods, the archaeological sites found in summer pastures are very few despite the increase in research and the improvement of research methodologies (Walsh 2005). Furthermore the interpretation of dry-stone structures as seasonal pastoral facilities is still doubtful and difficult to validate (Primas 1999). It seems that, rather than solving the problem of pastoral invisibility, this new evidence provided a new issue to tackle: the recognisability of the function of seasonal archaeological sites (Carrer 2012; see also Simms 1988).

Archaeological investigations and new archaeological data alone are therefore insufficient for the purposes of the archaeology of pastoralism. Ethnoarchaeology (see David and Kramer 2001), instead, may provide the interpretative tools necessary to change our perspective and to include the archaeological remains within a proper anthropological framework (Chang and Koster 1986). In this paper an ethno-archaeological study of upland pastoral strategies will be proposed. It will interpret how the activities and the goals of pastoral economy affect the composition of pastoral groups, their seasonal dwellings and their facilities. The study area will be a valley of the eastern Italian Alps.

Keywords: Pastoralism, archaeological visibility, upland sites, dairying, ethnoarchaeology, eastern Alps

Figure 7.1   The location of Val di Fiemme within the alpine region.

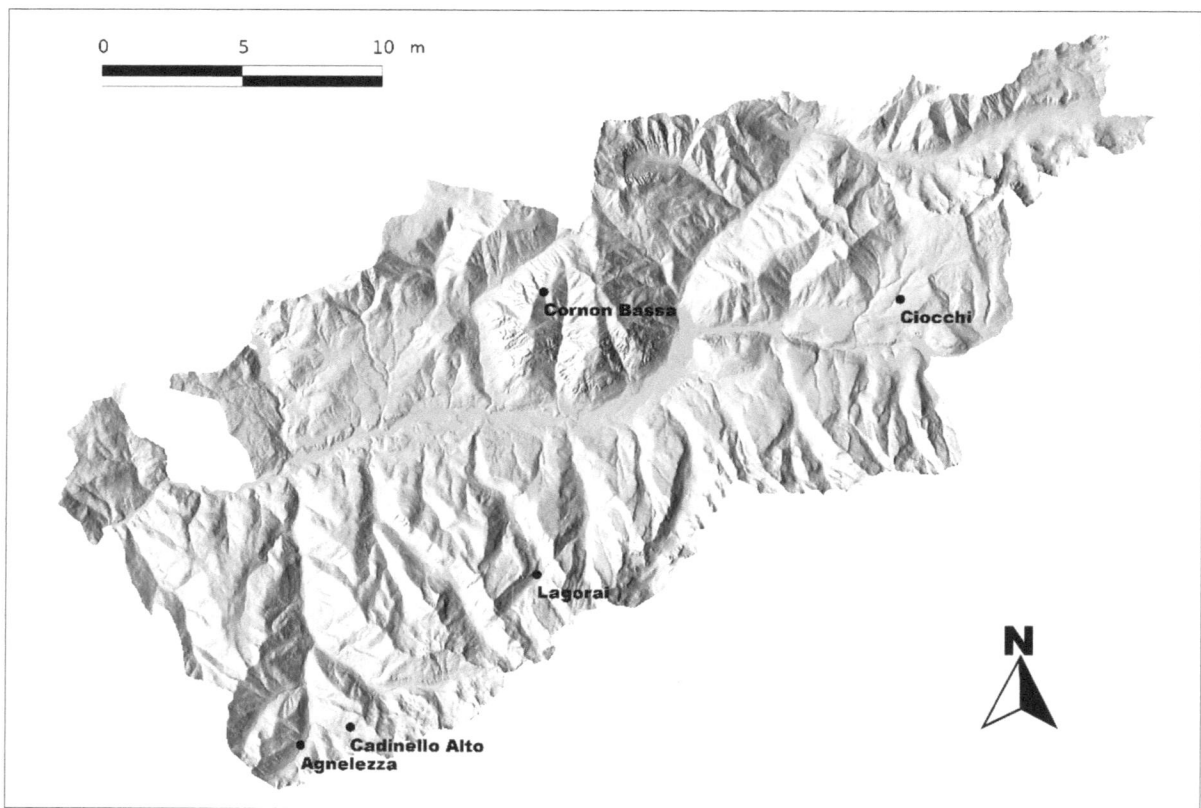

Figure 7.2   The location of the selected seasonal pastoral sites within the territory of Val di Fiemme.

An ethnoarchaeological approach to pastoral invisibility

This study starts with a basic question that will be detailed and answered in the next paragraph: what actually is pastoralism in the Alps? For Khazanov (1984:23) and Cribb (1991:19) there is only one type of middle-short range pastoral mobility in European mountains (and so in the Alps). However it is important to point out that their interest was on nomads, and that they believe transhumance to be a marginal manifestation of the more specialised nomadic pastoralism.

Other researchers (mainly geographers, anthropologists, historians, ecologists and ethno-archaeologists), instead, dealt specifically with the transhumance strategy and its internal variation, and they all agreed that there are two main types of movement that can be ascribed to transhumance: 'short' (or 'vertical' or 'Alpine'); and 'long' (or 'horizontal' or 'Mediterranean') (Gonzalez-Alvarez 2013; Baker 1999; Chang 1999; Clearly and Delano-Smith 1991; Boyazoglu and Flamant 1990; Braudel 1949:5–50; Pracchi 1942; Davies 1941; Frödin 1940/41). The first one is often undertaken by peasants who summer their livestock on small upland seasonal sites (*brañas* in the Asturias, *malghe* in the eastern Italian Alps, etc.) and winter them in the valley-bottom villages. The main goal of this strategy is to exploit the milk of sheep, goats and cattle to produce hard cheese, a fundamental food reserve for the winter months and, nowadays, a potential source of income (Viazzo and Wolf 2001). The second strategy, instead, is deeply related to trading: professional shepherds used to shift with their flock (of sheep and goats) between the plain and the high mountain plateaux in order to exploit the available pastures and to sell wool and lambs in the fairs; therefore these shepherds can be considered proper entrepreneurs rather than simple herders (Smerdel 1999). A secondary strategy might be also added to this framework; it has the same goals of 'long/horizontal' transhumance (namely wool and lambs) but has the same mobility range of 'short/vertical' transhumance, and hence it can be considered a sort of intermediate step within the spectrum of variation between the aforementioned short and long transhumance.

Oversimplifying this complex topic, it might be inferred that there are two different pastoral strategies in the European mountains (and consequently in the Alps). One has a limited mobility (from the mountain village to a mountain seasonal upland farm) and is focused mainly on dairying (initially for subsistence and then for trade); the other has a broader mobility (from fields in the plain to mountain pastures) and is focused mainly on trade in wool and lambs. Besides, there is an intermediate alternative version, characterised by limited mobility and the wool/lamb trade. The selected ethno-archaeological case-study has been analysed according to these theoretical categories.

The case-study: Val di Fiemme (Trentino province, eastern Italian Alps)

The sample area selected for this study is Val di Fiemme (Fig. 7.1), a 490km^2 valley in the eastern part of the Trentino province (eastern Italian Alps). It is part of the Avisio stream basin, and its orientation is northeast/southwest. The two sides of the valley have a significant geological difference: the north side is mostly made of limestone, while the south side is made of porphyric rocks. This dichotomy influences the quality of upland pastures (higher on the north side) as well as the quantity of surface watercourses (higher on the south side), but it affects also the local morphology. In fact the lower slopes of the north side are steeper than those of the south side, as the Pleistocene glacier that shaped local morphology eroded the limestone and the porphyric rocks in different ways. This morphologic feature, together with the abundance of surface watercourses, favours the pastoral exploitation of the southern side high plateaux (Morandini 1941).

But the main reason for selecting this valley as a case-study was not related to specific geological and morphological features, but to its particular historical evolution. In the Val di Fiemme, in fact, a closed corporate peasant community (Wolf 1957) is documented since the 12th century AD; it was named 'Magnifica Comunità di Fiemme' and managed most of the local woods and pastures, allowing all the peasants of the valley (*vicini*) to access these resources (Sartori Montecroce 2002). The socio-economic and political changes that took place in Europe from the end of the 18th and the beginning of the 19th century caused the disappearance of many of these communities, previously widespread within the alpine arc (Netting 1976). But the 'Magnifica Comunità di Fiemme' survived, and it keeps managing the woods and the high pastures on behalf of local peasants. This continuity enabled the survival of traditional pastoral strategies in this valley, which are carried out in the communal grazing areas during the summer (Croce 1972). Hence the Val di Fiemme has been selected as a case-study mainly for the important economic role that pastoralism still maintains in this area.

Five different herding strategies have been isolated within the pastoral economy of Val di Fiemme. Two of them correspond to the aforementioned 'short' transhumance, respectively of dairying cattle and dairying goats. One corresponds to the 'long' transhumance of non-dairying sheep; another corresponds to the cited intermediate variant, with short range mobility and wool/lamb trade (non-dairying sheep). A peculiar strategy is also undertaken by local herders: young cattle (that do not produce milk yet) are brought to the upland pastures to make them used to the mountain environment. In order to properly analyse all these different strategies, five cases-studies have been selected (Fig. 7.2). The collected data will be presented in the following paragraph.

Summer Farms: Seasonal exploitation of the uplands from prehistory to the present

Figure 7.3   The stable of Malga Cadinello Alta (Val di Fiemme, Trentino, Italy).

Figure 7.4   The hut of Malga Agnelezza, where the herders live during the summer (Val di Fiemme, Trentino, Italy).

Figure 7.5 Baito dei Ciocchi, a former hay-makers' site (now a transhumant site) close to the Lusia pass, between Val di Fiemme and Val di Fassa (Trentino, Italy).

The fieldwork

The fieldwork was carried out by the author of this paper during the summer of 2010. Therefore all the accompanying data refer to that time span.

The first ethno-archaeological field campaign was undertaken in the seasonal pastoral site called *Malga Cadinello Alta* (Val Cadino, southern side of the Val di Fiemme). It is a compound site with four masonry buildings: a dwelling; a stable (Fig. 7.3); an old hut used (in the past) to process the milk; and a former sty for pigs that contains the generator. The herd is composed of 66 cattle which are milked twice a day. The work group is an entire family (father, mother and two sons): the younger son herds the animals during the day, while the other members of the family just milk them, clean the activity areas (stable, milking area) and store the milk. The cheese-making activity is no longer allowed in this upland site, hence a van comes every morning to take the milk and take it to the communal dairy in the valley bottom.

Another dairying site called *Malga Agnelezza* has been analysed close to the site described above. Here almost 300 goats are summered by a work group of three people: two men (brothers) who milk the goats twice a day, clean the activity areas, store the milk and take the flock to the pastures; a woman (wife of one of the two brothers) who helps milking and cleaning the activity areas, and who also cooks and manages the seasonal dwelling. The entire site is composed of three structures: a masonry hut (Fig. 7.4), used as dwelling but in the past also as dairy and cellar for maturing the cheese; a masonry stable, with an annexe milking area; and a permanent wooden enclosure, next to the stable. As previously said, the goats are milked every morning and every evening, but the milk-processing activity is not allowed any more in this site, and a van comes every morning to take the milk to the town dairy.

The third site studied was a hut called *Baito dei Ciocchi* (Fig. 7.5), located in the Lusia Pass (close to the Val di Fassa territory) and occupied by a group of shepherds who have a 'long' transhumance strategy, and who used to winter their animals in the eastern Po plain (about 300km to the south) (see Perco 1982). The hut was built and exploited by hay-makers until the mid-20th century, and is now rented by these transhumant shepherds for the summer period. The animals (1500 sheep) are enclosed during the night inside an electrified mobile pen which is moved every 3–4 days (to have a uniform manuring of the pasture). The work group is composed of two shepherds who take the livestock to the pasture every morning and bring them back to the site every evening. Hence they have no hard work to do, and their workday is quite short. The wool is sheared in autumn and/or spring, and most of the lambs are sold before Easter (that is before the shepherds start to move from the plain to the mountains), so the upland summering is perceived by the shepherds as a rest period with limited financial income.

Figure 7.6   Malga Lagorai, a former dairying site, now a non-dairying pastoral site (Val di Fiemme, Trentino, Italy).

Another site was *Malga Lagorai* (Fig. 7.6) in the Val Lagorai (on the southern side of the Val di Fiemme). It is a former upland farm (*malga*) used to summer dairying cattle, but it is now exploited by shepherds of non-dairying sheep. The reason of this change of function is simple: as has been previously pointed out, the dairying activity is no longer allowed in the uplands of Val di Fiemme, and the milk from the seasonal sites has to be brought to the town dairies every morning; unfortunately this *malga* is difficult to reach with vehicles, and this inhibits the daily transportation of milk. Thus *Malga Lagorai* is no longer suitable for dairying purposes, and it has become a shelter for herders of non-dairying sheep. It is composed of two masonry buildings: a hut (former dwelling, milk-processing area and cellar for maturing the cheese) used as shelter; and a stable that is no longer used, as the current shepherds enclose their sheep (almost 1000) inside electrified portable pens. This site is used for a short time span (2–3 weeks), as these shepherds move very often from one site to another, searching for the most suitable pastures in different upland valleys.

The last site studied is the only one located on the northern side of the Val di Fiemme, in the high plateaux of Mount Cornon. It is a small wooden hut (called *Baito della Bassa*), built by hay-makers in the 20th century and currently used as a seasonal dwelling by a herder who looks after young non-dairying cattle (Fig. 7.7). These cattle are left free in the pastures, and their movements are constrained only by an electrified net that surrounds the entire grazing area. Hence the only activity this shepherd is required to do is controlling the animals in the morning and in the evening.

Ethno-historical overview

The fieldwork data showed that the way the shepherds exploit mountain landscapes has changed in the last century (see Carrer 2013a). The greatest transformations have taken place in the alpine arc since the Second World War and have caused a profound alteration of the traditional alpine economy (Netting 1981). An historical overview of seasonal pastoralism in the eastern Italian Alps is therefore needed to verify which were the traditional work groups and sites exploited in the uplands until the first half of the 20th century.

For the exploitation strategy of the dairying sites (*malghe*), the work group in the 19th – early 20th century was large (averaging 10 persons) and composed of herders and dairymen. The dairymen dealt with milk processing and cheese maturing, while the herders took the livestock to the pastures. Each *malga* had three fundamental activity areas: a milking area (a parlour or an enclosure); a milk-processing area (a hut, also used as a dwelling for the work group); and a cheese-storing area (a cellar or a cave) (Šebesta 1991). These were permanent structures, made of stones or wood (with a stone foundation), whose location was carefully selected according to the evaluation of specific environmental variables (Carrer 2013b).

Figure 7.7 Baito della Bassa, a former hay-makers' site, now a non-dairying pastoral site (Val di Fiemme, Trentino, Italy).

The flocks of non-dairying animals were herded by a few shepherds during the summer period. These shepherds did not carry out any productive activity in the uplands, and therefore they did not need to exploit permanent seasonal sites. Furthermore they needed to move within the high plateaux in order to find the better pastures for their animals, and so they used to have portable dwellings (tents, bark shelters, etc.) and pens (made of wooden sticks and leather thongs). They also used to exploit rock shelters as temporary refuges. It is also important to point out that the same ephemeral sites were used by local shepherds of non-dairying animals and by those shepherds who wintered in the plain (Pisoni 2013; Benetti 2007:70; De Diana 1997:173–174; Perco 1991:47; Da Roit 1991:87; Bagatella Seno 1982:45; Migliorini 1932:97), thus confirming that the main factor that influenced seasonal site typology was the type of livestock (dairying/non-dairying) rather than the range of mobility.

Discussion

Present-day case-studies and historical information contributed to the identification of the dichotomy between pastoralism of dairying and non-dairying animals, particularly evident within the upland seasonal landscapes.

In the aforementioned ethno-archaeological fieldwork two seasonal sites exploited by herders of dairying animals were studied: *Malga Cadinello* and *Malga Agnelezza*. Despite the difference of livestock (cattle and goats), there are several similarities between these sites. First of all, they are both exploited by entire families, whose daily activities are: taking the animals to the pastures; milking them; storing the milk; and cleaning the structures (milk-processing is no longer allowed in the uplands of the Val di Fiemme). Another similar feature is the typology of sites: they are composed of three to four permanent masonry buildings, each one with a specialised function originally related to a specific phase of dairying activity (milking, milk-processing, cheese-maturing). The ethno-historical data demonstrates that in the past (at least until the first half of the 20th century) the summer dairying activity was carried out by large (ten persons) male work groups, and that the *malghe* were identical to the current ones. These data suggest that a specific link exists between the complexity of work groups, the stability and specialisation of sites, and the seasonal dairying economy; milking, milk-processing and cheese-maturing are complex activities and need a large and specialised work group, as well as specific activity areas. These *malghe* can be described as proper summer farms used by herders and dairymen to exploit cattle and goat milk and produce cheese.

In Val di Fiemme three sample sites related to pastoralism of non-dairying animals were also taken into account. Two sites were former hay-makers' seasonal huts, exploited nowadays by local herders of young cattle (*Baito della Bassa*) and by shepherds carrying out 'long' transhumance (*Baito dei Ciocchi*). The third site was a former *malga* (*Malga Lagorai*) that was abandoned by 'dairying' pastoralists and is currently exploited by local shepherds (carrying out a 'short' transhumance). Otherwise in the past, until the first half of the 20th century, most of the shepherds used to dwell (during the summer months) in ephemeral shelters or under modified rock shelters, and they also had portable pens to enclose their livestock. This clearly indicates that herders of non-dairying animals do not need to use specialised sites during the summer as their activities are simpler than those of herders and dairymen dealing with dairying animals. The presence of small work groups (usually one or two persons) seems to confirm this assumption, as it is evidently related to the simplicity of daily activities (essentially taking the livestock to the pasture) and to the absence of specialised tasks (like milking, milk-processing, etc.). Therefore the use of ephemeral sites and the limited number of co-working herders can both be related to the absence of dairying specialisation.

Furthermore, due to these factors, the dairying activity seems to affect also the mobility range in the high mountain landscapes. As a matter of fact, those herders who have a dairying economy are less mobile within the uplands. They exploit for long periods the same specialised seasonal site as long as grass is available in the vicinity (that is, as long as the calories the livestock acquire and those they spend to reach the pastures are balanced: see Gremillon 2006). Then they move (uphill or downhill) to another *malga*. On the other hand, those herders who graze non-dairying animals move more often in order to find the most suitable pastures for their livestock. They do not need to use permanent sites to undertake dairying activities, so they can change their location every day according to the necessities of their herd.

These data suggest that dairying-focused pastoralists need to build different permanent sites in the uplands in order to achieve their goals. These sites are usually recognised as the main features (with paths) of the human landscapes in the high mountain territories. The herders of non-dairying animals, in contrast, do not create permanent sites, and so they hardly contribute to the shaping of seasonal alpine landscapes. Their presence in the uplands is less visible than the presence of 'dairying' pastoralists. From this point of view, the distinction between permanent dairying sites and ephemeral non-dairying sites can be generalised, and allows a distinction between 'invisible shepherds' and 'visible dairymen'. The dairymen are supposed to be more 'visible' in the uplands than the proper shepherds, and this concept tackles the archaeological invisibility issue that has been addressed in the opening paragraph. Therefore the ethno-archaeological and ethno-historical information that has been collected in Val di Fiemme will be valuable for archaeological interpretation.

Archaeological implications

The analysis detailed in the previous paragraphs allows us to distinguish between two different pastoral strategies in the alpine area: one focused on dairy products; and the other focused on wool and the 'primary products' of husbandry (see Sherrat 1983). Each strategy is related to specific site typologies, to specific settlement patterns (see Carrer 2013b) and to specific summer mobility ranges. The proper focus of this ethno-archaeological research is on site typology, and for this reason a link between dairying and seasonal upland sites has been identified; 'dairying' pastoralism is characterised by permanent and specialised sites, while 'non-dairying' pastoralism is characterised by ephemeral and mobile sites.

These inferences have fundamental archaeological consequences. The sites of 'non-dairying' pastoralists are traditionally made of perishable materials, like bark and wood; furthermore these shepherds move quite often, and therefore they take with them most of their material culture objects; for these reasons the simple pastoral sites leave few archaeological remains on the ground. On the other hand the seasonal dairying sites are usually permanent and compound structures, made of stone walls or foundations; besides the dairymen exploit the same seasonal sites every summer, and so they used to leave some objects on these sites for the following summer ('delayed curation': see Tomka 1993); the presence of walls and material culture enhance the visibility of these sites within the upland archaeological landscapes.

These interpretations challenge the simplistic suggestion that all pastoral groups are equally archaeologically 'invisible'. In fact, dairymen are more 'visible' than shepherds, and this means that those archaeological sites that have recently been identified in the European mountain uplands (essentially dry-stone huts and enclosures, as well as modified rock shelters) can be recognised as parts of seasonal 'dairying' sites rather than simple pastoral sites. Otherwise, the absence or dearth of sites in some upland areas and/or chronological periods may be attributed to the seasonal occupation of shepherds who did not exploit and process the milk of their livestock, rather than to the actual absence of herders in the high plateaus.

An awareness of the relationship between archaeological visibility and productive strategy has important consequences not only for the interpretation of archaeological evidence, but also for organising fieldwork activities in the uplands (see Carrer 2013b). The less visible 'non-dairying' sites require a more detailed survey strategy. Hence archaeological fieldwork aimed at filling the gap of 'non-dairying' pastoralism in a territory has to select specific scales and methodologies. On the other hand the visible 'dairying' sites may be analysed to identify indicators of dairying activities, and so shed new light on this topic. These examples suggest that the proposed ethno-archaeological study has the potential to improve the quality of archaeological surveys and excavations.

Besides, from a theoretical and methodological point of view, this new approach changes the current interpretative process of archaeology of pastoralism. In many studies of ancient pastoral groups, mobility was proposed as the key-factor which is used to explain other aspects of the studied group. In this research, instead, mobility has become a secondary feature of seasonal herding, as it apparently depends on another important feature, the productive goals (in this research, cheese on one hand, and wool/primary products on the other). Such a change of perspective will enable a new understanding of the 'invisibility' issue; mobility alone does not explain invisibility, but they are both deeply related to the productive strategy.

It is evident how ethnoarchaeology proved to be useful for tackling the axiom 'pastoral groups= invisible groups'. It will contribute to enhancing the potential of the archaeology of pastoralism, providing new interpretative tools. It will also contribute to the organisation of fieldwork in mountain landscapes, focusing on specific issues. Eventually, it will enable us to challenge the use of the concept of mobility as the exclusive key to explain pastoral strategies and patterns. Therefore a close interaction between ethnoarchaeology and archaeology of pastoralism is desirable in all the different phases of a research project: identification of the research topic, planning of the fieldwork and interpretation of the final results. This multidisciplinary approach has spread within European archaeological survey projects since the 1980s (Nandris 1985; Barker and Grant 1991), but only in the last years has it been adopted by some of the archaeologists who deal with the Alps (Carrer et al. 2013; Reitmaier 2012). Hence it can be claimed that the improvement of alpine archaeology is deeply related to the development of alpine ethnoarchaeology.

Acknowledgements

This research was conducted at the University of Trento (Italy), in collaboration with the Museo degli Usi e Costumi della Gente Trentina of San Michele all'Adige, and was funded by the APSAT Project (Provincia Autonoma di Trento – Bando 'Grandi Progetti' 2006; Delibera G.P. 2790/2006). Several people need to be acknowledged for their precious help and advice: Diego E. Angelucci, Graeme Barker, Marta Bazzanella, Andrea Bertagnolli, Fabio Cavulli, Cornelia Goss, Giovanni Kezich, Franco Marzatico, Annaluisa Pedrotti, Daniela Perco, Luca Pisoni and Kevin Walsh.

Bibliography

Angelucci, D.E., Carrer, F., Cavulli, F., Foradori, G., Medici, T., Pedrotti, A., Pisoni, D. and Rottoli, M. 2013. Primi dati archeologici da una struttura pastorale d'alta quota in Val di Sole: il sito MZ005S (Mezzana, Trento). In D.E. Angelucci, L. Casagrande, A. Colecchia and M. Rottoli (eds.). *Apsat 2. Paesaggi d'altura dalla preistoria all'età moderna: evoluzione naturale e aspetti culturali*. Pp. 141–162. Mantova: SAP.

Bagatella Seno, A. 1982. Tecniche tradizionali di allevamento e cura del gregge. In D. Perco (ed.) *La pastorizia transumante del feltrino*. Comunità Montana Feltrina, Centro per la Documentazione della Cultura Popolare, Quaderno n. 3:39–62. Feltre (BL): Regione Veneto.

Baker, F. 1999. The ethnoarchaeology of transhumance in the southern Abruzzi of central Italy: an interdisciplinary approach. In L. Bartosiewicz, and H.J. Greenfield (eds.) *Transhumant Pastoralism in southern Europe: recent perspectives from archaeology, history and ethnology*. Archaeolingua, Series Minor 11:99–109. Budapest: Archaeolingua.

Barker, G. and Grant, A. 1991. Ancient and modern pastoralism in central Italy: an interdisciplinary study in the Cicolano Mountains. *Papers of the British School at Rome* 59:15–88.

Benetti, D. 2007. Solingo o solastro. L'alpeggio: una tonalità della cultura alpina. In A. Beltrame (ed.) *La cultura delle malghe e il futuro dell'alpeggio. Atti dell'omonimo Convegno, Montebelluna (Treviso), Auditorium Centro Direzionale Veneto Banca, 21 ottobre 2006*:62–72. Piazzola sul Brenta (PD): Papergraf.

Boyazoglu, J. and Flamant, J.C. 1990. Mediterranean systems of animal production. In J.G. Galaty, and D.L. Johnson (eds.) *The World of Pastoralism: herding systems in comparative perspective*. Pp. 353–393. London: Guildford Press.

Braudel, F. 1949. *La Méditerranée et le monde méditerranéen à l'époque de Philippe II*. Paris: Colin.

Carrer, F., 2012. Upland sites and pastoral landscapes: new perspectives into the archaeology of pastoralism in the Alps. In G.P. Brogiolo, D.E. Angelucci, A. Colecchia and F. Remondino (eds.) *Apsat 1. Teoria e metodi della ricerca sui paesaggi d'altura*. Pp. 101–116. Mantova: SAP.

Carrer, F. 2013a. Paesaggi condizionati: un approccio ecologico ai sistemi insediativi stagionali dei pastori della Val di Fiemme. In M. Bazzanella and G. Kezich (eds.) *Apsat 5. Le scritte dei pastori: etnoarcheologia della pastorizia in Val di Fiemm*. Pp. 51–66. Mantova: SAP.

Carrer, F. 2013b. An ethnoarchaeological inductive model for predicting archaeological site location: a case-study of pastoral settlement patterns in the Val di Fiemme and Val di Sole (Trentino, Italian Alps). *Journal of Anthropological Archaeology* 32:54–62.

Carrer, F., Angelucci, D.E. and Pedrotti, A. 2013. Montagna e pastorizia: stato dell'arte e prospettive di ricerca. In D.E. Angelucci, L. Casagrande, A. Colecchia and M. Rottoli (eds.) *Apsat 2. Paesaggi d'altura del Trentino: evoluzione naturale e aspetti culturali*. Pp. 125–139. Mantova: SAP.

Cribb, R. 1991. *Nomads in Archaeology*. Cambridge: Cambridge University Press.

Chang, C., 1999. The ethnoarchaeology of pastoral sites in the Grevena Region of northern Greece. In L. Bartosiewicz and H.J. Greenfield (eds.) *Transhumant Pastoralism in Southern Europe: recent perspectives from Archaeology, History and Ethnology*. Archaeolingua, Series Minor 11:133–144. Budapest: Archaeolingua.

Chang, C., 2006. The grass is greener on the other side. a study of pastoral mobility on the Eurasian Steppe of southeastern Kazakhstan. In F. Sellet, R. Greaves and Pei-LinYou (eds.) *Archaeology and Ethnoarchaeology of Mobility*. Pp. 127–152. Gainesville: University of Florida.

Chang, C., and Koster, H.A. 1986. Beyond bones: toward an archaeology of pastoralism. In M.B. Schiffer (ed.) *Advances in Archaeological Method and Theory* 9:97–148. London: Academic Press.

Cleary, M.C., and Delano Smith, C. 1991. Transhumance reviewed: past and present practices in France and Italy. In R. Maggi, R. Nisbet and G. Barker (eds.) *Archeologia della pastorizia nell'Europa meridionale: atti della Tavola Rotonda Internazionale, Chiavari, 22–24 settembre 1989. Volume I*. Rivista di Studi Liguri, Anno LVI. Pp. 1–4, 21–38. Bordighera: Instituto Internazionale di Studi Liguri.

Croce, D. 1972. L'economia pastorale in Val di Fiemme. *Economia Trentina* 4:35–63.

Da Roit, C. 1991. Delle liti sui pascoli e sull'alpeggio dei bovini a la valle agordina. In D. Perco (ed.) *Insediamenti temporanei nella montagna bellunese*. Comunità Montana Feltrina, Centro per la Documentazione della Cultura Popolare, Quaderno n. 14:77–94. Feltre (BL): Libreria Pilotto.

David, N. and Kramer, C. 2001. *Ethnoarchaeology in Action*. Cambridge: Cambridge University Press.

Davies, E. 1941. The patterns of transhumance in Europe. *Geography* 26/4:155–168.

De Diana, E. 1997. Il pascolo e le casere nel Comune di Lozzo di Cadore. In D. Perco (ed.) *Insediamenti temporanei nella montagna bellunese*. Comunità Montana Feltrina, Centro per la Documentazione della Cultura Popolare, Quaderno n. 14:173–188. Feltre (BL): Libreria Pilotto.

Frödin, J., 1940/41. *Zentraleuropas Alpwirtschaft*. Oslo: Instituttet for Sammenlignende Kulturforskning.

Gassiot, E., Pèlachs, A., Bal, M.C., Garcia, V., Julià, R., Pérez, R., Rodrìguez, D. and Astrou, A.C. 2010. Dynamiques des activités anthropiques sur un milieu montagnard dans les Pyrénées occidentals catalanes durant la Préhistoire: une approche multidisciplinaire. In S. Tzortzis and X. Delestre (eds.) *Archéologie de la montagne européenne. Actes de la table ronde international de Gap, 29 septembre-1er octobre 2008*:33–44. Aix en Provence: Errance.

Gifford, D.P. 1978. Ethnoarchaeological observations of natural processes affecting cultural materials. In R.A. Gould (ed.) *Explorations in Ethnoarchaeology*. Pp. 77–101. Albuquerque: University of New Mexico.

González Álvarez, D. 2012. Traditional pastoralism in the Asturian mountains: an ethnoarchaeological view on mobility and settlement patterns. In F. Lugli, A. Assunta Stoppiello, and S. Biagetti (eds.) *Ethnoarchaeology: current research and field methods*. BAR S2472:202–208. Oxford: Archaeopress.

Gremillion, K.J. 2006. Central Place foraging and food production of the Cumberland plateau, eastern Kentucky. In D.J. Kennet and B. Winterhalder (eds.) *Behavioural Ecology and Transition to Agriculture*. Pp. 41–62. Los Angeles: University of California Press.

Hebert, F. and Mandl, F. (eds.) 2009. *Almen im Visier: Dachsteingebirge, Totes Gebirge, Silvretta*. Forschungsberichte der ANISA 2. Haus i.E. ANISA, Verein für alpine Forschung.

Hole, F. 1978. Pastoral nomadism in western Iran. In R.A. Gould (ed.), *Explorations in Ethnoarchaeology*. Pp. 127–167. Albuquerque: University of New Mexico Press.

Horvat, J. 1999. Colonizzazione preistorica e romana sulle Alpi di Kamnik (Slovenia). In S. Santoro Bianchi (ed.) *Studio e conservazione degli insediamenti minori romani in area alpina. Atti dell'incontro di studi, Forgaria del Friuli, 20 settembre 1997*. Pp. 63–69. Studi e Scavi 8. Bologna: Università di Bologna.

Khazanov, A.M. 1984. *Nomads and the Outside World*. Cambridge: Cambridge University Press.

Migliorini, E. 1932. *La Val Belluna. Studio antropogeografico*. Roma: Istituto di Geografia della R. Università di Roma.

Morandini, G. 1941. Notizie antropogeografiche sulla Val di Fiemme. *L'Universo*, 12/3:150–180, 12/4:231–268.

Nandris J., 1985. The Stina and the Katun: foundations of a research design in European Highland Zone ethnoarchaeology. *World Archaeology* 17:256–268

Netting, R. McC. 1976. What alpine peasants have in common: observation on communal tenure in a Swiss village. *Human Ecology* 4:135–146.

Netting, R. McC. 1981. *Balancing on an Alp: ecological change and continuity in a Swiss mountain community*. Cambridge: Cambridge University Press.

Perco, D. (ed.) 1982. *La pastorizia transumante del feltrino*. Comunità Montana Feltrina, Centro per la Documentazione della Cultura Popolare, Quaderno n. 3. Feltre (BL): Regione Veneto.

Perco, D. 1991. I malgari della Val Belluna. In D. Perco (ed.) *Malgari e Pascoli. L'alpeggio nella provincia di Belluno*. Comunità Montana Feltrina, Centro per la Documentazione della Cultura Popolare, Quaderno n. 10:39–60. Feltre: Comunità Montana Feltrina.

Pisoni, L. 2013. 'Leggevo Sandokan e i Pirati della Malesia': lavoro, oggetti e passatempi dei pastori del Monte Cornón (TN). In M. Bazzanella and G. Kezich (eds.) *Apsat 5. Le scritte dei pastori: etnoarcheologia della pastorizia in Val di Fiemme*. Pp. 85–112. Mantova: SAP.

Pracchi, R. 1942. *Il fenomeno della transumanza sul versante italiano delle Alpi*. Como: Marzorati.

Primas, M. 1999. From fiction to facts: current research on prehistoric human activity in the Alps. In P. Della Casa (ed.) *Prehistoric Alpine Environment, Society and Economy: papers of the international colloquium PAESE '97*. Universitätsforschungen zur Archäologie 55:1–10. Zurich: Rudolf Habelt.

Reitmaier, T. (ed.) 2012. *Letzte Jäger, erste Hirten: hochalpine Archäologie in der Silvretta*. Chur: Archäologischer Deinst Graubünden.

Rendu, C. 2003. *La montagne d'Enveig, un estive pyrénéenne dans la longue durée*. Cannet: Trabucaire.

Sartori Montecroce, T. 2002. *La Comunità di Fiemme e il suo diritto statutario*. Cavalese (TN): Magnifica Comunità di Fiemme.

Sherrat, A. 1983. The secondary exploitation of animals in the Old World. *World Archaeology*, 15/1:90–104.

Simms, S.R. 1988. The archaeological structure of a Bedouin camp. *Journal of Archaeological Science* 15:197–211.

Smerdel, I. 1999. The three 'sheepmaster'.transhumance in Pivka (Slovenia) from the middle of the 19th to the middle of the 20th. In L. Bartosiewicz, and H.J. Greenfield (eds.), *Transhumant Pastoralism in Southern Europe:*

recent perspectives from Archaeology, History and Ethnology. Archaeolingua, Series Minor 11:197–212. Budapest: Archaeolingua.

Šebesta, G. 1991. La via delle malghe. In G. Šebesta (ed.) *Scritti Etnografici* Pp. 466–500. San Michele all'Adige: Museo degli Usi e Costumi della Gente Trentina.

Tomka, S.A. 1993. Site abandonment behaviour among transhumant agro-pastoralists: the effects of delayed curation on assemblage composition. In C.M. Cameron, and S.A. Tomka (eds.) *Abandonment of Settlement and Regions: ethnoarchaeological and archaeological approaches*: 11–24. Cambridge: Cambridge University Press.

Viazzo, P.P. and Wolf, E. 2001 (eds.). *L'alpeggio e il mercato.* La Ricerca Folklorica, 43. Brescia: Grafo.

Walsh, K. 2005. Risk and marginality at high altitudes: new interpretations from fieldwork on the Faravel Plateau, Hautes-Alpes. *Antiquity* 79:289–305.

Walsh, K., Mocci, F., Court-Picon, M., Tzortzis, S. and Palet-Martinez, J.-M. 2005. Dynamique du peuplement et activités agro-pastorales durant l'âge du Bronze dans le massif du Haut Champsaur et de l'Argentierois (Hautes-Alpes). *Documents d'Archéologie méridionale* 28:25–44.

Walsh, K., Mocci, F. and Palet-Martinez, J. 2007. Nine thousand years of human/landscape dynamics in a high altitude zone in the southern French Alps. In P. Della Casa and K. Walsh (eds.) *Interpretation of Sites and Material Culture from mid-high Altitude Mountain Environments. Proceedings of the 10th annual meeting of the European Association of Archaeologists 2004. Preistoria Alpina* 42:9–22. Trento: Museo Tridentino di Scienze Naturali.

Wolf, E.R. 1957. Closed corporate peasant communities in Mesoamerica and central Java. *Southwestern Journal of Anthropology* 13/1:1–18.

Francesco Carrer, Department of Archaeology, University of York, King's Manor, York, YO1 7EP, UK
Email: francescokar@gmail.com

8. Going up the mountain! Exploitation of the Trentino highlands as summer farms during the Bronze Age: The Dosso Rotondo site at Storo (northern Italy)

Franco Nicolis, Elisabetta Mottes, Michele Bassetti, Elisabetta Castiglioni, Mauro Rottoli and Sara Ziggiotti

The Dosso Rotondo site (Storo, Trento) is located in the valley of the River Chiese at 1876 m.a.s.l. Investigations, still underway, began in 1998 and have involved a surface area of more than 50 m^2, although the occupied area, estimated using surveys and manual coring, is around 800–1,000 m^2. The archaeological deposits brought to light have made it possible to identify a stratigraphic sequence with four separate phases of occupation. In the area investigated it has currently been possible to identify 76 postholes, 70 of which can be related to four residential buildings constructed using load-bearing vertical posts planted in the ground, which can be attributed to three different settlement phases. During the fourth phase of occupation a fifth dwelling was built using a new construction technique with stone foundations, on which the wooden structure rested. Interdisciplinary research at the Storo Dosso Rotondo site has led to the conclusion that it was probably a seasonal settlement linked to mountain pasture activities. On the basis of the pottery and flint artefacts, occupation of the settlement can be dated to the early phase of the Middle Bronze Age.

1. Introduction

In the last few years mountain archaeology has excited considerable interest, as demonstrated by the numerous interdisciplinary research projects that have been started up in different European areas at high altitude.[1] The current framework of knowledge as regards occupation of the alpine environment by man in ancient times is no longer conditioned by the traditional image of poor, inhospitable areas inhabited by coarse, warlike peoples. On the contrary, the prevailing picture today is of regions with considerable economic potential, regularly exploited by groups of humans settling there, who played a role as cultural mediators between the populations settled to the north and south of the Alps (Bagolini and Broglio 1985; Cason 2002; AA.VV. 2002; Borrello 2013).

Mountain archaeology is not synonymous with excavations carried out at high altitude, nor does it have characteristics and peculiarities allowing it to be given a certain degree of autonomy (Nicolis 2006). The definition of archaeology, understood as the possibility of reconstructing ancient anthropogenic processes through material evidence, in relation to the passing of time and the succession of natural events, can indeed be applied to any context in which the discipline is active.

The natural and cultural evolution of mountain environments usually only takes place over a very long time. For this reason, there is a perception of stable natural elements that are not modified over man's lifetime. In actual fact, what we consider to be a natural landscape and a unified context is instead often the sum of human actions and natural events that have affected the area over a long period of time and given rise to genuine cultural landscapes, recording the superimposed and fading signs of the presence of man.

Our level of knowledge does not yet permit us to understand the real extent of ancient human occupation of mountain areas, above all in the pre-protohistoric period, or its effective nature.[2] The wealth of environments corresponds with the complexity of the relationship that man has always had with the mountains, his strategies for controlling, adapting or transforming them and the difficulty in finding the right balance between compatible use and uncontrolled exploitation.

Part of the problem related to a lack of knowledge comes from current conditioning as regards archaeological activities, which by now take place mostly in areas with a higher human density and thus give a distorted image of the settlement of our area in ancient times.

Thus the problem of mountain archaeology is to plan research in areas in which the current anthropogenic impact is less significant. Examples of this are the interdisciplinary research projects carried out in the southern French Alps (Walsh *et al.* 2007, and this vol-

---

1. See for example Tzortzis and Delestre 2010.
2. For a summary regarding the central southern alpine area see Marzatico 2007.

Keywords: Highlands, *Alpwirtschaft* economy, archaeobotany, use-wear analysis, Middle Bronze Age, Storo, Dosso Rotondo, Trentino, Central Alps.

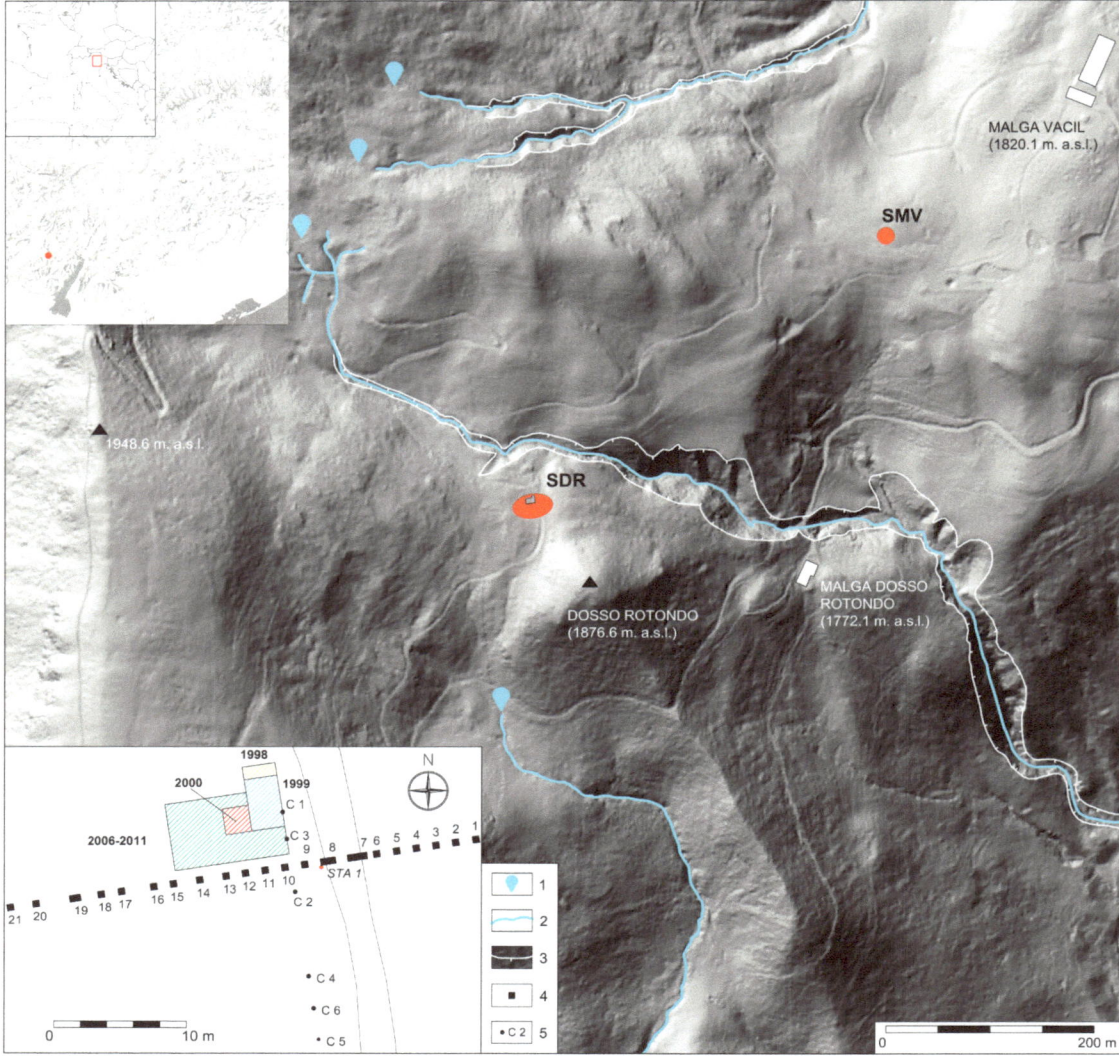

Figure 8.1 Storo Dosso Rotondo (Trento). Geographical and geomorphological setting. In the box bottom left the different excavation areas are outlined. SDR: Storo Dosso Rotondo site; SMV: Storo Malga Vacil site. 1: springs; 2: streams; 3: erosion escarpment; 4: archaeological test pits; 5: manual borings.

ume), the alpine region of Silvretta between the Lower Engadin (Switzerland) and the Paznaun Valley (Austria) (Reitmeier 2010; 2012; Reitmeier *et al.* 2013), the Schnidejoch area in the western Bernese Alps in Switzerland (Hafner 2009; 2012), the Montafon valley in Vorarlberg, Austria (Krause 2007), the Val di Sole area (Trentino) (Carrer *et al.* 2013; Angelucci *et al.* 2014) and the mountainous areas of Val Venosta/Vinschgau (Putzer 2011) and Val Senales/Schnals (Alto Adige), where the Similaun man (Ötzi the Iceman) was discovered in 1991 (Oeggl *et al.* 2009).

The research carried out by the Autonomous Province of Trento's Archaeological Heritage Office at the high altitude sites of Storo Dosso Rotondo (SDR) (Bassetti *et al.* 2003; 2008; Mottes and Nicolis 2004) and Malga Vacil (SMV) (Marzatico 2007:169–163) also falls within this framework of knowledge (Fig. 8.1).

2. The archaeological investigations

The Storo Dosso Rotondo site high in the mountains (Trentino–Alto Adige) is situated in the central section of the southern Alps, at an altitude of around 1,857 m.a.s.l. This mountainous area borders to the east with the Valle del Chiese, which is oriented NNE–SSW and along which the river of the same name flows, then flowing into Lake Idro (368 m.a.s.l.), and to the west with the Valle del Caffaro, running N–S.

The research and investigations have taken place starting from October 1998 and are still underway.

In the summers of 1999 and 2000 a trial trench was excavated over a surface area of 16.5m², which led to identification of important remains of dwellings (Bassetti *et al.* 2003; 2008; Mottes and Nicolis 2004).

Figure 8.2   Storo Dosso Rotondo (Trento). Excavation area.

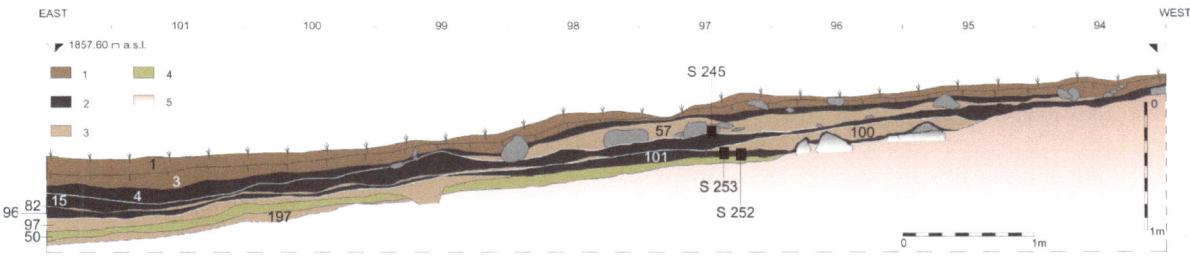

Figure 8.3   Storo Dosso Rotondo (Trento). Southern section. 1: colluvial soil; 2: charcoal layers; 3: colluvia; 4: prepared surface; 5: buried colluvial soils; S: sample.

Archaeological investigations began again in 2006, and currently concern a surface area of more than 50 m², while the occupied area, estimated using surveys and manual coring, is around 800–1,000m² (Fig. 8.2).[3]

The analysis of the stratigraphic sequence at Storo Dosso Rotondo has made it possible to document the presence of four separate structural phases, which can all be attributed chronologically to the beginning of the Middle Bronze Age.

In the area investigated it has currently been possible to identify 76 postholes, 70 of which can be related to four residential buildings constructed using load-bearing vertical posts planted in the ground. These can be attributed to three different settlement phases. During the fourth phase of occupation a fifth dwelling was built, using a new construction technique that insulated the foundations with stone blocks laid at right angles, on which the wooden structure must have rested. This evidence is covered by colluvial soil, which has sealed the archaeological deposits.

---

3. The archaeological research carried out and funded by the Autonomous Province of Trento's Archaeological Heritage Office was headed by Franco Nicolis and carried out by Cora Società Archeologica Srl, Trento with technical coordination by Michele Bassetti.

   The images in this paper are from the archives of the Autonomous Province of Trento's Archaeological Heritage Office. The graphics have been drawn up by Chiara Conci (Autonomous Province of Trento's Archaeological Heritage Office), Marco Grosso, Chiara Maggioni and Livia Stefan (Cora Società Archeologica Srl, Trento). Translation by Vivienne Frankell.

Figure 8.4 Storo Dosso Rotondo (Trento). General view of the area subject to archaeological research (arrow indicates the Dosso Rotondo site).

3. Geological and geomorphological background

The Storo Dosso Rotondo site lies on the western slopes of the Rio S. Barbara valley, between the Col Perpetue Mountains (2,016 m.a.s.l.) to the north and Monte Tonolo (1,533 m.a.s.l.) to the south.

The zone in question is situated within the geological sector known as the Lombard Southern Alps, The site is characterised by red sandstone outcrops alternating with yellow siltstone (the Val Gardena sandstone formation of the Middle–Upper Permian, Dal Piaz 2008). The area in question was not covered by glaciers in the last glacial expansion (ALGM – Alpine Last Glacial Maximum, Penk and Brückner 1901–1909; Ravazzi *et al.* 2007), remaining an environment subject to periglacial processes, dominated by cycles of freezing and thawing. Processes of selective erosion were mainly active on the substrate. Indeed, due to the variability in the thickness of the layers, their different arrangements and mechanical resistance, these types of rock are subject to weathering. Specifically, the site is located on a saddle between a gentle slope (5°) uphill (Fig. 8.3), and the high ground represented by Dosso Rotondo (local name Doredont) (1,876.6 m.a.s.l.), whereas to the north and south it is delimited by the head of two small coombes created by fluvial erosion (Fig. 8.1). The types of stone making up the substrate have a low level of permeability and the infiltration of rainwater is concentrated in the thin eluvial and sedimentary coverage. In watershed areas the aquifer emerges on the surface, giving rise to springs which feed ephemeral watercourses of a seasonal nature, flowing into the Rio S. Barbara. The site currently falls within an alpine grassland environment covering around 160 hectares, still used for alpine pasture (Malga Vacil, Malga Dosso Rotondo) and characterised by plant species typical of acid soils and rich in the humus of the subalpine belt, between the present timberline and alpine grasslands (Rhododendron ferrugineum, Vaccinium myrtillus) (Fig. 8.4).

4. Stratigraphy

4.1. Pedostratigraphic description of section 1[4]

The anthropogenic area has mainly been subject to accumulation phenomena, as demonstrated by the sequence of colluvial deposits. The morphology has therefore allowed the conservation of the stratigraphy through aggradation and burial of the pedogenetic horizons. Specifically, in section 1 buried podzolic soil is evident, from the bottom up, evolving on parent material made up of weathered siltstones and quartz sand. In the alpine and subalpine area podzols developed on acid substrates, under forest coverage made up

---

4. Section 1 is the eastern limit of excavation. For the descriptions of pedostratigraphic units, reference has been made to Sanesi (1977), whereas the symbology used to classify soils and define the horizons follows the criteria of *Soil Taxonomy*, Soil Survey Staff-U.S.D.A. 1999, *Keys to Soil Taxonomy*, Soil Survey Staff-U.S.D.A. N.R.C.S. ninth edition, 2003. The colours have been codified using the *Munsell® Soil Color Charts* and were determined while wet (*Munsell Soil Color Charts, 1992 Revised Edition*. Munsell® Color, New Windsor). As regards geoarchaeological problems, reference was made to Waters 1992; Goldberg and Macphail 2006, while for the sedimentological description reference was made to Ricci Lucchi 1980; Bosellini *et al.* 1989; Marchetti 2000.

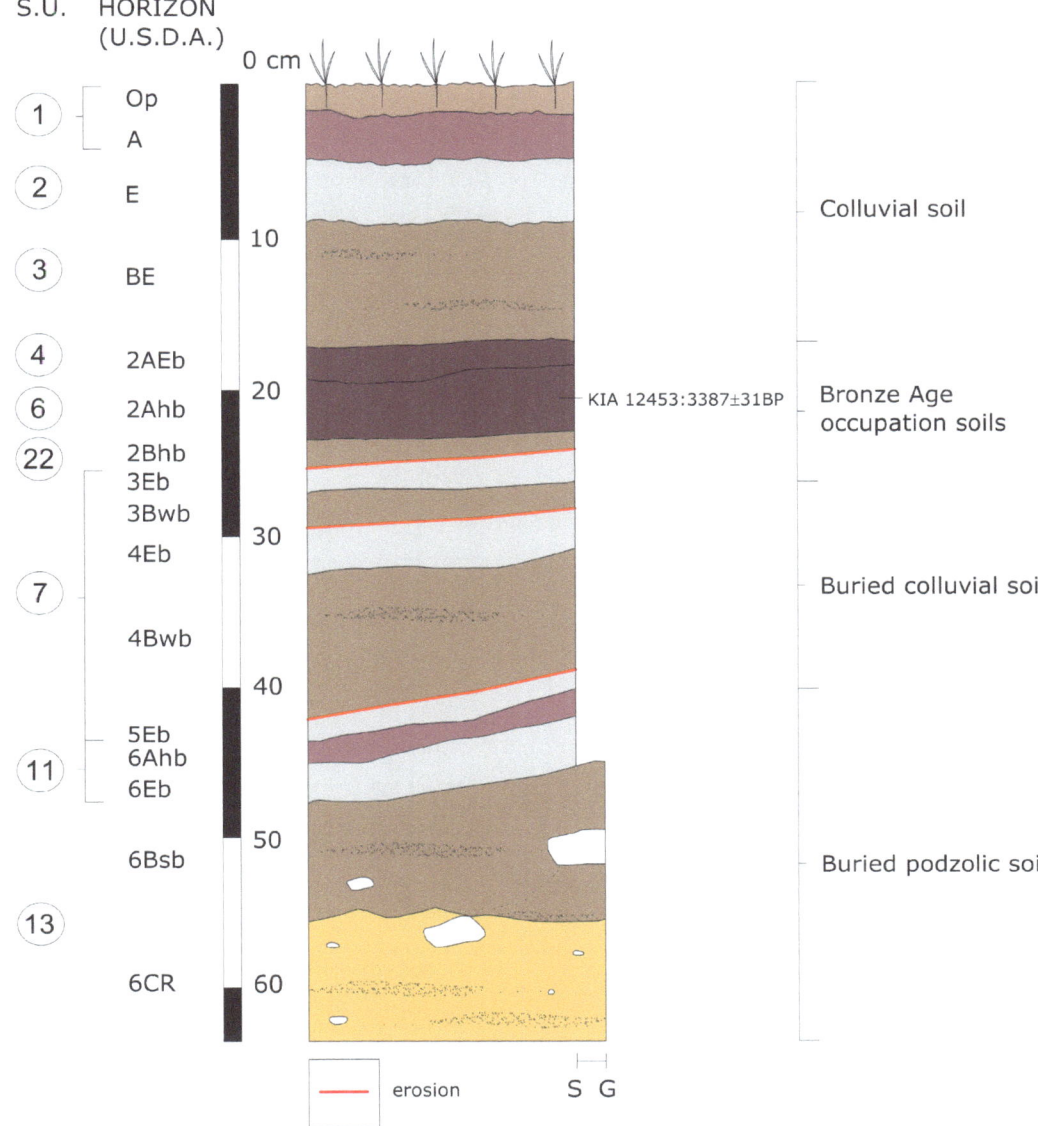

Figure 8.5 Storo Dosso Rotondo (Trento). Stratigraphic sequence of section 1. S: sand; G: gravel.

of conifers and/or ericaceous shrubs, vegetation capable of producing a sufficient biomass of acid litter to spark off the podsolization process (Duchaufour 1998; Stützer 1999). In the alpine area podzols were limited to the belt between an altitude of 1.000m (Cremaschi and Rodolfi 1991) and 2,620 m.a.s.l. (D'Alessio and Previtali 1988). The podzolic soil was buried by detrital-colluvial sediment, subsequently pedogenised (buried colluvial soils) (Fig. 8.5). The development of polycyclic soils can be interpreted as a geomorphological response to the phase of instability that manifests itself with erosion of the soil on the overlying slopes, possibly due to deforestation. On these fine layers of deposits, the soil formation gave rise to the weak podsolization typical of ochric brown soils (Duchaufour 1976), with intermittent, little pronounced (E) eluvial horizons, and underlying horizons characterised by a limited accumulation of sesquioxides (Bw).

Three sequences of soils truncated at the top by erosive surfaces have been distinguished in the section. During the Bronze Age the systematic occupation of the area began in this environmental context (Bronze Age occupation soils), with the creation of semi-permanent structures. In general, the succession of horizons highlights the progressive substitution of conifer forest, represented by podzolic soil, by alpine grasslands, following degradation linked to human activities (colluvial soil) (Table 8.1).

S.U.	Horizons	Description of the Dosso Rotondo sequence
Colluvial soil:		
1	Op	0–2cm, sandy loam, dark brown (7,5YR 3/2), weak very fine crumb structure, very few very fine pores. Abrupt linear boundary to:
1	A	2–5cm, sandy loam, very dark grey (5YR 3/1), weak very fine crumb structure, very few fine pores. Clear linear boundary to:
2	E	5–9cm, sandy to sandy loam, dark reddish brown-brown (5YR 3/2–7,5Y4/2), common charcoal <1 mm, moderate very fine crumb primary structure, moderate medium subangular blocky secondary structure, common very fine pores. Clear linear boundary to:
3	BE	9–17cm, sandy, very dark grey (5YR 3/1), common clasts at the bottom, common charcoal <1 mm, moderate medium subangular blocky structure, common very fine pores. Clear linear boundary to:
Bronze Age occupation soils:		
4	2AEb	17–19cm, sandy, black (7,5YR 2,5/1), frequent subangular to subrounded weathered clasts (<7cm) common charcoal (<1mm), moderate medium subangular blocky structure, common fine pores. Clear linear boundary to:
6	2Ahb	19–23cm, charcoal layer (fragments <1cm), and occasional sandy laminae (colluvium), black (10YR 2/1), moderate fine platy structure, common fine pores. Clear linear boundary to:
22	2Bhb	23–25cm, sandy, brown (7,5YR 5/3), common subangular/subrounded siltstone and sandstone clasts (from 0.5 to 3cm), frequently with heat alteration, common charcoal <1 mm, moderate medium subangular blocky structure, common organic coating on the clasts, common fine pores. Clear linear boundary to:
Buried colluvial soils:		
7	3Eb	25–27cm, sandy-loam, brown (7,5YR 4/3), few subangular/subrounded siltstone and sandstone clasts, from 0.5 to 3cm), moderate medium subangular blocky structure, common fine pores. Clear linear boundary to:
7	3Bwb	27–29cm, sandy, brown (7,5YR 4/4), common iron mottles (strong brown, 7,5YR 5/6), few subangular/subrounded siltstone and sandstone clasts (from 0.5 to 3cm), weakly medium subangular blocky structure, common fine pores. Clear linear boundary to:
7	4Eb	29–32cm, sandy, brown (7,5YR 4/3), few subangular/subrounded siltstone and sandstone clasts (from 0.5 to 3cm), weakly medium subangular blocky structure, common fine pores. Clear linear boundary to:
7	4Bwb	32–41.5cm, sandy, brown (7,5YR 4/4), common iron mottles (strong brown, 7,5YR 5/6), few subangular/subrounded siltstone and sandstone clasts (from 0.5 to 3cm), weakly medium subangular blocky structure, common fine pores. Clear linear boundary to:
7	5Eb	41.5–43cm, sandy loam, brown (7,5YR 5/3) few subangular/subrounded siltstone and sandstone clasts (from 0.5 to 3 cm), weakly medium subangular blocky structure, common fine pores. Clear linear boundary to:
Buried podzolic soil:		
11	6Ahb	43–44.5cm, silty loam, very dark gray (7,5YR 3/1), abundant subangular/subrounded siltstone and sandstone clasts (from 0.5 to 3 cm), few charcoal fragments <1 mm, well developed fine granular structure, abundant very fine pores. Clear linear boundary to:
11	6Eb	44.5–47cm, sandy, brown (7,5YR 5/2), very few subangular/subrounded siltstone and sandstone clasts (from 0.5 to 3cm), moderate fine subangular blocky structure, tubular pores filled with yellowish brown (10YR 5/6) silty coating, abundant very fine pores. Clear linear boundary to:
13	6Bsb	47–55cm, sandy, reddish brown (5YR 5/3), common strong brown (7,5YR 5/6) iron mottles, few subangular/subrounded siltstone and sandstone clasts (from 0.5 to 3cm), well developed very fine subangular blocky structure, common very fine pores. Clear linear boundary to:
13	6CR	55–63cm, heterometric clast supported saprolite, weathering of the siltstones bedrock surface. Lower boundary not reached.

Table 8.1   Storo Dosso Rotondo (Trento). Pedostratigraphic description (section 1).

Sample N°	SU	East (m.)	North (m.)	Elevation (m.a.s.l.)
107	15, 57	100.00	46.30	1856.87
108	15, 23	100.00	46.30	1856.80
203	94	98.88	44.34	1856.75
211	15, 82	98.55	45.05	1856.83
212	15, 82	98.55	46.05	1856.86
218	73	95.87	46.88	1857.18
219	73	95.76	47.11	1857.16
231	82, 96	99.00	46.86	1856.72
235	96	98.50	45.90	1856.81
252	50, 100, 101	97.80	43.95	1856.82
253a	15, 1	97.65	43.95	1856.88
253b	50, 100, 101	97.65	43.95	1856.88
254	15, 57	97.55	43.95	1856.98

Table 8.2 Storo Dosso Rotondo (Trento). List of the thin sections.

microfacies	MS	GM		anthropogenic components			burned clasts
		b-f	rdp	charcoal			
				<100ìm	>100ìm < 1 mm	>1mm	
mf_1a	v, p	ss	op	*			
mf_1b	m, sb	ss	ssp	*	*	**	
mf_2a	p, c	u	op	**	**	****	
mf_2b	g	u	op	*	**	****	
mf_2c	g, sb	ss	op	***	**		
mf_3a	sb	ss	ssp	*	*		**
mf_3b	sb	ss	dsp	*	*		***
mf_3c	sb, p	ss	ssp	*			***
mf_4	m, c	ss	op	*	**		****

Table 8.3 Storo Dosso Rotondo (Trento). Synthetic micromorphological description of microfacies anthropogenic components. Abundance of fabric units expressed according to Stoops (2003): * very rare (<5%), ** rare (5–15%), *** common (15–30%), **** frequent (30–50%), ***** abundant (>50%). MS (microstructure): v = vughy, sb = subangular blocky, g = granular, p = platy, m = massive, c = crack. GM (groundmass): b-f (b-fabric): u = undifferendiated, ss = stipple-speckled; rdp (related distribution pattern): ssp = single spaced porphyric, dsp = double spaced porphyric, op = open porphyric.

4.2. Micromorphological analysis[5]

Microstratigraphic analysis and micromorphological aspects of the archaeological deposits have contributed to achieving an accurate interpretation of the Storo Dosso Rotondo site, although this is preliminary and in the process of definition.

Recognition of anthropogenic components within thin sections provided information about human activities (e.g. the use of fire), but also about subsequent anthropogenic changes (e.g. trampling) and post-depositional phenomena of natural origin (e.g. erosion and colluvial deposits) (Goldberg and Macphail 2006; Goldberg and Berna 2010; Mallol et al. 2013) (Table 8.2). The detailed stratigraphic subdivision carried out during the archaeological excavations was further refined through analysis of the microfacies. This approach made it possible to identify changes to the sedimentary and pedological features within the individual thin sections (Fig. 8.6).

Analysis of the microfacies (mf), recognisable within the units identified in the field (SU), made it possible to interpret the archaeological sediment and attribute it to specific anthropogenic activities carried out in different areas of the site (Table 8.3) (Courty 2001; Goldberg et al. 2009). In this way it was possible to understand better the use of space over the course of the

---

5. The thin sections were analysed using a microscope with polarised light at 20×, 40×, 100×, 400×, described and interpreted according to the criteria adopted in the following reference texts: Bullock et al. 1985; Courty et al. 1989; Stoops 2003; Stoops et al. 2010. The thin sections were prepared by the Laboratorio Servizi per la Geologia di Massimo Sbrana, Piombino (Livorno).

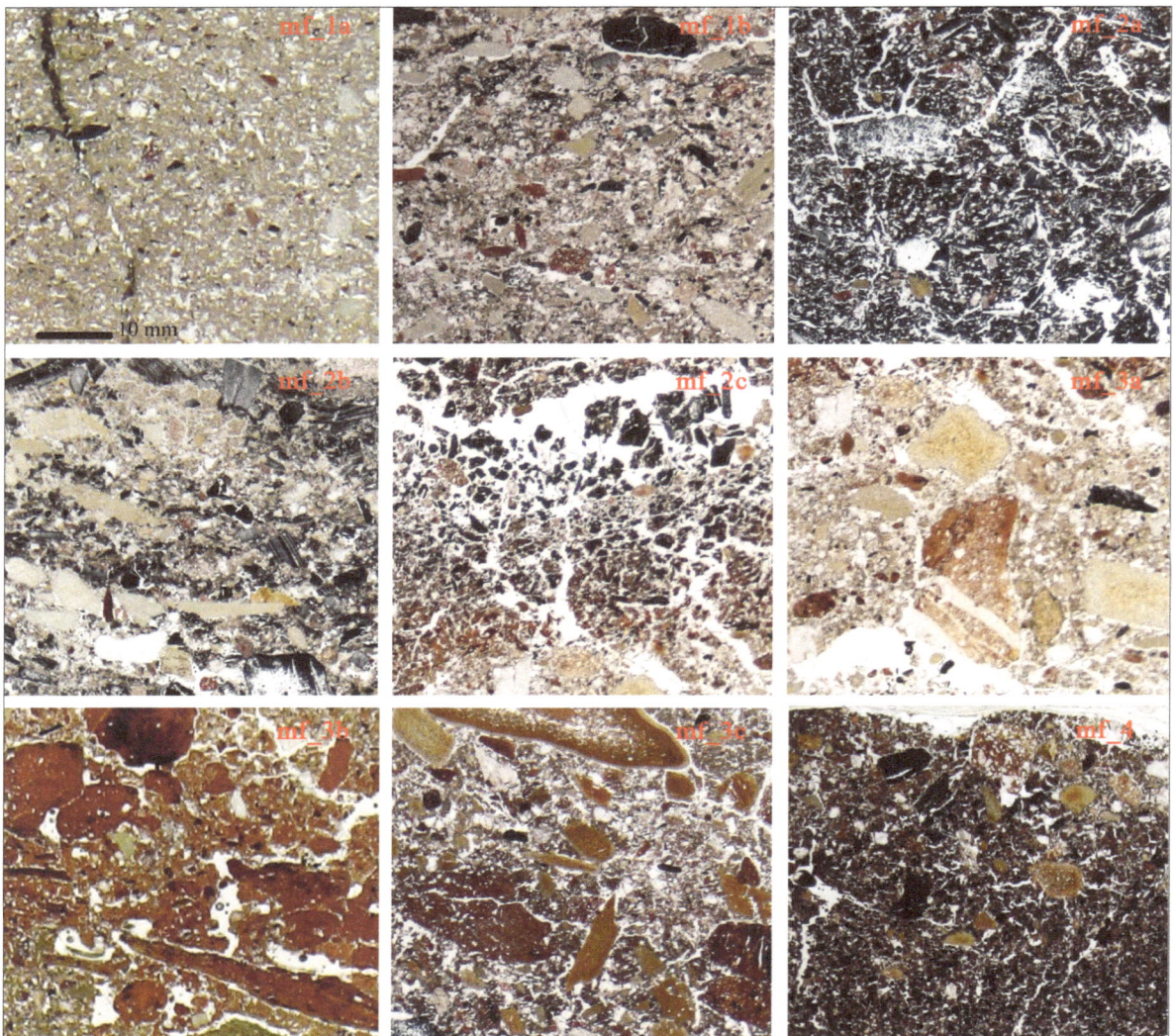

Figure 8.6   Storo Dosso Rotondo (Trento). Synoptic table of microfacies (mf) (description in the text).

various construction phases. This work concentrates in particular on the analysis of two vertical associations of microfacies (samples 107–108 and 252–253b–254) and their relationship with construction phases 3 and 4. The following types of microfacies were distinguished, with a thickness ranging from a few millimetres to several centimetres (Figs. 8.7, 8.8).

mf_1: colluvial deposits

Colluvial deposits generally made up of heterogeneous sand-size detritus, from well sorted to poorly sorted, with angular grains. These can be classified in the context of laminated colluvial deposits and non-laminated colluvial deposits (Mücher *et al.* 2010). They occur cyclically alongside carbonate deposition (mf_2) and are structured into lenses or laminae with a thickness varying from a few millimetres to several centimetres. The thickness is conditioned by the topographic gradient, which tends to concentrate the sediment downhill (SE corner of the excavation sector) and in hollows located in the holes of posts that have been removed in antiquity or rotted in situ, The macroscopic aspect of the unit is generally massive. Two subtypes were distinguished:

mf_1a (Fig. 8.6)

The texture of the sediment is sandy, homogeneous, with moderate granulometric selection, generally of massive aspect and well-sorted. Micro-charcoal is scarce and anthropogenic components are absent. The sediment comes from processes of sheet erosion of the slope. The microstructure is platy, linked to freezing and thawing cycles (Van Vliet-Lanoë 2010). Sedimentation of this subtype is conditioned above all by natural factors and could therefore be indicative of phases of periodic abandon (seasonal?).

mf_1b (Fig. 8.6)

Colluvial deposits of natural and anthropogenic sediment with a heterogeneous texture, little or rarely selected, with common charcoal fragments up to 5mm, common to frequent texture made up of sandstone and siltstone grains. The stratification shows iso-oriented grains and bedding with deformation structures, produced by moderate trampling (human or by animals) in conditions with an oversaturation of water.

These characteristics suggest that the microfacies is linked to a phase of erosion of the slopes, taking place during phases of occupation. The structure has a massive aspect, while there is a subangular blocky microstructure.

mf_2: charcoal layers

Charcoal lenses extend over almost all the area investigated. The units are characterised by charcoal frustules with a size of up to 2cm. As in the case of the colluvial microfacies, the thickness is conditioned by the topographic gradient, which tends to concentrate sediment towards the downhill area (SE corner of the excavation sector) (Fig. 8.3). The unit is characterised by the abundant presence of fragments of burnt wood with an easily recognisable cell structure of very variable size, from 50 micron to several centimetres, mostly related to conifer branches. Some fragments are made up of cells filled with decomposition products of a brownish-yellowish colour, probably cellulose and/or phlobaphene, a colouring substance characterised by intense brownish-yellowish tones, often tending towards red, resulting from tannins. In some cases, transfer from the altered cellulose to the inside of half-burnt wood cells was observed (mf_2b). In the mf_2c microfacies, decomposed material was in contact with the ground, while carbonisation increased gradually towards the top and was interpreted as in situ combustion of conifer branches with residues of charred needles (see sample 253B). No residues of burnt herbaceous plants were found, also due to the poor preservation of the cell structure. Furthermore, traces of phytolites, depositions of amorphous silica of plant cells present mainly in herbaceous plants, were not found in the thin section. This data would consequently exclude the periodic burning of animal dung and vegetal remains, commonly interpreted as the product of pastoral activities, such as residues of forage or animal bedding. Archaeological sediment of this type is known starting from the Neolithic and is conserved in caves and rock shelters (Brochier 1983; Courty et al. 1992; Macphail et al. 1997; Boschian and Montagnari-Kokelj 2000; Miracle and Forenbaher 2005; Boschian and Miracle 2007; Angelucci et al. 2009). The charcoal layer was generally compact, adherent and often directly superimposed over sandy laminae of colluvial origin related to microfacies mf_1. The charcoal fibres were often iso-oriented and subhorizontal, with elements showing deformation and fractures due to stress from compacting and/or trampling. The archaeological record has not conserved any evidence of ash resulting from the burning of wood, being made up of microcrystalline aggregates of calcite cell pseudomorphs, due to pedogenesis of an acidic nature.

The wood ash was made up of fine-grained crystals of calcite (15–20μ), and may display relic plant shapes due to replacement of plant material by calcite cell pseudomorphs (Wattez and Courty 1987; Courty et al. 1989; Wattez et al. 1990; Brochier 2002). It is presumed that for the same reason possible excrement deposits, made up in particular of spherulitic calcite produced by the digestion of herbivores, have not been conserved. Indeed, in open-air sites the conservation of dung and excrement in general is rare and it is well-

microfacies		SU	thin sections n°	interpretation
type	subtype			
mf_1 colluvial deposits	mf_1a	50-96	235-252	Sheet erosion
	mf_1b	15-82-96-100-101	107-108-211-231-235-252-253b-254	Sheet erosion, rain-wash and trampling
mf_2 charcoal layers	mf_2a	15-82-96-101	107-108-211-231-253b	Secondary deposit of charcoal (burned wood)
	mf_2b	15-50-82-96-100-101	108-211-231-235-252-253b-254	Secondary deposit of charcoal (burned wood), trampling and dumping
	mf_2c	15	108	Primary deposit of burnt and humified wood
mf_3 occupation surfaces	mf-3a	23-57-50	108-254	Prepared surfaces and dumping
	mf_3b	57	254	Prepared surfaces with rubified bedding and dumping
	mf-3c	15-57	107-108	Prepared surfaces with reworked and residual burned components
mf_4 hearth		73	218-219	Combustion structure

Table 8.4 Storo Dosso Rotondo (Trento). Synthetic description of microfacies (mf).

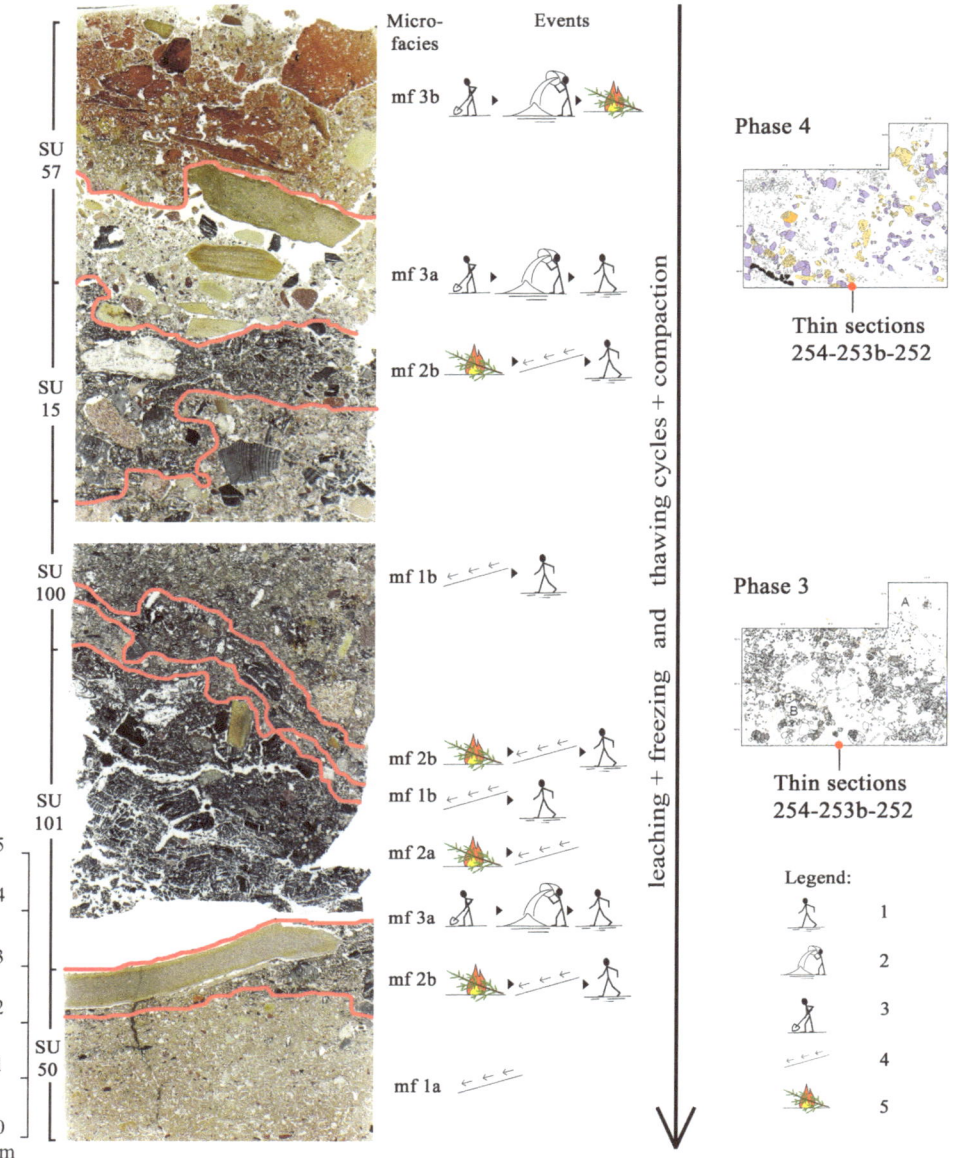

Figure 8.7 Storo Dosso Rotondo (Trento). Samples 252, 253b, 254. 1: trampling; 2: dumping; 3: prepared surface; 4: sheet erosion; 5: use of fire.

known that spherulites deteriorate rapidly, even after a single winter, if they are not buried rapidly (Brochier 2002; Brochier *et al.* 1992). Following the analysis carried out to date, the archaeological record shows no presence of siliceous algae (diatoms and chrysophyceae) at the site. These components can indicate both stabling areas, as the dung of domestic herbivores is present, and dwelling areas (Brochier 2002). The deposition of dark red clay coatings (mf_2b) may indicate enrichment with organic matter and phosphates, suggesting activities linked to the stabling of animals or in any case to their temporary concentration on the topsoil. Unfortunately, in the case of the Storo Dosso Rotondo site, animal stabling and activities linked to animal management in general cannot be identified through analysis of the organic phosphate deposited in the form of dung (Courty *et al.* 1989; Macphail and Cruise 2001), as the sediment has been polluted by the effects of intensive pasture currently underway. This unit has experienced colluvial phenomena to a limited extent. The sediment is located just a few metres away from the original position. The microstructure generally is platy, tending towards subangular blocky, in some cases granular. The minute fragmentation of the charcoal can be attributed to freezing and thawing phenomena and/or trampling, which have caused the

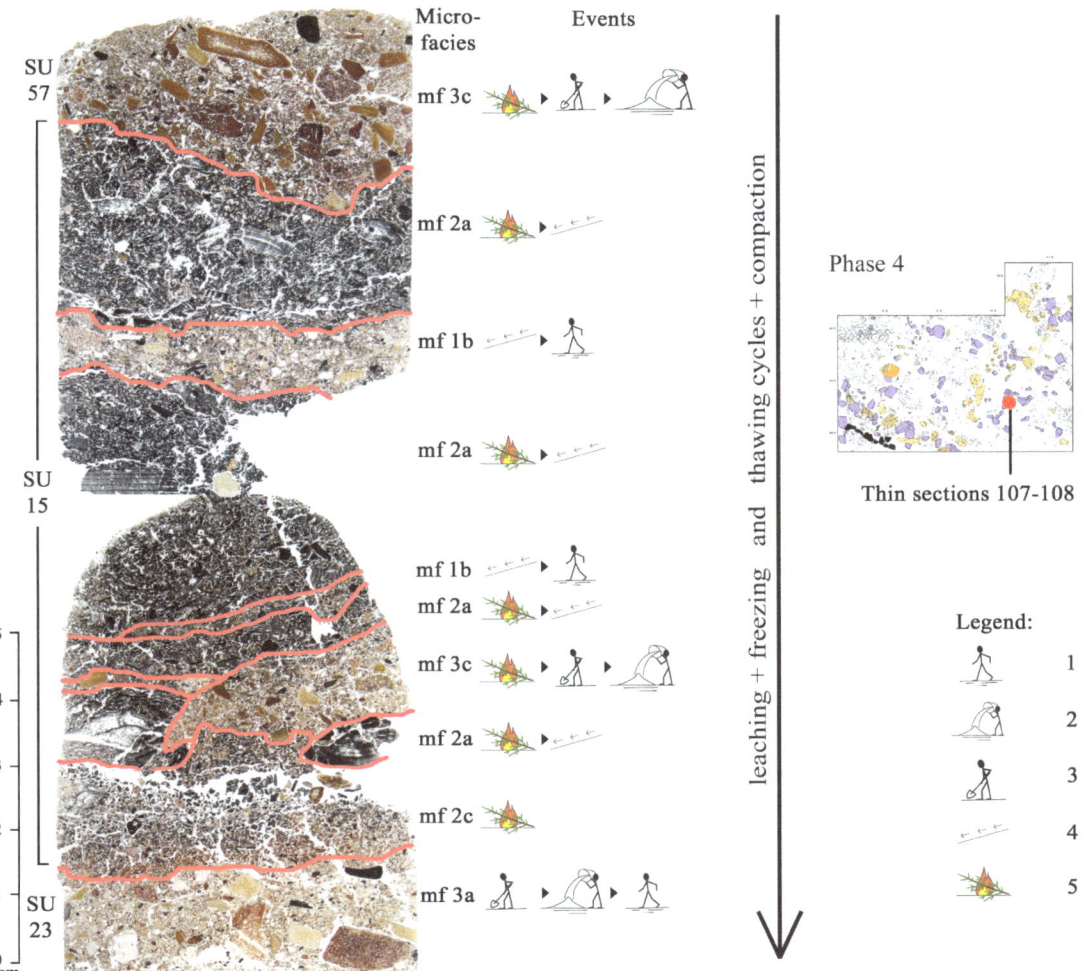

Figure 8.8 Storo Dosso Rotondo (Trento). Samples 107–108. 1: trampling; 2: dumping; 3: prepared surface; 4: sheet erosion; 5: use of fire.

collapse of the cellular structure and the subsequent compacting of the spaces in the microstructure. Three sub-types have been distinguished:

mf_2a (Fig. 8.6)

Subtype made up almost entirely of charcoal frustules (>80%). It has a massive aspect, but laminated structures can often be noted, made up of iso-oriented, sub-horizontal fibres.

mf_2b (Fig. 8.6)

Subtype made up of at least 50% charcoal frustules. The remaining component is of colluvial origin, represented by clastic sediments, made up of grains of sandstone and tabular clast of siltstone, in a sandy matrix. Overall, the unit is made up of heterogeneous unsorted material, which is interpreted as the result of mixing up of the stratification due to trampling in humid conditions.

mf_2c (Fig. 8.6)

Unit made up of abundant wooden plant remains, partially burnt, decomposed and humified. Cell residues made up of a decomposition product of a brownish-yellowish colour are evident. One can note a gradual transition from charred wooden remains to uncharred organic residues. The latter are often in direct contact with the surface of the original soil. The effects of partial combustion of woody materials in situ can be noted. This unit is around 2cm thick. The unit is characterised by an evident granular microstructure, following evolution of the pedogenetic horizon on the surface of the soil (organic horizon, O). The development of this very porous microstructure is symptomatic of limited trampling or its total absence (Gè *et al.* 1993; Goldberg and Macphail 2006).

mf_3: occupation surfaces

Deposits of anthropogenic origin made up of abundant sandy matrix support clastic material of heterogeneous size. Tabular siltstone clasts were mainly selected, with surface areas covering various m^2. In the thin section horizontal stratification is visible as a result of the alternation of different textural clasts. There is a subangular blocky and platy microstructure.

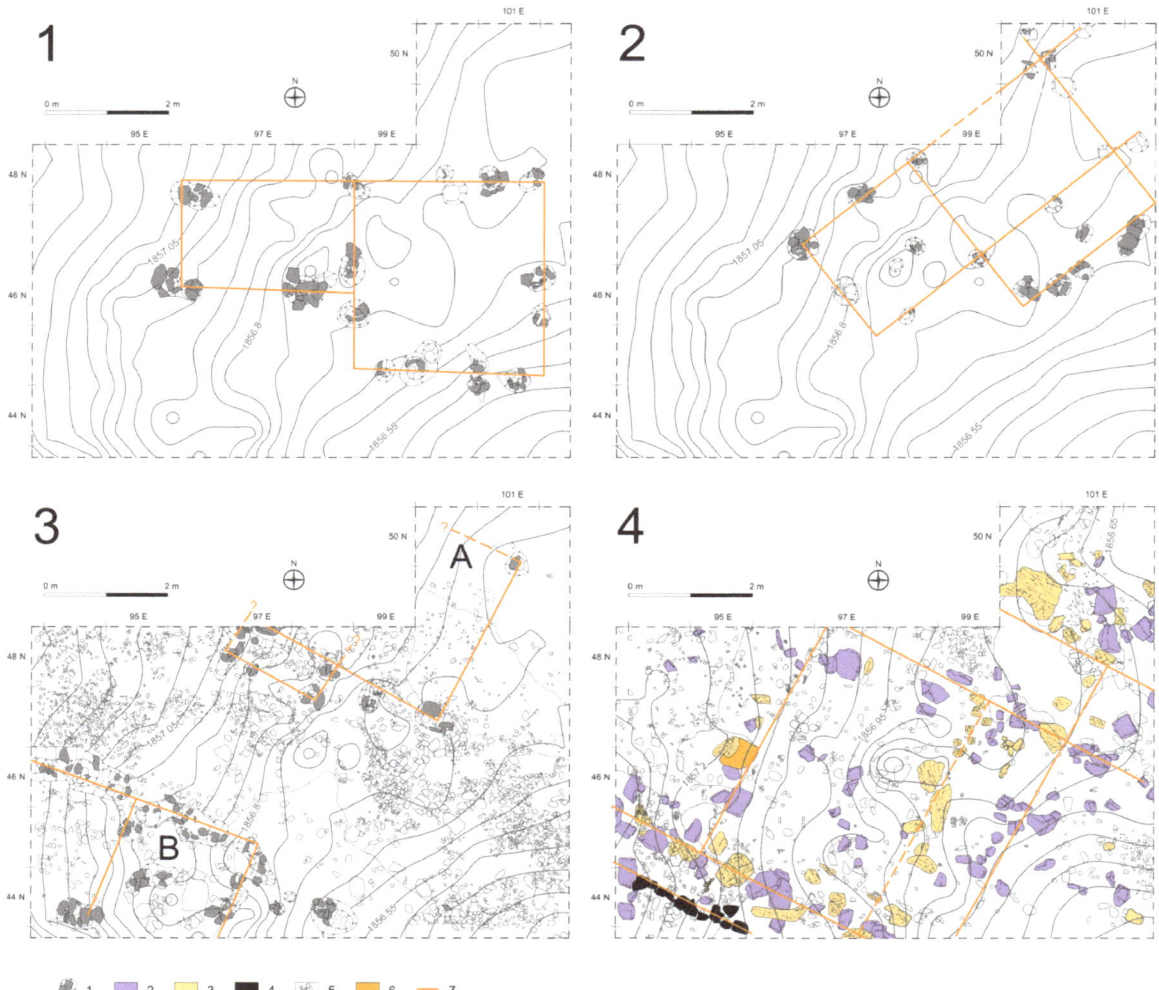

Figure 8.9 Storo Dosso Rotondo (Trento). Building phases 1–4. 1: post-hole with stone wedges; 2: block of sandstone; 3: siltstone slab; 4: structural line of stones; 5: stone clasts and blocks; 6: combustion structure (US 73); 7: structural perimeter.

mf_3a (Fig. 8.6)

Lenses of anthropogenic origin made up of clasts and grains, without evidence of alteration due to combustion. They come from carryovers of selected sediment removed from the local substrate.

mf_3b (Fig. 8.6)

As compared to the previous subtype, the surface area shows traces of combustion activities in situ. One can note rubefication in the clasts (sometimes very deep and intense, concerning in particular siltstone lithotypes. This thermal impact can be obtained with temperatures of around 450° – 500°C (Canti and Linford 2001).

mf_3c (Fig. 8.6)

Deposit of intentional anthropogenic origin with thermally altered clasts due to combustion. The occupation surface was obtained with sediment having traces of rubefication, deriving from the dismantling of combustion structures (mf_4) or the levelling of surfaces with traces of combustion (mf_3b).

mf_4: hearth (Fig. 8.6)

Combustion structure constructed on the ground made up of a rectangular cooking surface measuring 50×45cm, delimited by a border of stones. The cooking surface is significantly rubefied and was constructed by laying fine sediment.

4.3. Discussion (Table 8.4)

Interpretation of microfacies associations shows a cyclical sequence of anthropogenic activities and post-depositional modifications, providing for (Figs. 8.7, 8.8):

1. The use of clasts and sediment to create the surface used (mf_3a), also coming from the dismantling (mf_3c) or restoration of combustion areas (mf_3b). In this phase it is believed that an occupation surface was prepared close to the dwellings/production areas,

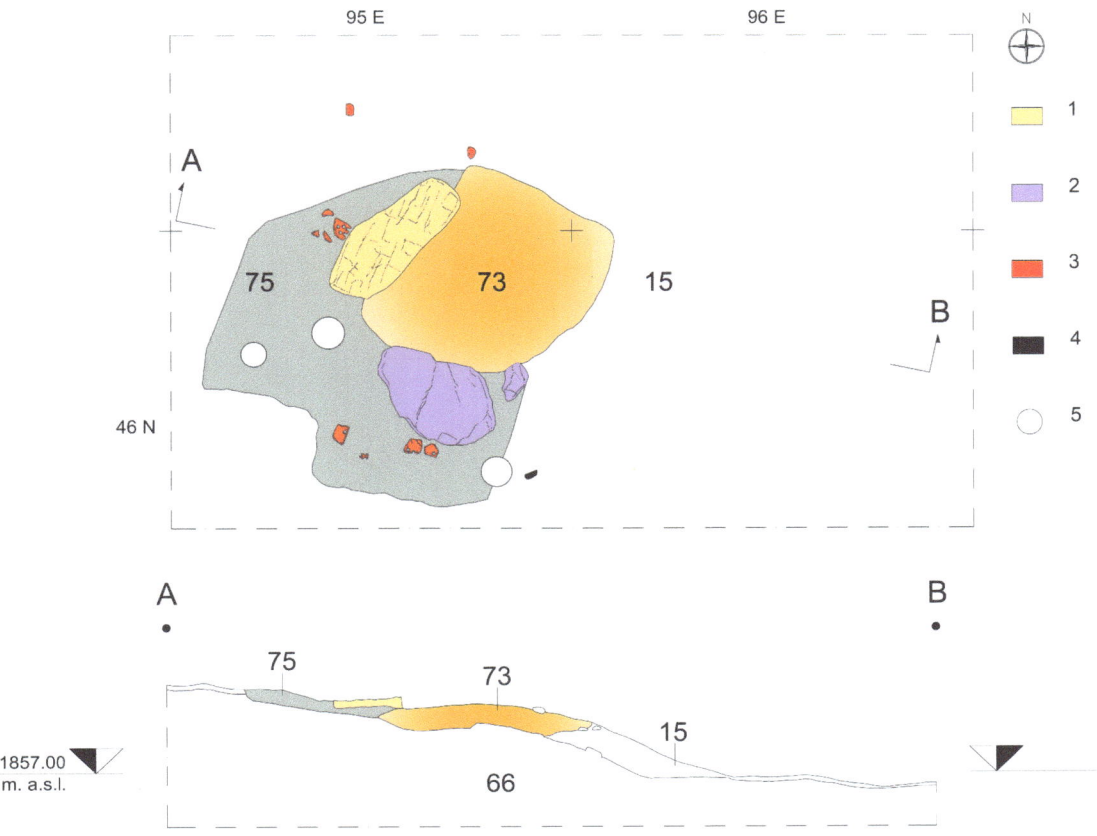

Figure 8.10 Storo Dosso Rotondo (Trento). Plan of hearth US 73 (mf_4). 1: siltstone slab; 2: block of sandstone; 3: fragments of pottery (some of them are strainers); 4: flint; 5: post-hole.

creating an insulating base on a sandy substrate of colluvial origin (mf_1a) or on the unit abandoned during the previous phase of occupation.

2. Combustion activities in situ on used surfaces (mf_3b) with accumulation of charcoal deriving from the burning of conifer branches (mf_2c). The latter can be interpreted as residues of animal bedding, or waste produced as the result of deforestation activities (mf_2c). It can also be surmised that the periodic reconstruction of the structures usually provided for the elimination of wooden residues (such as branches) by burning, whereas the load-bearing elements were reused or maintained. In the archaeological record there are indeed only faint traces of the remains of large wooden elements. In this case renovation probably provided for the elimination of less important wooden elements that must have been represented by walls, roofs and furnishings that could not be reused. It is believed that the products of combustion, exclusively charcoal which we may recall extended over a distribution area of around 800 – 1,000m², may result partly from the use of structured hearths (mf_4, US 73) but above all from the lighting of fires on surface areas (mf_3b and mf_2c). Over the course of the occupation, these products of combustion were subject to erosion and limited transport in the form of colluvial deposits (mf_2a), in addition to changes taking place during and after deposition due to trampling (mf_2b).

3. In the phase of seasonal abandonment the bare earth would have been subject to erosion. Colluvial phenomena represented by massive lenses (mf_1b) are interposed with the products of combustion, which are also displaced further downhill (mf_2a).

5. The structural phases

The area currently investigated has brought to light 76 postholes, 70 of which can be related to four residential buildings constructed using load-bearing vertical posts planted in the ground. These can be attributed to three separate stratigraphic phases (Fig. 8.9).

An initial structure oriented E–W was constructed in phase 1, made up of two rectangular modules placed side by side, delimited by postholes around the perimeter. The first module measures 2.80×1.75m and the second 3.20×3.05m, for a total surface area of around 14.80m².

A similar, two-module structure of slightly different size (2.15×1.95m and 2.75×3.3m) was constructed in phase 2, with an overall surface area of 12.5m², but with the longer side oriented in a NE–SW direction.

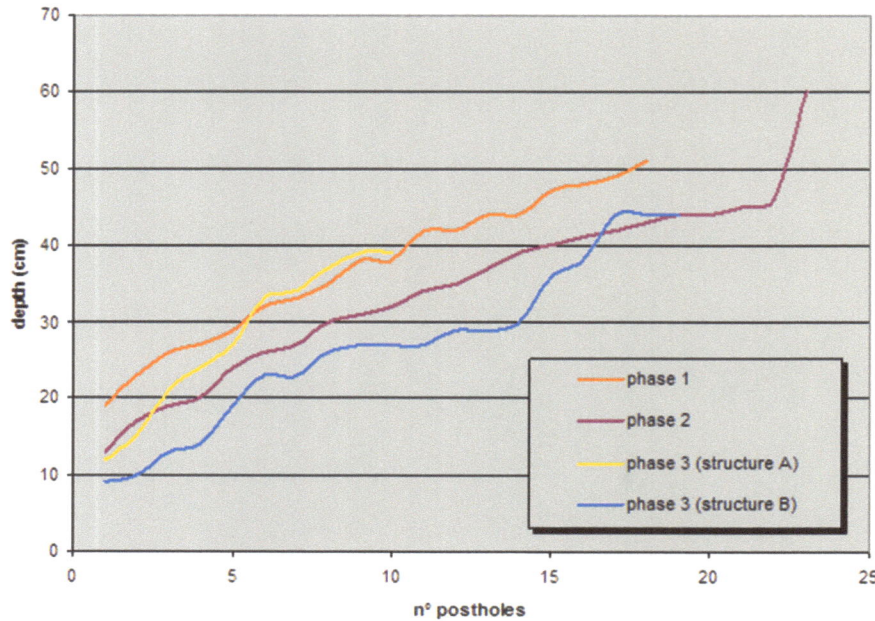

Figure 8.11 Storo Dosso Rotondo (Trento). Depth of post-holes in the different building phases.

Two structures oriented WNW–ESE and only partially included in the excavation area were constructed in phase 3. The southern structure (structure 3B) is characterised by a different construction technique, adopting perimeter foundation ditches within which vertical posts were positioned using stone wedges. The internal partitioning of the ditches seems to suggest that the diameter of the foundation posts was smaller as compared to that documented in previous phases.

The north-eastern perimeter of structure 3B was investigated for a length of 3.9m. Further north, the limits of the second structure (structure 3A) are less clear and with the current state of knowledge a quadrangular structure measuring 3.15×2.95m can be surmised. The area concerned by the two structures is characterised by an intentional distribution of clasts and flakes, indicating work to ensure drainage for the settlement space.

Phase 4 was characterised by further modification of the construction technique. At this point a fifth dwelling was built, with an insulated foundation using stone elements made up of siltstone slabs alternating with blocks of sandstone positioned at right angles. It is possible to surmise that this structure represented the base for a wooden structure with horizontal beams, very similar to types of alpine buildings in the protohistoric and historic times (Reitmeier et al. 2013; Putzer 2011; Bassi and Cavada 1994; Cataldi 1986; Aspesi and Cataldi 2013). In the northern area the line of stones brought to light retraces the perimeter of structure 3A from the previous phase 3, whereas in the southern area the line is parallel but shifted around a metre south as compared to the older 3B structure. The two lines of stones documented in the southern area in phase 4 are particularly well conserved. The one further south, which could also be referred to a sixth dwelling, is made up of blocks of sandstone and was documented for a length of 4.45m, whereas the more internal parallel line was constructed using slabs of sandstone and siltstone and was documented for a length of 2.55m. The blocks of sandstone are many-sided and have a maximum size of 60×50cm, whereas the slabs, albeit subject to rapid changes given the lithological characteristics, have nevertheless remained of considerable size, measuring up to 80×70cm. The finding of a considerable number of charred wooden fibres, brought to light along the lines of stones identified, could represent a further indication of the original presence of posts laid horizontally and at right angles.

A combustion structure (US 73) constructed on the ground, made up of a rectangular cooking area measuring 50×45cm, delimited by a sandstone rock and a slab of siltstone can be referred to the more recent phase of occupation (Fig. 8.10). The cooking surface is highly rubefied and was created by laying a layer of fine sediment. A series of pottery fragments belonging to at least two strainers (Fig. 8.18) and remains of calcined bones were found close to the combustion structure.

In the different phases of occupation the dwellings were constructed on a gentle slope (around 5°) (Fig. 8.3) and it is therefore possible to surmise that the floors were constructed by laying protruding decks.

Analysis of the morphometric parameters of the postholes has shown that in phases 1, 2 and 3 the most significant aspect is the variability in the depth of the holes in which the posts were planted. This was never constant over the various phases and would not appear

Figure 8.12 Storo Dosso Rotondo (Trento). Post-hole (US 98) filled with charcoal and colluvial lenses (US 99).

to be related to their position on the slope. It is therefore likely that the depth of the hole was related to the static function of each individual post and diminished gradually over time, from phase 1 to phase 3 (Fig. 8.11).

The fact that the depth at which the posts were planted decreased over the course of time could be interpreted as the effect of a progressive change in the construction technique. It was indeed observed that over the course of the first three phases of occupation the load-bearing capacity of the terrain (reduction in depth) and the individual wooden elements (different depths), took on increasingly less importance, ultimately arriving at the fourth phase, without the adoption of load-bearing posts planted in the ground, but rather with the presence of stone supports to bear the weight of the structure.

As far as the shape of the holes is concerned, only in phase 2 was there a trend for a depth greater than the diameter.

In the different construction phases it is possible to observe that some of the posts had very probably been removed in ancient times and reused as construction material.[6] This interpretation was supported by the following evidence:

- the backfill of the postholes was homogeneous and made up of sandy sediment, without signs of a postpipe. At all events, there were charcoal frustules dispersed in a heterogeneous manner in the matrix;
- the presence of wedging stones was documented in an anomalous position, namely in the upper part of the backfill instead of at the sides of the hole;
- in some cases it is possible to observe perimeter deformation of the postholes, which could suggest the removal of the post by rocking;
- the posthole hollow was filled with an alternation of charcoal and colluvial lenses (Fig. 8.12). The backfill was present alongside the charcoal and colluvial units covering the excavation area. These data has been interpreted as the result of a post-depositional sedimentary process that can occur both when there has been deterioration of the post in situ and as a result of its intentional removal. No traces due to organic deterioration of the post were found and only in one case was a charred post identified in situ.

To summarise, it is possible to surmise a series of construction phases providing for the dismantling of the previous building, with the removal and perhaps recovery of the load-bearing elements before subsequent reconstruction. The cause of these repeated construction phases can probably be sought in the specific environmental conditions of the site, which are not particularly favourable for the conservation of wooden structures. The low hydraulic conductivity of the

---

6. Similar observations were made as regards some structures in a Neolithic settlement dating back to the 5th millennium BC cal brought to light in Trentino (Mottes and Degasperi 2014).

taxon	US 3	US 4	US 57	US 61	US 63	US 65	sond. 14	sum nr	%
CONIFERS									
Abies alba		7	1	1			1	10	3.9
cf. *Abies alba*		2	3	3				8	3.1
Picea excelsa	18	32	19	2	12	4	1	88	34.0
cf. *Picea excelsa*	1	6	1			1	1	10	3.9
Abies/Picea	3	4	1		1		1	10	3.9
cf. *Larix decidua*							1	1	0.4
Pinus sylvestris/mugo				1				1	0.4
cf. *Pinus cembra*		1						1	0.4
Coniferae		2				1		3	1.2
Coniferae (indeterminate)		44	9			10		63	24.3
BROAD-LEAVED TREES									
Acer sp.	3	18	9	7	1	2		40	15.4
cf. *Acer* sp.			1			1		2	0.8
Alnus viridis		5			1			6	2.3
Alnus sp.		2						2	0.8
Cornus/Viburnum	1							1	0.4
cf. *Corylus avellana*	1							1	0.4
Fagus sylvatica				1		1		2	0.8
Laburnum sp.	2	2						4	1.5
cf. *Laburnum* sp.	1							1	0.4
Pomoidea		3						3	1.2
Tilia sp.		1						1	0.4
indeterminate (bark)		1						1	0.4
sum	30	86 (+44)	35 (+9)	15	15	10 (+10)	5	259	100.0

Table 8.5 Storo Dosso Rotondo (Trento). Anthracological remains.

substrate indeed favours the creation of persistently humid and waterlogged conditions, which in areas at high altitude are accentuated during the period of winter snow. These aspects probably influenced the conservation of the structures, made up vertical load-bearing posts in conifer wood that must also have been subject to rapid biodegradation due to ligniferous fungi and xylophagous insects. Consequently it would have been necessary to proceed with the frequent removal and substitution of wooden components. It can be noted that none of the 76 holes brought to light cut across other holes, probably because there was contemporary removal and reconstruction of the structures in a displaced position.

In phase 4, the abandoning of the construction method with vertical load-bearing posts, used in particular in settlements on the valley floor or flat land characterised by peaty or clayey substrates, in favour of insulated foundations in stone, on which the wooden structure rested, could be explained as an adaptation to conditions imposed by the mountain environment by the groups of men seasonally occupying the Dosso Rotondo site.

6. Botanical remains

6.1. Materials and methods

The material for archaeobotanical analysis came from 8 stratigraphic units and one test trench, with a total of 29 samples. These related to charcoal levels belonging to US 3 (phase of abandonment modified by colluvial processes), US 4 (Bronze Age flooring), US 57 (drainage material delimiting a utilisation space), US 73 (combustion structure), US 75 (charcoal lenses around the combustion structure), UUSS 61, 63 and 65 (backfill of postholes) and test trench 14, section 8.

In order to recover carpological remains, all the samples were examined carefully using a binocular microscope, including those already subjected to the flotation process by archaeologists. During the separation of carpological remains the charcoal fragments for anthracological analysis were also separated. The presence of scoriaceous material difficult to interpret was found in almost all the samples. In many cases one could surmise that this is conifer wood deformed by combustion, because homoxylous wood parts were still visible at some points. In other cases they could be other types of remains, in particular of prepared food 'bursting' during combustion. However, no anatomical structures belonging to kernels or other parts of the ears that could confirm this hypothesis were observed within these scoriaceous fragments.

The clear prevalence of conifers and the difficulties in identifying them led to the decision to rapidly examine part of the charcoal to distinguish those with homoxylous wood (conifers) from those with heteroxylous wood (broad-leaved trees), to increase the number of broad-leaved tree species identified. The conifers observed in this way were not determined at the level of genus or species but were in any case included in the table under the item 'undetermined conifer', whereas identification of the broad-leaved trees was carried out.

The number of charcoal fragments analysed varied for each stratigraphic unit, on the basis of the quantity of samples and material available. When possible, the original sizes of the parts from which the fragments came were recorded (twigs, branches or trunks and their diameter), along with other dendrological characteristics (growth anomalies, hyphae, parasite attack, etc.).

6.2. Sample results

6.2.1. Anthracological remains

Conifers represented around 75% of individual identifications (Table 8.5), the most commonly documented species being the spruce (Picea excelsa), followed by the silver fir (Abies alba). The Scots pine/mountain pine (Pinus sylvestris/mugo) was documented by a single fragment, as was the larch (cf. Larix decidua) and the Swiss stone pine (cf. Pinus cembra); the evidence regarding the last two species is however uncertain. Of the broad-leaved trees, there is significant documentation of the maple (Acer sp.), whereas there are only a few remains of other species. These are the green alder (Alnus viridis), Cornelian cherry/guelder rose (Cornus/Viburnum), beech (Fagus sylvatica), laburnum (Laburnum sp.), lime (Tilia sp.) and Maloideae; the maloideae include various genera (apple, Malus; pear, Pyrus etc.), also including the sorbus (Sorbus spp.), documented by a few carpological remains, albeit in an uncertain manner. Evidence regarding the common hazel (cf. Corylus avellana) is uncertain, while for two elder charcoal fragments it was not possible to specify the species (Alnus sp.).

6.2.2. The vegetation around the site

A substantial part of the charcoal came from branches or twigs and no elements clearly relating to the structures were identified. A significant number of fragments of maple charcoal came from trunks or large branches, but the traces of working observed on these relate to the production of objects.

Spruce was by far the most prevalent species, needle-shaped leaves also being found for this species. The presence of small branches – which sometimes have knots, cracking, tension wood and a scoriaceous appearance, occasionally resiniferous channels of traumatic origin and in rare cases hyphae – confirms that the charcoal analysed did not come from the structure of the building but were rather the remains of wood used for fuel, collected on the ground, or by removing the lower branches of trees in the woods, or alternatively by removing the branches from trunks destined for other uses. Another possibility is that this material was employed to construct wattle structures to shelter livestock or as partitions in the buildings. The high number of charcoal fragments and the type of use surmised suggest that spruce formations dominated at the site altitude.

The silver fir charcoal also belongs almost exclusively to branches and twigs. The presence of this species is also indicated by the finding of needle-shaped leaves. It is possible that fronds of silver fir were used for bedding, as well as for firewood. As compared to spruce, which has diamond-shaped needles, the leaves of the silver fir are softer, flatter and with a blunt point. The silver fir today is generally found in mixed formations (often with the beech), at an altitude of up to 1,600m, but isolated individuals, mixed with spruce trees, can be found at much higher altitudes. The more limited presence of silver fir as compared to spruce, and the even scarcer presence of beech, could suggest that the upper limit of the spruce and beech formation – currently more extensive on the opposite side of the Valle del Chiese due to the exposure – was situated at an altitude just below the site. The limited presence of beech charcoal (at least one of which comes from a small piece) suggests that they came from a bundle of twigs or an artefact.

Beech, together with silver fir, represents the species most widely used at Fiavé Carera (Trento) for the production of objects, being adopted in particular in order to make handles and tools with an elongated shape (Perini 1988, Fig. 1; 1987: 374–378, Fig. 182). Young beech twigs were used at the site to feed livestock at the beginning of the spring (Haas et al. 1998).

Of the species of maple, the sycamore (Acer pseudoplatanus) is the one arriving at the highest altitude (1,500 metres, exceptionally arriving at 1,900 m.a.s.l.). From an anatomical point of view, the particular width of the rays, up to 8–10 cells wide, observed in the charcoal from Dosso Rotondo, would also seem to be characteristic of the Norway maple (Acer platanoides), a tree that grows at lower altitudes, in respect to sycamore. Maple charcoal was documented in almost all the stratigraphic units. In most cases, as already mentioned, the fragments came from large pieces and were mostly tabular, with a

taxon		US 3	US 4	US 57	US 61	US 63	US 65	US 73	US 75	s.t. 14
CEREALIA										
Hordeum vulgare/distichum	grain	1	1						1	
Triticum monococcum	glume base				1					
Triticum dicoccum	grain		1							
Triticum dicoccum	spikelet fork				2					
Triticum dicoccum	glume base	1			1					
cf. "nuovo frumento vestito"	spikelet fork				1					
cf. "nuovo frumento vestito"	glume base				1					
cf. *Cerealia*	grain			7		1		3		
FRUITS										
Cornus mas	stone	5	11		1		4		3	
Corylus avellana	shell		15	5	2	1	3	1	5	3
cf. *Juniperus* sp.	seed		2							
Quercus sp.	receptacle		1							
cf. *Quercus* sp.	fruit				1					
Rosacea tipo *Sorbus*	fruit								5	
OTHER PLANTS										
Lamiaceae	fruit								1	
cf. *Luzula* sp.	fruit		1							
Poaceae	grain		1					5		
indeterminate	small shell/seed?		5			2	2			
CONIFERS NEEDLE-SHAPED LEAVES										
Abies alba	needle-shaped leaf	17	4						6	
Picea excelsa	needle-shaped leaf		9	3				2		
Coniferae	needle-shaped leaf		2							

Table 8.6   Storo Dosso Rotondo (Trento). Carpological remains.

thickness of at least 3mm. In some cases, perfectly polished worked surface areas were evident: these characteristics can be related to flakes coming from artefacts. Maple wood lends itself to fine working (for bowls or kitchen utensils) and this tradition is confirmed by data coming from the pile-dwelling settlement of Fiavé Carera (Trento) (Perini 1987; 1988; 1990). In this context it can be noted that in some alpine valleys, maple wood was used to make objects and tools linked to cheese production right up to the modern age (Sordi 1979).

As far as pines are concerned, the Swiss pine could have been present at higher altitudes, but it is a species that is and was more widespread in internal areas of the Alps. With the essentially gentle slopes such as those around the site, the mountain pine and/or the Scots pine would not have found the most suitable environmental conditions and substrate.

High in the mountains, on north-facing slopes and along watersheds, green alder is widespread, generally preferring siliceous substrates. This is a plant that currently tends to invade pastureland, with compact and impenetrable formations that have to be uprooted in order to eliminate them. The wood of green alder would not seem to have any specific uses. Among the twigs analysed there was one which was still covered with bark, with the last ring complete, suggesting cutting in the late summer or subsequently.

Laburnum, probably Laburnum alpinum which grows at higher altitudes (from 500 to 1,600m, up to 1,900 m.a.s.l.), can be used for wattle work and is a reasonably good fuel.

The only lime charcoal, probably coming from a large branch, is likely to be linked to the working of an artefact, given that neither Tilia platyphyllos nor Tilia cordata – which cannot be discriminated on the basis of the anatomical characteristics of the wood – grow at an altitude of over 1,000–1,200 metres. Finally, there was a fragment from a twig identified as Cornelian cherry or guelder rose (Cornus/Viburnum): once again this would appear to be material collected at lower altitude as compared to the site. Both species have flexible branches that can be used for wattle work.[7]

6.2.3. Carpological remains

In total 107 remains of cereals, fruits and other plants of environmental significance were identified (Table 8.6). There was documentation of barley (Hordeum vulgare/distichum), Emmer wheat (Triticum dicoccum), Einkorn wheat (Triticum monococcum) and uncertain identification of 'new glume wheat'. The remains include both kernels and parts of the ears. The barley kernels were too fragmentary to identify the

---

7. As regards this see the extraordinary example of head-covering made using interwoven plant fibres (*Viburnum, Picea, Phragmites* (?) found in the pile-dwelling at Fiavé Carera (Trento) (Perini 1987:187–192, 360–361, Figs. 75–79; 1990:261–262, Figs. 14–16).

morphological type. Fragments of hazelnuts (Corylus avellana), Cornelian cherry berries (Cornus mas), acorns (Quercus sp.) and perhaps juniper berries (cf. Juniperus sp.) and sorbus berries (Rosaceae tipo Sorbus sp.) were recognised among the fruit. Other plants were documented by fragmentary remains of grasses (Poaceae) and a Lamiacea not precisely identifiable and perhaps one example of a rush (cf. Luzula sp.). During the screening, 43 fragments of conifer leaves belonging to the spruce and silver fir were also recognised.

6.2.4. Plants used for food

Barley is a plant that can cope with particularly unfavourable growing conditions. According to Dalla Fior (1981), in the last century, it was cultivated "up to the upper limit of crops in general", namely in the lower mountain horizon, situated at an altitude of between 1,000 and 1,500 m.a.s.l. The same author also states that two-row barley can mature at an altitude of up to 1,800 metres. Other archaeobotanical and ethnobotanical data have shown that barley was cultivated at an altitude of up to 2,000 metres, above all in more internal valleys in the Alps (Valais, Zermatt; Schmidl *et al.* 2007).

Einkorn wheat has not been cultivated in the central-eastern Alps for some time, so we cannot know at what altitude it was sown in recent times. In general it is considered to be a very frugal plant and due to this characteristic it is possible that, like barley, it was grown up to the upper cultivation limit.

Emmer wheat is considered to be slightly more demanding than Einkorn wheat, so it is only likely to have been cultivated at lower altitudes. In the Apennines it is still used in mountain areas and a variety sown in spring is also grown as an emergency cereal crop, in the event of very bad harvests of winter cereals. We do not know whether these characteristics can also be extended to varieties cultivated in the past.

As is well-known in the archaeobotanical field, there is generally caution when assigning a precise specific attribution to finds identified as 'new glume wheat' (Jones *et al.* 2000). In general reference is made to Triticum timopheevii, but the question is still open. For this reason it is difficult to evaluate the agronomical aspects. Triticum timopheevii is currently limited to Georgia, in a steppe/mountain environment. It would therefore have been a cereal with similar requirements to those of Emmer wheat and Einkorn wheat. Spelt, another type of wheat frequently found in mountain areas in the Bronze Age, was not documented in the carpological record.

The climatic and environmental conditions, the characteristics of podzol type forest soils and the data obtained from anthracological tests nevertheless lead us to exclude the idea that there were fields at the site, although it is possible that cereal crops were cultivated on better exposed slopes in the valley, up to relatively high altitudes. It is not however possible to establish the origin of the human group occupying Storo Dosso Rotondo and to establish a location for fields.

The fruits present at the site were also picked at lower altitudes: the hazelnut grows at an altitude of up to 1,400m in Trentino, while the Cornelian cherry grows at a maximum altitude of up to around 1,300 m.a.s.l. It is possible that the practice of encouraging the expansion and growth of the Cornelian cherry, by now ascertained in the Bronze Age (Rottoli 1997), made it possible to extend cultivation to slightly higher altitudes, but it is unlikely that this was possible at altitudes of over 1,500 metres.

Of the different species of oak present in Trentino, according to Dalla Fior (1981), most varieties do not grow above 1,300 m.a.s.l. The only fruit picked close to the site could have come from the sorbus: the dwarf whitebeam (Sorbus chamaemespilus), the mountain ash (S. aucuparia) and the common whitebeam (S. aria) can indeed grow at altitudes of up to 2,000 m.a.s.l. and beyond. The juniper could also have grown close to the site; the berries, as is known, were and still are used for nutritional and medicinal purposes.

The botanical remains come from various stratigraphic units, the concentrations being relatively limited, and are represented by remains accidentally burnt during the preparation of meals or waste eliminated in the fire. On the basis of these data, it would appear that during their occupation of the site, the inhabitants of Dosso Rotondo had a certain quantity of food supplies transported from settlements situated further down the mountain, with a range very similar to the area of origin. The presence of hulled wheat glumes, implying the hulling of cereals day by day on site, suggests that in order to improve conservation, cereal crops were transported from the valley still encased in their glumes and stored in the form of ears, a system also adopted in settlements on the valley floor.

Assuming summer occupation of Storo Dosso Rotondo, the presence of cereals reaching their maturity starting from the beginning of summer is not problematical, whereas the species of fruit documented can raise some questions. Bearing in mind the ripening period for the maturing of Cornelian cherry fruit, starting from the second half of August and throughout September, the presence of burnt stones could indicate the transport of just ripe fruit to high altitude, which would however assume more lengthy occupation of the site. Another hypothesis is that the fruit was instead conserved at length and taken to the site dried or toasted. In the case of hazelnuts, it is again possible that those from the previous year were consumed. Such lengthy conservation is possible and undoubtedly simpler as compared to the preserving of Cornelian cherry berries, especially if the hazelnuts were partially toasted or smoked. It is however possible that the hazelnuts were picked freshly – even carrying out daily picking at lower altitudes – in the last part of the season in the mountains (September–October). Long-term conservation of acorns poses greater problems; however early maturing would have made it possible to have access to them already at the beginning of the summer. Of the mountain species of Sorbus, the only one with edible fruit is the common whitebeam, whereas the mountain ash is still planted today in order to attract birds. The service tree (Sorbus domestica), which has the most highly appreciated fruit, belongs to the same genus. However, this tree grows at an altitude of up to around 800 metres and, in any case, the fruit ripens in the late summer or autumn.

There were very few remains belonging to spontaneous species and what is more, these were not clearly identifiable. The wood rush belongs to the Juncaceae family, which includes different species reaching an average height of 20–40cm, many of which live in mountain environments, spruce forests and heaths. The impossibility of better identifying the remains of Gramineae and Lamiaceae meant that it was not possible to establish whether they belonged to local plants, or instead represented remains that were accidentally transported together with food supplies.

7. Preliminary results of functional analysis on flint tools

A number of samples of flint artefacts coming from the Storo Dosso Rotondo site were subjected to functional studies, through microscope observation with a low magnification level (using an Optech stereo–microscope, up to 40×) and with high magnification (using a reflected light microscope – up to 200×), according to the commonly adopted protocols (Keeley 1980; Vaughan 1985).

The samples were made up of 33 finds, including 22 sickle blades (Fig. 8.13) and 11 arrowheads (Fig. 8.14).

Despite the presence of post-depositional scars on some of these finds and changes to the surface (particularly of a thermal nature), the conservation of the artefacts, mostly intact, was very good.

Almost all the sickle blades, of sub-rectangular shape, showed traces of use-wear (20 out of 22), both at macroscopic and microscopic level. In general it was possible to see the presence of a single active functional area, located on one of the two long edges, with a straight or concave shape, contrasting with a convex-shaped edge, where there were no traces of use-wear and which probably represented the part of the tool fastened into the haft; this hypothesis is supported by traces of use-wear attributable to contact with a handle (bright spots, micro-chipping or rounding of the edge and the ribs) on some of the convex edges or close to them.

The traces of use-wear observed, above all under the microscope, showed characteristics attributable to the treatment of siliceous plants, with a micro-polish from bright to opaque in appearance, which often concerned not only the edges of the tools but also the more internal part, showing a certain degree of development and hence intensive and prolonged use of the tools. In at least a couple of cases, the distribution of the traces of use-wear made it possible to recognise the recovery

Figure 8.13 Storo Dosso Rotondo (Trento). Lithic industry: sickle blade.

Figure 8.14 Storo Dosso Rotondo (Trento). Lithic industry: arrowhead.

Figure 8.15 Storo Dosso Rotondo (Trento). Polish micro-wear on the sickle blade RR 136 (original enlargement 200×).

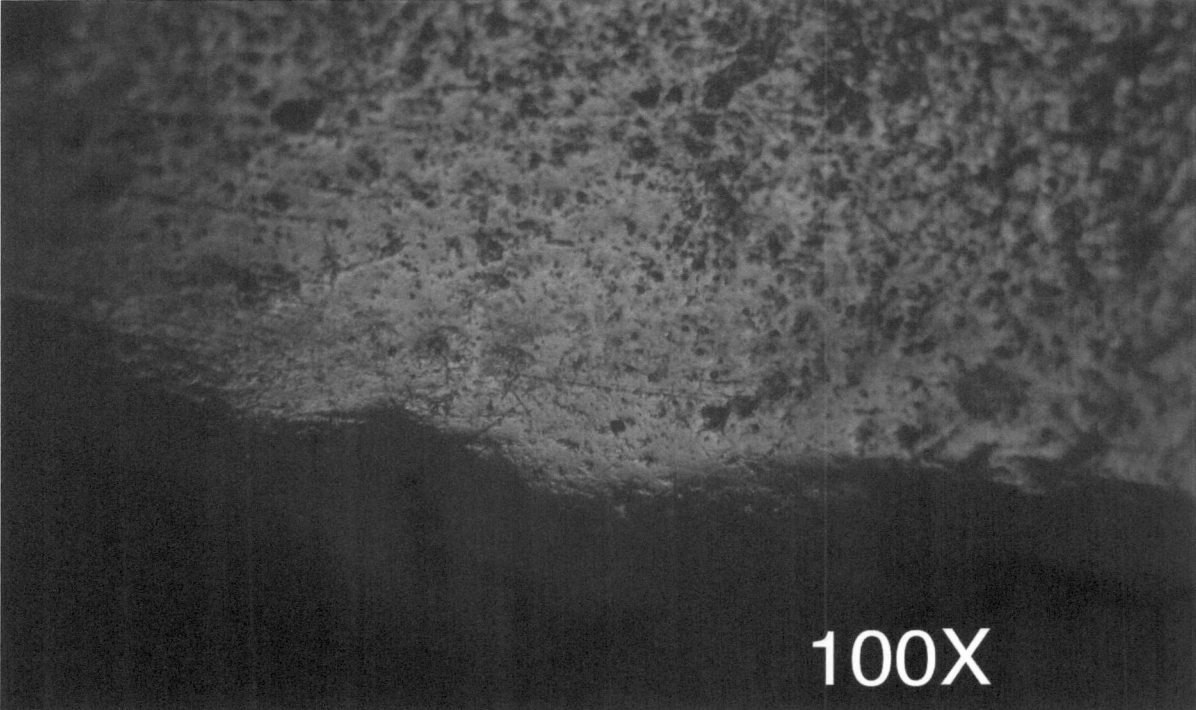

Figure 8.16 Storo Dosso Rotondo (Trento). Polish micro-wear on the sickle blade RR 329 (original enlargement 100×).

of broken hafts and their reuse, probably after having been reinserted in the handle.

Given the position of the site, at high altitude, one possible hypothesis is that these tools were linked to haymaking activities, involving the harvesting of spontaneous grasses. In particular it was possible to observe the presence of a very bright micro-polish, with a smooth texture and flat or corrugated topography. However, this was often characterised by the presence of rough streaks of differing widths, sometimes oriented in a specific manner and on other occasions in a less organised way (Figs. 8.15, 8.16). In the literature it has been observed by several authors that extensive formation of streaks takes place with the treatment of domestic cereal crops, whereas the phenomenon would seem to be rare in the case of harvesting of spontaneous grasses (Unger-Hamilton 1985; Van Gijn 1989; Juel Jensen 1994). The type of trace of use-wear observed on the sickle blades from Storo Dosso Rotondo would therefore be more compatible with the treatment of domestic cereals, which we would imagine to have taken place further down the mountain. In this case we would have to conclude that the human groups moved around with a range of equipment, used to harvest cereal crops at lower altitudes, perhaps with a view to their sporadic use for other types of plants at high altitude, or perhaps also to see to their repair, with the substitution or repositioning of damaged flint blades. However, given the current state of knowledge, it is not possible to altogether exclude the hypothesis that these tools were used for haymaking, given that at the moment there has not been specific experimentation with local forage plants, which could provide further elements assisting with interpretation.

Furthermore, in some cases the traces observed would seem to have characteristics attributable to contact with hide (marked rounding of the edges, extensive presence of streaks and the rough appearance of the micro-polish), so contact of the sickle blades with this material can therefore be surmised, probably as a secondary activity.

As regards the arrowheads, in at least three cases they had traces of use-wear as projectiles, although in terms of type and size, the traces observed did not indicate a particularly violent impact, perhaps as a result of the size of the prey hunted and/or the hunting strategies.

8. Discussion

Comparative analysis of the archaeological data, together with pedostratigraphic interpretation and the results of analysis of anthracological and carpological remains, makes it possible to suggest a reconstruction of the vegetation in the Middle Bronze Age and certain dynamics related to the organisation and utilisation of this high altitude site.

The pottery materials (Figs. 8.7, 8.8) and flint artefacts found to date (belonging exclusively to the categories of bifacial side scrapers and arrowheads) establish that utilisation of the site dates back to the beginning of the Middle Bronze Age. Radiocarbon dating carried out on a sample of charcoal at Leibniz Labor in Kiel provided the following measurement: KIA12453: 3387±31 BP corresponding to 1751–1616 BC cal (95.4% probability) (Mottes and Nicolis 2004),[8] which places it in the period between the end of the Early Bronze Age and the beginning of the Middle Bronze Age.

Due to the characteristics of the soil at Dosso Rotondo, archaeozoological remains have not survived and sifting of the samples for botanical analysis only made it possible to recover small fragments of unidentifiable bones, almost all charred, which did not allow data regarding animal husbandry to be obtained.

Functional analysis of the arrowheads found at the site, which in some cases have traces of use-wear as throwing weapons, confirms the practice of hunting or the defence of livestock from predatory animals.

The elements collected suggest that the dwellings were located in the midst of Norway spruce forest, probably still compact, whereas today the area is characterised by pastureland with an almost total absence of arboreal vegetation. The existence of well-structured Norway spruce forest is confirmed among other things by the presence of podzolic soil, which typically develops under this formation.

The limited and dubious documentation of the larch and stone pine among the charcoal suggests that the limit for arboreal vegetation in this period stood at around 2,200–2,400 m.a.s.l. (Tinner and Vescovi 2007),[9] in agreement with the palynological and pedological data, so above the altitude of the surrounding peaks (which are just over 2,000 metres high). Thus there are not likely to have been formations characterised by these two conifers in the immediate vicinity of the site, but only solitary individuals as a secondary component of Norway spruce forest.

The absence of the larch, a pioneer species that establishes itself spontaneously in open areas and on more degraded soils at high altitude, also depending on the openness of pastureland, could be a clue suggesting that deforestation practices had only just begun. It is likely that only the area concerned by the settlement was cleared, as suggested by the formation of colluvial soil covering a podzolic type of forestry palaeosoil.

The limited documentation of the green alder, juniper and dwarf pine can be linked to the morphology of the landscape, mainly characterised by rounded forms, due to the type of substrate (sandstone and siltstone), with a limited presence of rocky outcrops, watersheds, escarpments, etc., where these bushes prefer to grow.

Relatively substantial documentation of the silver fir, accompanied by the mountain maple, beech and laburnum, could suggest easy access to spruce woods with some beech trees, forest formations situated at an altitude below Norway spruce forests and partially cutting into these, although this type of forest currently finds more appropriate conditions (Pedrotti *et al.* 1998) on the slopes situated on the left-hand bank of the River Chiese.

The substantial documentation of wooden finds coming from the stratigraphic sequence at the pile-dwelling site at Fiavé Carera (Trento) shows that from the formative phase of the Middle Bronze Age I to the later phase of the Middle Bronze Age III (according to the chronology proposed by R. Perini) there was a considerable development in woodworking, and production of objects with a greater variety of types and forms (Perini 1988:60).

The presence of worked splinters of maple wood at Dosso Rotondo, together with the regularity with which the species is documented among the charcoal, can be explained as the result of anthropogenic selection destined for some artisan activity. The charcoal could be a result of the elimination of broken artefacts or the offcuts of working in hearths, assuming that the production of these artefacts was carried out on site, or of objects burnt during a fire. As already mentioned, maple wood is particularly suitable for working and would have been available around the site or at slightly lower altitude. The presence of charcoal of other broad-leaved trees (beech, sorbus, Cornelian cherry/guelder rose, elder, hazelnut and lime) can be related to artefacts (objects, wickerwork, ropes. etc.), brought from settlements situated at lower altitude, or constructed on site using wood from different altitudes. Although most of the charcoal can be linked to species commonly used for carpentry (above all spruce, silver fir and beech), elements clearly relating to the structures of the buildings were not found. The numerous branches could come from wood collected to feed fires or from fences or partitions with interwoven branches, subsequently burnt.

For environmental reasons (altitude, soil and forest coverage) it would seem possible to exclude the possibility that any form of agricultural cultivation was carried out at Dosso Rotondo; the cereals and fruit must have been transported from settlements on the valley floor or at all events picked at lower altitudes.

The traces of use-wear observed on parts of a sickle, which can be referred to the treatment of domestic cereal crops, probably related to activities carried out further down the mountain with the same type of tools.

The presence of hulled cereals and parts of ears suggests that the cereals were transported in the form of sheaves of ears, in order to ensure longer conservation.

The species documented at Storo Dosso Rotondo are barley, Emmer wheat and Einkorn wheat (and perhaps 'new glume wheat'), species constantly present in alpine settlements.[10] Spelt and millet were not among the cereals documented and there is no trace of leguminous plants, but it is possible that the incompleteness of the

---

8. The dating was calibrated using OxCal 4.2 software, on the basis of the atmospheric data of Reimer *et al.* 2013.

9. The tree line descended to this altitude due to climatic recrudescence known as the Löbben oscillation (Patzelt, and Bortenschlager 1973; Patzelt 1977), namely between 3,800 and 3,400 cal BP (Ivy-Ochs *et al.* 2009), which saw the advance of numerous glaciers in the Alps.

10. As an example, see Fiavé Carera (Trento) (Jones and Rowley-Conwy 1994; Karg 1998) and Ledro (Trento) (Bellintani *et al.* in press).

picture depends on the limited number of remains found.

The elements helping to determine the period of utilisation for the settlement are provided by a combination of observations:

- the hulled cereals were transported up the mountain after harvesting, so presumably at the time of the transfer to pastures at high altitude (June–July);
- the presence of acorns and some juniper berries shows that the site was utilised in the summer months;
- as it is believed to be unlikely that the hazelnuts, Cornelian cherry and sorbus berries were from the previous year's harvest, it is surmised that they were picked during the later period of utilisation of the site (late August–September), during excursions at lower altitudes or when going down to settlements in the valley to take alpine products there (cheese? milk? game? timber?) or to get new food supplies.

One can thus suggest that even in the Bronze Age, the permanence in the structures situated at around 1,850 m.a.s.l. corresponded to the classic '100 days' (Dematteis 1996), whereas there is no evidence showing whether the practice of using lower alpine pasture, half-way up the mountain, normally taking place from April to November, was adopted in this period. As is known, in Bronze Age settlements on the valley floor (for example at Fiavé Carera, Karg 1998) the permanence of animals in the village is documented during the winter months (between January and March), fed with the branches and shoots of early flowering arboreal species (beech twigs, hazelnut, elder and beech shoots, etc.). It is not however clear where the animals were taken in the spring months and whether they remained close to the village or whether the practice of using any available low mountain pasture had already begun.

With the current state of research, reliable interpretation of the real meaning and nature of the activities carried out at the Storo Dosso Rotondo site can only be surmised.

It is likely to have been a seasonal settlement linked to mountain pasture activities, in a similar way to the contemporary site at Malga Vacil (1,820.1 m.a.s.l.) (Marzatico 2007:169–173), which is only a few hundred metres away (Fig. 8.1).

For the moment, archaeobotanical data and functional analysis of flint tools have not provided elements making it possible to clarify the production activities carried out during the permanence in the Alps in terms of the rearing of livestock, nor whether a genuine haymaking system had been started up. In this context, a specific experiment is planned on local forage plants, which could provide further and more certain elements for interpreting the nature of the traces of use-wear.

The presence of pottery fragments riddled with holes, belonging to at least two sieves interpreted as 'cheese-strainers' (Bogucki 1984) (Fig. 8.18), most of which were brought to light in the area of the combustion structure (US 73) and linked to the fourth structural phase in the utilisation of the site, makes it possible to surmise in situ processing of milk and its derivatives. However, it should be noted that this type of pottery container, documented in European archaeological contexts from the Early Neolithic, can be used not only to process milk, namely to separate fat-rich milk curds from lactose-containing whey (Salque et al. 2013), but also to strain broth, in order to separate out the meat (Salque et al. 2012:58). Only specific analysis of possible organic residues present on the strainer fragments will be able to provide more precise information as regards this.[11]

The impression gained from comparison of the data available for the Storo Dosso Rotondo site is that the system of exploitation of areas at high altitude was still at an early stage, involving only a restricted number of people and livestock, and had not yet led to major modifications in the natural environment, which was characterised by relatively compact forest with limited pasture areas.

The presence of plant foods very similar to those found at sites on plains also confirms the idea of a lack of specialist alpine pasture, characterised in recent times by the almost exclusive consumption of products of animal origin.

As the lack of specialisation would not have allowed any surplus in terms of production, the presence of a settlement high in the mountains in this era would seem to have had the scope of extending the areas subject to some form of exploitation. This diversification of activities, still at an early stage, tended to separate cultivated areas from those destined for pasture.

To summarise, at the Storo Dosso Rotondo site there are a series of clues suggesting pioneering colonisation of the high mountains in the period between the end of the Early Bronze Age and the beginning of the Middle Bronze Age by a community with a cultural and technological background typical of areas on the valley floor or plains. This is documented by the extensive use of pottery containers and the same range of flint tools used in settlements in the valley (sickles), the transporting of ears of cereals to high altitude and the use of a construction technique typical of plain areas (with posts planted in the ground). The latter was gradually abandoned in favour of technology more suitable for environmental conditions high in the mountains (with insulating foundations in stone).

Processing of the data obtained from the investigations at Storo Dosso Rotondo has made it possible to reconstruct, albeit not yet in a clearly defined manner, the embryonic stage of a process that was to lead to a new model of exploitation and economic development, linked to a different approach to areas at high altitude by communities in the alpine valleys.

---

11. On this subject see: Evershed et al. 2008; Šoberl et al. 2008; Salque 2012; Salque et al. 2013; Curry 2013.

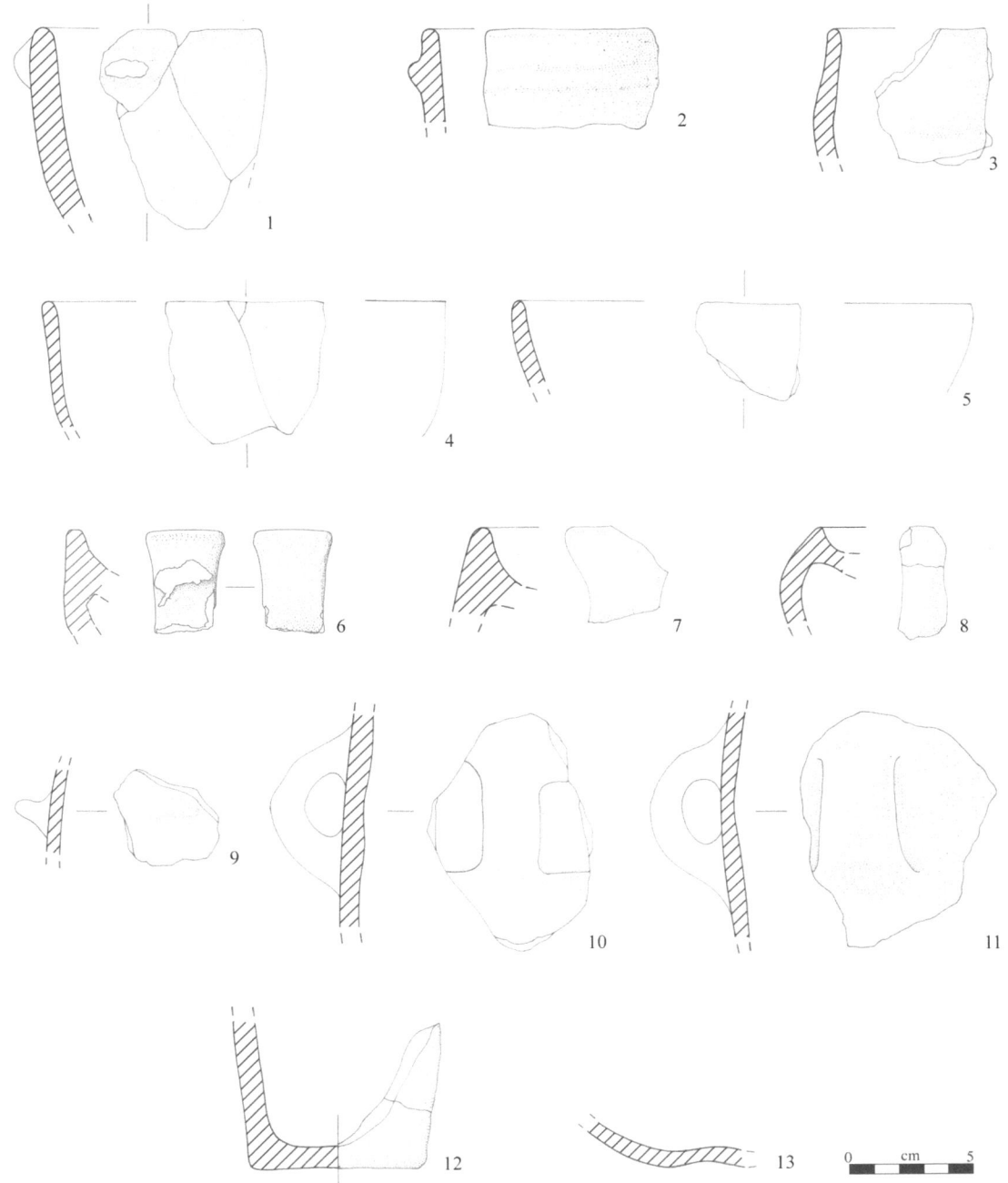

Figure 8.17 Storo Dosso Rotondo (Trento). Pottery (Drawings: Chiara Conci, Livia Stefan).

At Storo Dosso Rotondo one can note the effects of this 'economic conversion' on the mountain environment, but it is clear that this transformation must also have had economic, technological and social implications in the cultural contexts of origin.

The importance of the results obtained at Storo Dosso Rotondo is due to the fact that structural and analytical data comes from a site which has been excavated regularly, albeit not in an exhaustive manner, and the use of these data has enabled interpretation of economic and technological phenomenon on an objective basis. This makes it possible to avoid interpretations based simply on the application of purely theoretical models, often at sites characterised by a few materials taken out of context, trusting only to generic references to types of exploitation of mountain areas which can be referred to chronologically well-established periods and included in clear economic and social dynamics.

Figure 8.18 Storo Dosso Rotondo (Trento). Fragments of strainers (Drawings: Livia Stefan).

As regards the Storo Dosso Rotondo site, the reference made by us in the past to the model of exploitation of mountain environments going by the name of the 'mountain dairy economy' ('economia di malga') (Bassetti et al. 2003:929–930; Mottes and Nicolis 2004:84) was based on clear analogies and similarities, albeit with an awareness of its specific historical nature, and not merely due to the fact that they share the same context of reference in the mountains. As regards this, while seeking to avoid any nominalist considerations, it would not seem to be by chance that following careful analysis of all the data, the permanence and economic activities in the Alps at Storo Dosso Rotondo have been shown to correspond to the classic '100 day' period of mountain dairies.

Bibliography

AA.VV. 2002. *AttraVerso le Alpi; uomini, vie, scambi nell'antichità.* Archäologischen Landesmuseum Baden-Württemberg. ALManach 7/8. Stuttgart: Konrad Theiss Verlag.

Angelucci, D.E., Boschian, G., Fontanals, M., Pedrotti, A. and Vergès, J.M. 2009. Shepherds and karst: the use of caves and rock shelters in the Mediterranean region during the Neolithic. *World Archaeology* 41/2:191–214.

Angelucci, D., Carrer, F., Cavulli, F. and Pedrotti, A. 2014. Antichi pastori in Val di Sole (Trento, Italia); primo bilancio del progetto ALPES, 2010–2013. In M. Avanzini and I. Salvador (eds.) *Antichi pastori; sopravvivenze, tradizione orale, storia, tracce nel paesaggio e archeologia.* Atti della Tavola Rotonda, Boscochiesanuova (Vr) 26–27 ottobre 2013, Museo delle Scienze di Trento. Pp. 53–66.

Aspesi, G.M. and Cataldi, G. 2013. La casa alpina in tronchi/blockbau, varianti locali ed evoluzione tipologica. *Quaderni di cultura alpina.* Scarmagno (Torino): Priuli and Verlucca editori. Pp. 102–103.

Bagolini, B. and Broglio, A. 1985. Il ruolo delle Alpi nei tempi preistorici (dal Paleolitico al Calcolitico). In M. Liverani, A. Palmieri and R. Peroni (eds.). *Studi di Paletnologia in Onore di Salvatore M. Puglisi,* Università di Roma La Sapienza. Pp. 663–705.

Bassetti, M., Dalmeri, G., Mottes, E. and Nicolis, F. 2003. Nuovi dati sulle modalità di sfruttamento dei territori di alta quota nell'età del Bronzo: il sito di Storo – Dosso Rotondo in Valle del Chiese (Trentino sud-occidentale). Atti della XXXV Riunione Scientifica dell'IIPP. Le comunità della Preistoria italiana studi e ricerche sul Neolitico e le età dei metalli, Castello di Lipari, 2–7 giugno 2000. Pp. 927–931.

Bassetti, M., Dalmeri, G., Mottes, E. and Nicolis, F. 2008. La frequentazione delle alte quote nell'età del Bronzo; il sito di Storo – Dosso Rotondo. In E. Mottes, F. Nicolis and G. Zontini (eds.) *Archeologia lungo il Chiese; nuove indagini e prospettive della ricerca preistorica e protostorica in un territorio condiviso fra Trentino e Lombardia.* Atti del 1° convegno interregionale, Storo 24–25 ottobre 2003. Trento: Tipografia Effe e Erre. Pp. 107–127.

Bassi, C. and Cavada, E. 1994. Aspetti dell'edilizia residenziale alpina tra l'età classica e il medioevo: il caso trentino. In G.P. Brogiolo (eds.) *Edilizia residenziale tra V e VIII secolo; 4° Seminario sul Tardoantico e l'Altomedioevo in Italia centrosettentrionale.* Documenti di archeologia 4:115–134.

Bellintani, P., Bassetti, M., Bettinardi, I., Degasperi, N., Magny, M., Martinelli, N., Pignatelli, O. and Rottoli, M. in press. *Nuove Ricerche per la Tutela del Sito palafitticolo di Molina di Ledro (Trento).* Oxford: British Archaeological Report.

Bogucki, P.I. 1984. Ceramic sieves of the Linear Pottery Culture and their economic implications. *Oxford Journal of Archaeology* 3/1:15–30.

Borrello, M.A. 2013 (ed.). *Les hommes préhistoriques et les Alpes.* Oxford: BAR International Series 2476.

Boschian, G. and Miracle, P.T. 2007. Shepherds and caves in the Karst of Istria (Croatia). *Atti della Società Toscana di Scienze Naturali*, Memorie, Serie A 112:173–180.

Boschian, G. and Montagnari-Kokelj, E. 2000. Prehistoric shepherds and caves in the Trieste Karst (Northeastern Italy). *Geoarchaeology* 15/4:331–71.

Bosellini, A., Mutti, E. and Ricci Lucchi, F. 1989. *Rocce e Successioni sedimentarie.* Torino: UTET.

Brochier, J.E. 1983. Bergeries et feux de bois néolithiques dans le Midi de la France; caracterisation et incidence sur le raisonnement sedimentologique. *Quartar* 33–4:119–35.

Brochier, J.E. 2002. Les sediments anthropiques; mèthodes d'ètude et perspective. In J.-C. Miskovsky (ed.) *Géologie de la Préhistoire; méthodes, techniques, applications.* Association pour l'étude de l'environnement géologique de la Préhistoire, Paris, Géopré, Presses universitaires de Perpignan. Pp. 453–477.

Brochier, J.E., Villa, P. and Giacomarra, M. 1992. Shepherds and sediments; geo-ethnoarchaeology of pastoral sites. *Journal of Anthropological Archaeology* 11:47–102.

Bullock, P., Fedoroff, N., Jongerius, A., Stoops, G.J. and Tursina, T. 1985. *Handbook for Soil Thin Section Description.* Albrighton: Waine Research Publications.

Canti, M.G. and Linford, N.T. 2001. Geophysical evidence for fires in antiquity; preliminary results from an experimental study. Paper given at the EGS XXIV General Assembly in The Hague, April 1999. *Archaeological Prospection* 8/4:211–225.

Carrer, F., Angelucci, D.E., Pedrotti, A. 2013. Montagna e pastorizia; stato dell'arte e prospettive di ricerca. In D.E. Angelucci, L. Casagrande, A. Colecchia and M. Rottoli (eds.) APSAT 2, *Paesaggi d'altura del Trentino; evoluzione naturale e aspetti culturali.* Progetti di archeologia, Mantova, SAP Società Archeologica. Pp. 125–139.

Cason, E. (ed.) 2002. *Uso dei Valichi alpini orientali dalla Preistoria ai Pellegrinaggi medievali.* Fondazione Giovanni Angelini, Centro Studi sulla Montagna, Udine; Forum.

Cataldi, G. 1986. *All'Origine dell'abitare.* Studi e documenti di architettura 13. Firenze: Alinea.

Courty, M.A. 2001. Microfacies analysis assisting archaeological stratigraphy. In P. Goldberg, V.T. Holliday and C.R. Ferring (eds.) *Earth Sciences and Archaeology.* New York: Kluwer. Pp. 205–239.

Courty, M.A., Goldberg, P. and Macphail, R.I. 1989. *Soils and Micromorphology in Archaeology.* Cambridge Manuals in Archaeology: Cambridge University Press.

Courty, M.A., Macphail, R. and Wattez, J. 1992. Soil micromorphological indicators of pastoralism, with special reference to Arene Candide, Finale Ligure, Italy. In R. Maggi, R. Nisbet and G. Barker (eds.) *Archeologia della Pastorizia nell'Europa Meridionale.* Rivista di Studi Liguri 2 (57):127–150.

Cremaschi, M. and Rodolfi, G. (eds.) 1991. *Il suolo. Pedologia nelle scienze della terra e nella valutazione del territorio.* NIS: Roma.

Curry, A. 2013. The milk revolution; when a single genetic mutation first let ancient Europeans drink milk, it set the stage for a continental upheaval. *Nature* 500:20–22.

D'Alessio, D. and Previtali, F. 1988. I podzoli della Valle Camonica (Alpi meridionali bresciane). *Annali del Museo Civico di Scienze Naturali*, Brescia 24 (1987):47–73.

Dal Piaz, G.V. 2008. *Note illustrative della Carta Geologica d'Italia alla scala 1:50.000. Foglio 058 Monte Adamello.*

Dalla Fior, G. 1981. *La nostra Flora.* Monauni: Trento.

Dematteis, L. 1996. *Il Fuoco di Casa nelle Tradizioni dell'Abitare alpino.* Quaderni di Cultura Alpina. Torino: Priuli e Verlucca Editori.

Duchaufour, P. 1976. *Atlas écologique des Sols du Monde.* Paris, Massons.

Duchaufour, P. 1998. *Handbook of Pedology; soils, vegetation and environment.* Lisse, The Netherlands: A.A. Balkema.

Evershed, R.P., Payne, S., Sherratt, A.G., Copley, M.S., Coolidge, J., Urem-Kotsu, D., Kotsakis, K., Özdoğan, M., Özdoğan, A.E., Nieuwenhuyse, O., Akkermans, P.M., Bailey, D., Andeescu, R.R., Campbell, S., Farid,

S., Hodder, I., Yalman N., Özbaþaran, M., Býçakci, E., Garfinkel, Y., Levy, T. and Burton, M.M. 2008. Earliest date for milk use in the Near East and southeastern Europe linked to cattle herding. *Nature* 455:528–531.

Gè, T., Courty, M.A., Matthews, W. and Wattez, J. 1993. Sedimentary formation process of occupation surfaces. In P. Goldberg, D.T. Nash and M.D. Petraglia (eds.). *Formation Processes in Archaeological Contexts*. Monographs in World Archaeology 17:149–163.

Goldberg, P. and Berna, F. 2010. Micromorphological and context. *Quaternary International* 214:56–62.

Goldberg, P. and Macphail, R.I. 2006. *Practical and Theoretical Geoarchaeology*. Oxford: Blackwell.

Goldberg, P., Miller, E. C., Schiegl, S., Ligouis, B., Berna, F., Conard, N.J. and Wadley, L. 2009. Bedding, hearths, and site maintenance in the Middle Stone Age of Sibudu Cave, KwaZulu-Natal, South Africa. *Archaeological Anthropological Sciences* 1:95–122.

Haas, J.N., Karg, S. and Rasmussen, P. 1998. Beech leaves and twigs used as winter fodder; examples from historic and prehistoric times. *Environmental Archaeology* 1:81–86.

Hafner, A. 2009. Geschichte aus dem Eis; Archäologische Funde aus alpinen Gletschern und Eismulden. *Mitteilungen der Naturforschenden Gesellschaft in Bern* 66:159–171.

Hafner, A. 2012. Archaeological discoveries on Schnidejoch and at other ice sites in the European Alps. *Arctic* 65, suppl. 1:189–202.

Ivy-Ochs, S., Kerschner, H., Maisch, M., Christl, M., Kubik, P.W. and Schluchter, C. 2009. Latest Pleistocene and Holocene glacier variations in the European Alps. *Quaternary Science Reviews* 28:2137–2149.

Jones, G. and Rowley-Conwy, P. 1994. Plant remains from the north lake dwellings of Fiavé (1400–1200 BC). In R. Perini (ed.) *Scavi archeologici nella Zona palafitticola di Fiavé-Carera. Parte I. Campagne 1969–76, situazioni dei depositi e dei resti strutturali*. Trento: Patrimonio Storico Artistico del Trentino 8. Pp. 323–355.

Jones, G., Valamoti, S. and Charles, M. 2000. Early crop diversity: a 'new' glume wheat from northern Greece. *Vegetation History and Archaeobotany* 9:133–146.

Juel Jensen, H. 1994. *Flint Tools and Plant Working – hidden traces of stone age technology; a use wear study of some Danish Mesolithic and TRB implements*. Aarhus University Press.

Karg, S. 1998. Winter- and spring-foddering of sheep/goat in the Bronze Age site of Fiavé-Carera, Northern Italy. *Environmental Archaeology* 1:87–94.

Keeley, L.H. 1980. *Experimental Determination of Stone Tools Uses; a microwear analysis*. Chicago and London: The University of Chicago Press.

Krause, R. 2007. The prehistoric settlement of the inner alpine valley of Montafon in Vorarlberg (Austria). *Preistoria Alpina* 42:119–136.

Macphail, R., Courty, M.A., Hather, J., Wattez, J., Ryder, M., Cameron, N., and Branch, N.P. 1997. The soil micromorphological evidence of domestic occupation and stabling activities. In R. Maggi R. (ed.) *Arene Candide: a functional and environmental assessment of the Holocene sequence (excavations Bernabò Brea-Cardini 1940–50)*. Memorie dell'Istituto Italiano di Paleontologia Umana, n.s. 5, Roma, Il Calamo. Pp. 53–88.

Macphail, R.I. and Cruise, G.M. 2001. The soil micromorphologist as team player: a multianalytical approach to the study of European microstratigraphy. In P. Goldberg, V. Holliday and R. Ferring (eds.) *Earth Science and Archaeology*. New York: Kluwer Academic/Plenum Publishers. Pp. 241–267.

Mallol, C., Hernández, C.M., Cabanes, C., Machado, J., Sistiaga, A., Pérez, L. and Galván, B. 2013. Human actions performed on simple combustion structures: an experimental approach to the study of Middle Palaeolithic fire. *Quaternary International* 315:3–15.

Marchetti, M. 2000. *Geomorfologia Fluviale*. Bologna, Pitagora.

Marzatico, F. 2007. La frequentazione dell'ambiente montano nel territorio atesino fra l'età del Bronzo e del Ferro: alcune considerazioni sulla pastorizia transumante e 'l'economia di malga'. *Preistoria Alpina* 42:163–182.

Miracle, P.T. and Forenbaher, S. 2005. Neolithic and Bronze-Age herders of Pupiæina Cave, Croatia. *Journal of Field Archaeology* 30/3:255–281.

Mottes, E. and Degasperi, N. 2014. Analisi interpretativa delle strutture di abitato neolitiche del Trentino. In P. Bellintani, M. Cunaccia and M. Riggio (eds.) *Conoscere e ricostruire edifici in legno: dalle palafitte preistoriche all'età contemporanea*. Abstract delle giornate di Studio, Larido (Bleggio Superiore, Trento) 17–18 ottobre 2014, Centro Duplicazioni della Provincia autonoma di Trento.

Mottes, E. and Nicolis, F. 2004. Storo – Dosso Rotondo (Trento): un sito di alta quota dell'età del Bronzo in Valle del Chiese. *Annali del Museo* 19 (anni 2001–2002). Gavardo: Civico Museo Archeologico della Valle Sabbia. Pp. 81–88.

Mücher, H., Van Steijn, H. and Kwaad, F. 2010. Colluvial and mass wasting deposits. In G. Stoops, V. Marcelino and F. Mees (eds.) *Interpretation of Micromorphological Features of Soil and Regoliths*. Amsterdam, Elsevier. Pp. 37–46.

Nicolis, F. 2006. Archeologia di montagna tra tutela, ricerca e valorizzazione. L'esperienza della Provincia autonoma di Trento. In *Alpis Graia; archéologie sans frontières au col du Petit-Saint-Bernard*. Projet Interreg IIIA, Quart (AO), Musumeci S.p.A. Pp. 373–380.

Oeggl, K., Schmidl, A. and Kofler, W. 2009. Origin and seasonality of subfossil caprine dung from the discovery site of the Iceman (Eastern Alps). *Vegetation History and Archaeobotany* 18:37–46.

Patzelt, G., 1977. Der zeitliche Ablauf und das Ausmass post-glazialer Klimaschwankungen in den Alpen. In B. Frenzel (ed.) *Dendrochronologie und postglaziale Klimaschwankungen in Europa*. Wiesbaden, Steiner Verlag. Pp. 248–259.

Patzelt, G. and Bortenschlager, S. 1973. Die postglazialen Gletscher- und Klimaschwankungen in der Venedigergruppe (Hohe Tauern, Ostalpen). *Zeitschrift für Geomorphologie N.F. Supplementband* 16:25–72.

Pedrotti, F., Gafta, D. and Minghetti, P. 1998. *Carta della vegetazione potenziale del Trentino*. Firenze; S.EL.CA.

Penk, A. and Brückner, E. 1901–1909. *Die Alpen in Eiszeitalter*. V. III; Leipzig.

Perini, R. 1987. *Scavi archeologici nella zona palafitticola di Fiavè-Carera. Parte II. Campagne 1969–1976. Resti della cultura materiale metallo – osso – litica – legno*. Trento: Patrimonio storico e artistico del Trentino 9.

Perini, R. 1988. La suppellettile lignea fra i resti degli antichi abitati di Fiavè e Lavagnone. In R. Perini (ed.) *Archeologia del Legno; documenti dell'età del Bronzo dall'area sudalpina*. Quaderni della Sezione Archeologica Museo Provinciale d'Arte 4:65–94.

Perini, R. 1990. Manufatti in legno dell'età del Bronzo nel territorio delle Alpi meridionali. In AA.VV. *Die ersten Bauern*, Band 2, Schweizerisches Landesmuseum Zürich. Pp. 253–265.

Putzer, A. 2011. Eine prähistorische Almhütte auf dem Schwarzboden im Maneidtal, Südtirol/Vinschgau. *Archaeologia Austriaca* 93 (2009):33–43.

Ravazzi, C., Peresani, M., Pini, R. and Vescovi, E. 2007. Il Tardoglaciale nelle Alpi e in Pianura Padana: evoluzione stratigrafica, storia delle vegetazione e del popolamento antropico. *Il Quaternario, Italian Journal of Quaternary Sciences* 20/2:163–184.

Reimer, P.J., Bard, E., Bayliss, A., Beck, J.W., Blackwell, P.G., Bronk Ramsey, C., Buck, C.E., Cheng, H., Edwards, R.L., Friedrich, M., Grootes, P.M., Guilderson, T.P., Haflidason, H., Hajdas, I., Hatté, C., Heaton, T.J., Hoffmann, D.L., Hogg, A.G., Hughen, K.A., Kaiser, K.F., Kromer, B., Manning, S.W., Niu, M., Reimer, R.W., Richards, D.A., Scott, E.M., Southon, J.R., Staff, R.A., Turney, C.S.M. and van der Plicht, J. 2013. IntCal13 and Marine13 radiocarbon age calibration curves 0–50,000 years cal BP. *Radiocarbon* 55/4:1869–1887.

Reitmaier, T. 2010. Auf der Hut; Methodische Überlegungen zur prähistorischen Alpwirtschaft in der Schweiz. In F. Mandl and Stadler H. (eds.) *Archäologie in den Alpen: Alltag und Kult*. ANISA, Haus im Ennstal. Pp. 219–238.

Reitmaier, T. (ed.) 2012. *Letzte Jäger, erste Hirten: hochalpine Archäologie in der Silvretta*. Archäologie in Graubünden, Sonderheft 1, Chur.

Reitmaier, T., Lambers, K., Walser, C., Zingman, I., Haas, J.N., Dietre B., Reidl, D., Hajdas, I., Nicolussi, K., Kathrein, Y., Naef, L. and Kaiser, T. 2013. Alpine Archäologie in der Silvretta. *Archäologie Schweiz* 36/1:4–15.

Ricci Lucchi, F. 1980. *Sedimentologia*. Bologna: Clueb.

Rottoli, M. 1997. I resti botanici. In P. Frontini (ed.) *Castellaro del Vhò; campagna di scavo 1995*. Comune di Milano, Settore Cultura e Spettacolo, Raccolte Archeologiche e Numismatiche. Pp. 141–158.

Salque, M. 2012. Was milk processed in these ceramic pots? Organic residue analyses of European prehistoric cooking vessels. In F. Feulner, P. Gerbault, R. Gillis, H. Hollund, R. Howcroft, M. Leonardi, A. Liebert, M. Raghavan, M. Salque, O. Sverrisdottir, M. Teasdale, N. van Doorn and C. Wright (eds.) *May Contain Traces of Milk; investigating the role of dairy farming and milk consumption in the European Neolithic*. LeCHE, York: University of York. Pp. 127–141.

Salque, M., Bogucki, P.I., Pyzel, J., Sobkowiak-Tabaka, I., Grygiel, R., Szmyt, M., and Evershed, R.P. 2013. Earliest evidence for cheese making in the sixth millennium BC in northern Europe. *Nature* 493:522–525.

Salque, M., Radi, G., Fabbri, C., Tagliacozzo, A., Pino Uria, B., Wolfram, S., Stäuble, H., Hohle, I., Whittle, A., Hofmann, D., Pechtl, J., Schade-Lindig, S., Eisenhauer, U. and Evershed, R.P. 2012. New insights into the early Neolithic economy and management of animals in southern and central Europe revealed using lipid residue analyses of pottery vessels. *Anthropozoologica* 47/2:45–61.

Sanesi, G. (ed.) 1977. *Guida alla Descrizione del Suolo*. C.N.R. Progetto Finalizzato Conservazione del Suolo 11, Firenze.

Schmidl, A., Jacomet, S. and Oeggl, K. 2007. Distribution patterns of cultivated plants in the Eastern Alps (Central Europe) during Iron Age. *Journal of Archaeological Science* 34/2:243–254.

Šoberl, L., Žibrat Gašparič, A., Budja, M. and Evershed, R.P. 2008. Early herding practices revealed through organic residue analysis of pottery from the early Neolithic rock shelter of Mala Triglavca, Slovenia. *Documenta Praehistorica* 35:253–260.

Sordi, I. 1979. Il mondo degli oggetti. In G. Bertolotti, I. Melli, E. Minervini, G. Sanga, P. Sassu, and I. Sordi (eds.) *Mondo popolare in Lombardia*. 10, Premana: Ricerca su una comunità artigiana. Milano: Silvana editoriale. Pp. 547–599.

Stoops, G., 2003. *Guidelines for Analysis and Description of Soil and Regolith Thin Sections*. Soil Science Society of America. Wisconsin, USA, Inc. Madison.

Stoops, G., Marcelino, V. and Mees, F. 2010. *Interpretation of Micromorphological Features of Soil and Regoliths*. Amsterdam: Elsevier.

Stützer, A. 1999. Podzolisation as a soil forming process in the alpine of Rondane, Norway. *Geoderma* 91:237–248.

Tinner, W. and Vescovi, E. 2007. Ecologia e oscillazioni del limite degli alberi nelle Alpi dal Pleniglaciale al presente. *Studi trentini di Scienze naturali. Acta Geologica* 82 (2005):7–15.

Tzortzis, S. and Delestre, X. 2010. *Archéologie de la montagne européenne*. Actes de la table ronde internationale de Gap, 29 septembre–1er octobre 2008. Bibliothèque d'archéologie méditerranéenne et africaine 4. Paris: Errance, Centre Camille Jullian.

Unger-Hamilton, R. 1985. Microscopic striations on flint sickle-blades as an indication of plant cultivation: preliminary results. *World Archaeology* 17:121–126.

Van Gijn, A.L. 1989. *The Wear and Tear of Flint; principles of functional analysis applied to Dutch Neolithic assemblages*. Analecta Præhistorica Leidensia 22.

Van Vliet-Lanoë, B. 2010. Frost action. In G. Stoops, V. Marcelino and F. Mees (eds.) *Interpretation of Micromorphological Features of Soils and Regoliths*. Amsterdam, Elsevier. Pp. 81–108.

Vaughan, P. 1985. *Use-wear Analysis of Flaked Stone Tools*. Tucson: University of Arizona Press.

Walsh, K., Mocci, F. and Palet-Martinez, J. 2007. Nine thousand years of human/landscape dynamics in a high altitude zone in the southern French Alps (Parc National des Ecrins, Hautes-Alpes). *Preistoria Alpina* 42:9–22.

Waters, M.R. 1992. *Principles of Geoarchaeology; a North American perspective*. Tucson, University of Arizona Press.

Wattez, J. and Courty, M.A. 1987. Morphology of ash of some plant materials. *Soil Micromorphology* 25:677–683.

Wattez, J., Courty, M.A. and Macphail, R.I. 1990. Burnt organo-mineral deposits related to animal and human activities in prehistoric caves. *Soil Micromorphology* 45:431–440.

Franco Nicolis, Provincia autonoma di Trento, Soprintendenza per i beni culturali, Ufficio beni archeologici, Via Manatova 67, I-38122 Trento.
Email: franco.nicolis@provincia.tn.it

Elisabetta Mottes, Provincia autonoma di Trento, Soprintendenza per i beni culturali, Ufficio beni archeologici, Via Mantova 67, I–38122 Trento.
Email: elisabetta.mottes@provincia.tn.it

Michele Bassetti, Cora Società Archeologica S.r.l., Via Salisburgo 16, I–38121 Trento.
Email: michele@coraricerche.com

Elisabetta Castiglioni, Laboratorio di Archeobiologia dei Musei Civici di Como, Piazza Medaglie d'Oro, 1, I–22100 Como.
Email: archeobotanica@alice.it

Mauro Rottoli, Laboratorio di Archeobiologia dei Musei Civici di Como, Piazza Medaglie d'Oro, 1, I–22100 Como.
Email: archeobotanica@alice.it

Sara Ziggiotti, Via Matteotti 62/a I–35010 Villafranca Padovana (Padova).
Email: sara.ziggiotti@gmail.com

9. Pastoral land use and climate between the 17th and 19th century in the Italian Southern Alps (Pasubio Massif, Trento): A preliminary report

Marco Avanzini and Isabella Salvador

Upland population levels are strongly correlated to environmental dynamics such as morphology, exposure and climate. A temperature fall leads to a shortening of the plant growth season, which can lead to lower pasture productivity hence shorter periods spent by the livestock in the mountains. The aim of this research is to correlate natural climate constraints with variations in post-medieval human settlements of the Pasubio Plateau, located between 1500 and 1800m, in the Italian southern Alps (Trento). A survey of a 630ha area has identified the remnants of 145 structures used for milk processing, which date from the 17th to the 19th century. A 'building density curve', grouped into fifty year intervals, shows oscillations over time. The high population density that characterises the first half of the 17th century is followed by an almost total abandonment of pastures in the second half of the century. Dairyman-shepherds returned to the high pastures only after the mid-18th century. The correlation of this trend with climatic oscillations in the same area derived from speleothems, reveals that the decrease in highland exploitation between 1650 and 1750 was linked to the Maunder temperature fall. The subsequent temperature rise corresponds to the resettlement of the highland pastures.

Introduction

Buildings linked to dairy production in the Trentino Pre-Alps (Italian southern Alps) were already documented during the 13th century (Giacomoni 1998; 2001). These structures are commonly described as wooden mobile units, similar to those in the neighbouring area of Lessinia (Verona Province). The raised wooden structures were easy to dismantle, making them highly mobile and suitable for the maintenance and fertilisation of the pastures (Varanini 1991).

During the 16th and 17th centuries, wooden structures were progressively replaced by masonry ones and the *malghe* (highland pasture and dairy buildings) specialised in the production of butter and cheese (the *baiti*) and in the storage of the products (the *casere*).

Between the 18th and 19th centuries, while the production buildings did not change much, the storage ones became bigger, permanent and connected to several production buildings.

At the beginning of the 20th century, in the post-First World War period, the traditional function and spatial organisation of the pasture buildings was redefined and their architecture was standardised: the *baiti*'s dimensions increased and storage was included within the structure. The animal shelters changed from open air fenced areas to proper stables.

Not many traces remain of the early infrastructure of the highlands. The only surviving buildings belong to the 20th century but recent studies in sectors of the venetian-tridentine Pre-Alps, found traces of older buildings. These traces are recognisable over wide areas of pasture and they cover a period from the 15th to the 20th century (Carrer 2012; Migliavacca 2012; Migliavacca *et al.* 2013).

Studies of the settlement mode and production chain of these pre-industrial activities are currently in progress (Carrer 2012; Migliavacca et al. 2013). However, studies on the relationship between environmental factors and the settlement modes as well as the exploitation of pastures, are still very scarce.

Climate and Population

It cannot be overlooked that the population of the highlands is always related to the morphological and climatic conditions of the territory, especially when the permanence at high altitudes is strongly connected with activities of production, which require a stable system of buildings and facilities. In the territories placed at the upper limit of vegetation, fluctuations in rainfall and temperature have direct effects on the vegetative cycle of plants. It is therefore clear that in economies based on the exploitation of pastures, climatic oscillations may have played a role in the mode of use of the place.

The interaction between climatic factors, agro-forestry systems and ecosystem productivity is complex (Baglioni *et al.* 2009; Bosello and Zhang 2005; Roson 2003; Hegerl *et al.* 2009; Solomon *et al.* 2007). Due to the extreme variability of the environment and different management techniques, the alpine pastures are one of the most complicated case studies. The productivity and quality of forage is strongly linked to en-

Keywords: Pasubio pastures, climatic oscillations, dairy production, *malghe*

vironmental factors such as temperature, fertility and soil moisture (Menzel and Fabian 1999). The pastures are also characterised by a rapid growth in production during spring and summer, followed by a period of gradual decline and drop in quality due to changes in the structure of the vegetation and the deterioration of the nutritional characteristics of individual species (Ziliotto and Scotton 1993; Orlandi *et al.* 2004; Gusmeroli *et al.* 2005; Orlandi and Clementel 2007).

The pastures' productivity and the economy of the highlands are strongly linked to natural constraints, which in those places exercise, more than elsewhere, a direct effect on the vegetative cycle and productivity of grass. The productivity of these territories is variable over time and consequently the resulting economic component fluctuates. The link between climate change and resource availability is often overlooked; an increase or decrease in temperature of a few tenths of a degree can result in an increase or decrease of resources and thus contribute to the formation or loss of wealth. A climatic deterioration may in fact lead to a series of extreme weather events that contribute to making it difficult to live at high altitude, but above all, the fall in temperature and a shortening of the growing season lead to a shortening of the time of attendance and use of the mountain pastures, with the consequent decrease in the value of the pasture and radical changes in the use of mountain areas even over short time frames.

The aim of this work is to define the extent to which natural climate constraints are related to the phases and modes of exploitation of the highlands in a Pre-Alps sector of the eastern and central Alps.

Pasture as an indicator of climatic variation

The Pasubio Massif appears as a wide plateau separated by two deep valleys (Leno and Terragnolo Valleys); it extends southeast of Rovereto at altitudes mainly between 1500 and 2000m. The morphological structure is that of a large homoclinal structure crossed by two systems of tectonic lineaments oriented north-northeast–southwest and northwest–southeast, which allow the development of parallel valleys and ridges.

The summit sector is characterised by a sub-flat morphology from which originates a series of little valleys which cut the slopes. The structural and carbonate nature of the mountain led to the development of surface karst landforms. They drain rainwater and melting snow at deep levels, making the summit territory quite arid.

In areas like this, the environmental component is of paramount importance in the control of production and welfare (Pfister and Brazdil 2006; Dearing 2006; Fraser 2009; Paavola and Fraser 2011).

The link between the availability of energy (*sensu lato*) of former agricultural societies and the change in climate is often overlooked (Bozhong 1998; Anfodillo 2007).

The Medieval Warm Period (MWP) (Medieval Climate Optimum or Medieval Climatic Anomaly) was probably more important to the economic welfare than technical discoveries and changes in institutions (Malanima 2006). It is in fact associated with an increase in population in the whole of Europe. With the beginning of what is known as the Little Ice Age (LIA), the average temperature decreases significantly. In the series of temperatures obtained for Europe (Luterbacher *et al.* 2004), there is a clear downward trend, culminating in the second half of the 17th century in what is known as the 'Maunder minimum' (1675–1715). This is a period of stagnation, if not fall, in the population of the whole of Europe. The first phase of demographic recovery took place as a result of an increase in temperature, as observed in England, France and Italy from the end of the 17th century to about 1750. Between the end of 18th century and 1820, the temperature returned for a few decades to that found in the 17th century. During this period, the glaciers advanced again until they reached their maximum expansion in 1816, with significant economic consequences (Michaelowa 2001).

In the central area of Pasubio, at an altitude of 1025m above sea level, a large cavity opens, called the Cogola of Giazzera. The cavity has been the object of palaeoclimatic studies for nearly a decade (Frisia *et al.* 2007). The climatic curves reconstructed for the last 4500 years are comparable with those obtained from archives, dendrochronology studies and glacier data of other alpine and European areas (Frisia 2007).

The evolution of the local climate over the last millennium marks the end of the Medieval Warm Period (MWP) with a drop in temperatures of -0.5° less than the reference average from the middle of the 15th century. The decrease continued down to -0.6° in the second half of the 17th century. Although less pronounced, this negative peak corresponds with the 'Maunder minimum' of solar activity – the coldest period of the Little Ice Age (LIA) (Xoplaki *et al.* 2011). A gradual and steady rise in temperature occurred in the Pasubio massif from 1720 to 1880, with a long period of temperatures close to those of today, which reached their highest values in the forty years around the transition from 1700 to 1800 and around 1860.

Sandwiched between these two periods is a short cold phase between 1820 and 1840 (the maximum of LIA in the Alps is placed around 1850). This negative oscillation is reflected in Trentino, in the thermometric series recorded in Trento (Rea *et al.* 2003). The data highlighted a collapse of temperatures around 1840. The negative oscillation is also evident in the dendroclimatic series of *Larix decidua* and *Pinus cembra* (Leonelli *et al.* 2012) ('Dalton minimum' of solar activity. cf. Büntgen *et al.* 2006). Between 1880 and 1890 the temperature decreases with an evident negative peak, although less intense than the ones found in the Alps (the Trento thermometric series define a decrease of about 1.5°, Rea *et al.* 2003). From 1890 the tempera-

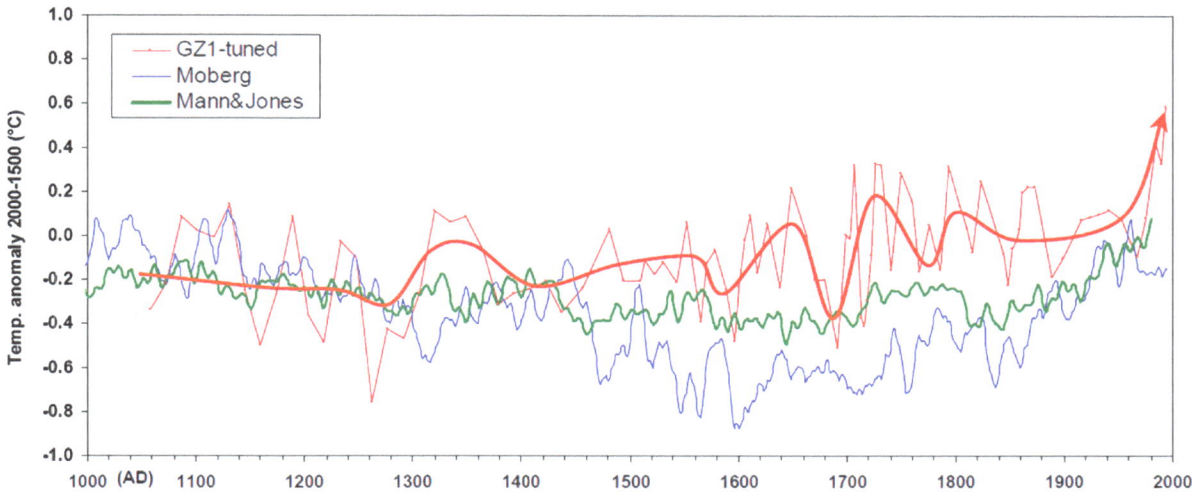

Figure 9.1 Comparison between the temperature anomaly calculated from the $\delta^{18}O_c$ of Giazzera Cave (GZ1) for the last 1000 years (tuned record) with other temperature reconstructions in the Northern Hemisphere: Mann and Jones 2003; Moberg et al. 2003. In all the reconstructions the 7-year average values are shown (from: Frisia et al. 2007 mod.).

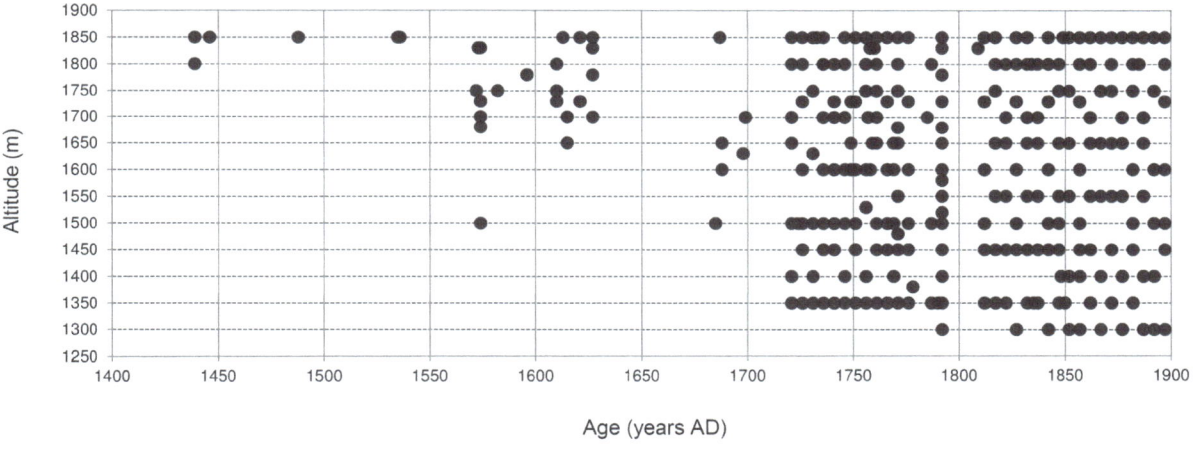

Figure 9.2 Chronology of pasture exploitation related to altitude in the studied area. Until the 16th century, only natural pastures at high altitudes were used as grazing land. In the second half of the 16th century, intensive felling of forest and land clearance for agriculture allowed the use of pastures at lower altitudes. From the 18th century a continuous exploitation of the highlands is attested.

ture increases (with a brief cold period at the end of the first decade of the 20th century) until 1950. After a colder period between 1960 and 1970, the temperature is now on a clear and steady rise (Fig. 9.1).

The Pasubio *malghe* from historical documents

The Pasubio territory, located in southern Trentino, is a good example of the complex layering of traces of people dedicated to the exploitation of pasturelands. Since the 16th century, the independence achieved by local communities allowed them to self-regulate the use of pastures in a 'rational' and preservative way and to optimise dairy production.

Documents in historical archives of local municipalities illustrate well the rules and timing of grazing, animals bred in the mountains, structures related to dairy production and to conservation.

While the use of natural pastures at high altitudes started in the 13th century, the intensive forest harvesting and land clearance for agriculture at middle altitudes driven by the need to obtain new land for farming, dates back to the 16th century (Fig. 9.2). At the same time, precise regulations are drawn up for the use of mountain resources and high altitude buildings are described in detail. The typological evolution of the latter, which went from mobile wooden structures to solid masonry buildings, can be precisely correlated to the rules contained in leasing con-

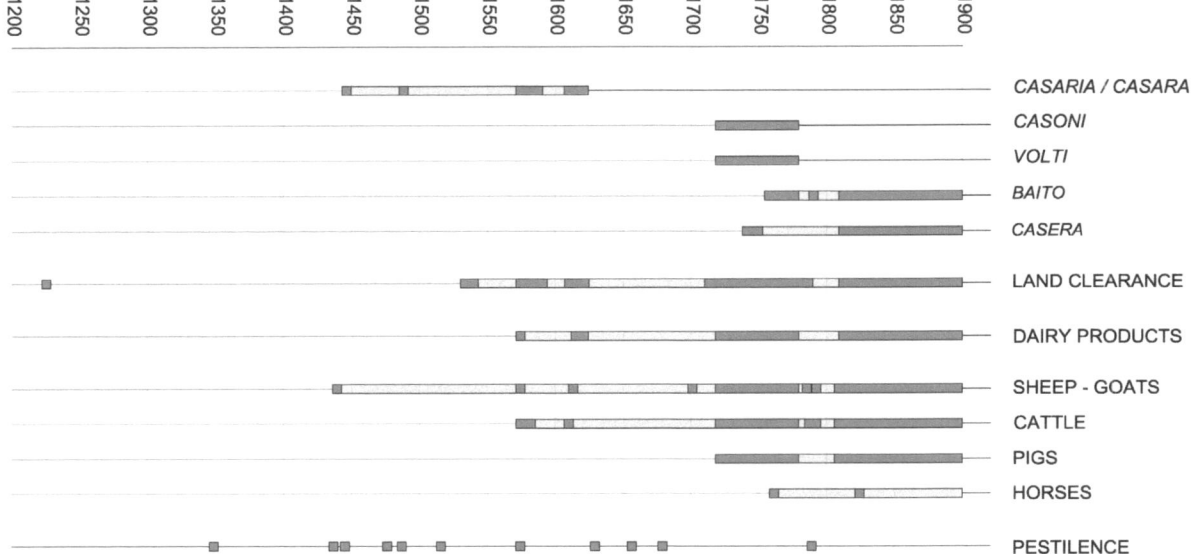

Figure 9.3　The evolution of highland exploitation related to dairy production from historical documents.

tracts, practices for the restoration and the concessions of use of water and wood.

Quotes about the oldest building structures date back to the 15th century (Fig. 9.3). The documents describe them as *cascine* or *casare* (alpine dairy farms) in which cheese was produced. Structures for maturing of cheese or animal shelters are not mentioned. These structures are generally described as mobile and wooden.

Only in the 18th century was the difference between buildings specified: the *casoni*, temporary huts with a stone base and a wooden upper part are used for the processing of milk (that had to be moved at each new leasing, usually every five years); and the *volti*, vaulted storage buildings in which cheese and butter were kept. During the 19th century the *casoni* came to be called '*baiti*' and the *volti* '*casere*'.

The documents never mention shelters for cattle; it was normal practice to leave a wooded area in the pastures under which the cows could find shelter.

The first type of structures (*casoni* or *baiti*) corresponds to rectangular low buildings; one half was partially terraced into the slope of the pasture and reinforced by large stones, the other half was built with a wooden trellis, completely above ground and positioned on a small embankment. The interior space was divided into three parts: an entrance with wooden walls separated the milk parlour (*camera del latte*, where the fresh milk was allowed to cool down and where cream was skimmed); the fire room (*camera del foco*), which housed the great fire boiler where cheese was produced; and where the *malgari* (dairyman) lived. The milk cooling was secured by air circulation, by the presence of water or, where the presence of water was rare due to the karst phenomenon, through the use of snow and ice stored in appropriate structures. A thatched roof made of rye straw or conifer bark covered the building. The new location of the *baito* was decided by the municipality before the cattle were brought to the pasture. This construction and management model remained unchanged until the early 20th century (Fig. 9.4).

The cheese and butter produced in the *baiti* were kept in the *volti* or *casere*, storage buildings that had two minimum requirements: insulation (to maintain a constant humidity suitable for the proper storage and maturing of the cheese); and protection against theft. Both requirements were met thanks to buildings consisting of a single masonry covered room, which was protected, in turn, by a covering of timber and straw. These storage buildings are mentioned in the 18th century as small semi-subterranean and easy to dismantle structures, which were functionally connected to each *baito*.

During the 19th century they become municipal properties and are described as more and more impressive buildings with thick stone walls and limestone slab roofs. They become the only permanent buildings of the *malghe*. A single main door provided access and a small window at the opposite end guaranteed air circulation. The *casere* were a mark of distinction and a sign of wealth for the community, the huts were built and managed directly by the municipality, who were also responsible for their maintenance (Fig. 9.5). Some of these artifacts have survived to the present day.

The Pasubio *malghe* from material traces

Several traces of similar morphological structures, although different in dimensions and state of preservation, were identified by the study of the detailed Habsburgic Cadastre (the first cadastral maps of Habsburg territories) of 1859, together with an investigation and exploration of the territory with the help of vertical

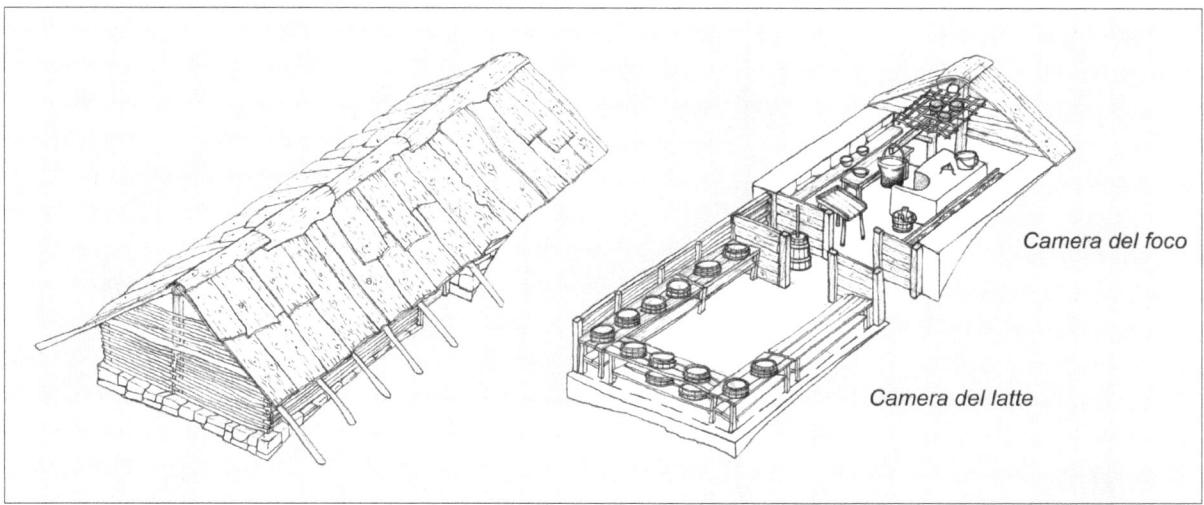

Figure 9.4 Constructive scheme of a '*baito*' inferred from historical documents. *Camera del latte*: 'milk parlour' with wooden walls where the fresh milk was allowed to cool down; *camera del foco*: 'fire room' with stone walls which housed the fire boiler and where the dairymen lived.

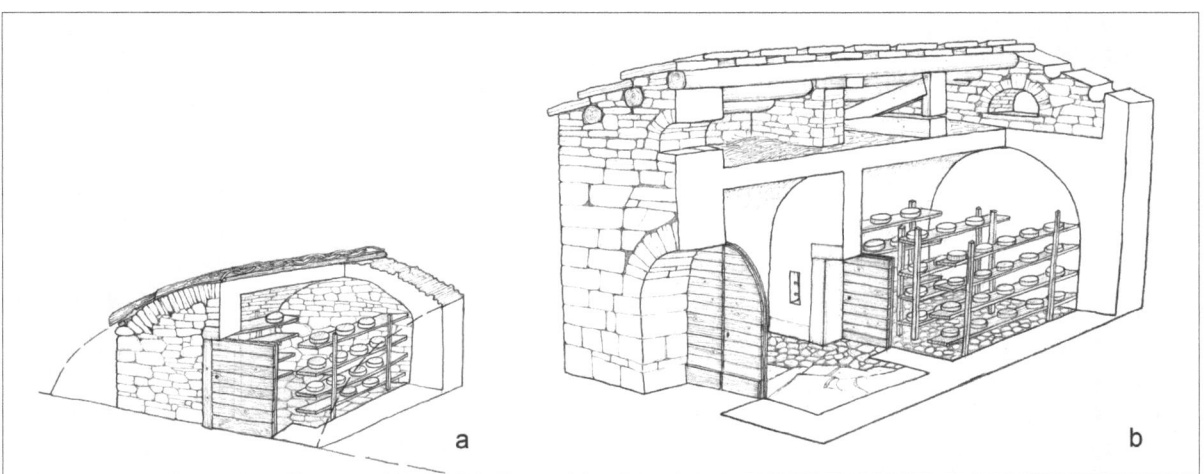

Figure 9.5 Construction scheme of the storage building. a: 18th century dairy store, a small semi-subterranean building; b: 19th century '*casera*', a solid building with thick stone walls and limestone slab roof.

photographs, historical aerial photos and LiDAR (digital terrain model) (Fig. 9.6). Some appear as depressions not very dissimilar from natural cavities, others respected the perimeter walls.

Overall the survey identified 145 recorded and georeferenced structures, mostly concentrated in pasture areas not yet reached by the expanding reforestation of recent decades (Fig. 9.7).

The traces are mainly concentrated in the summit area of grassy ridges along the watershed between two small valleys (Fig. 9.6d). This consistent location is not only due to the greater stability of the slope which, in turn, insures the solidity of the structure, but also to the mode of migration to summer pastures; the choice of the highest part of the territory was fundamental to control herd movements and trespassing, that often occurred between neighbouring territories.

The traces are mainly found along slopes leading towards the valley; the longitudinal axis is usually perpendicular to the slope. The orientation of the structures follows the land morphology, especially for those located on very steep slopes.

On the basis of a series of morphological and historical elements, it is possible to separate the remains into two groups.

a. structures without walls and almost completely obliterated;

b. structures with wall elements or remnants of spatial organisation.

Through the analysis of plano-altimetric distribution, preliminary lichenometric surveys, morphological-constructive characteristics, state of degradation and materials identified during and investigations, it was

Summer Farms: Seasonal exploitation of the uplands from prehistory to the present

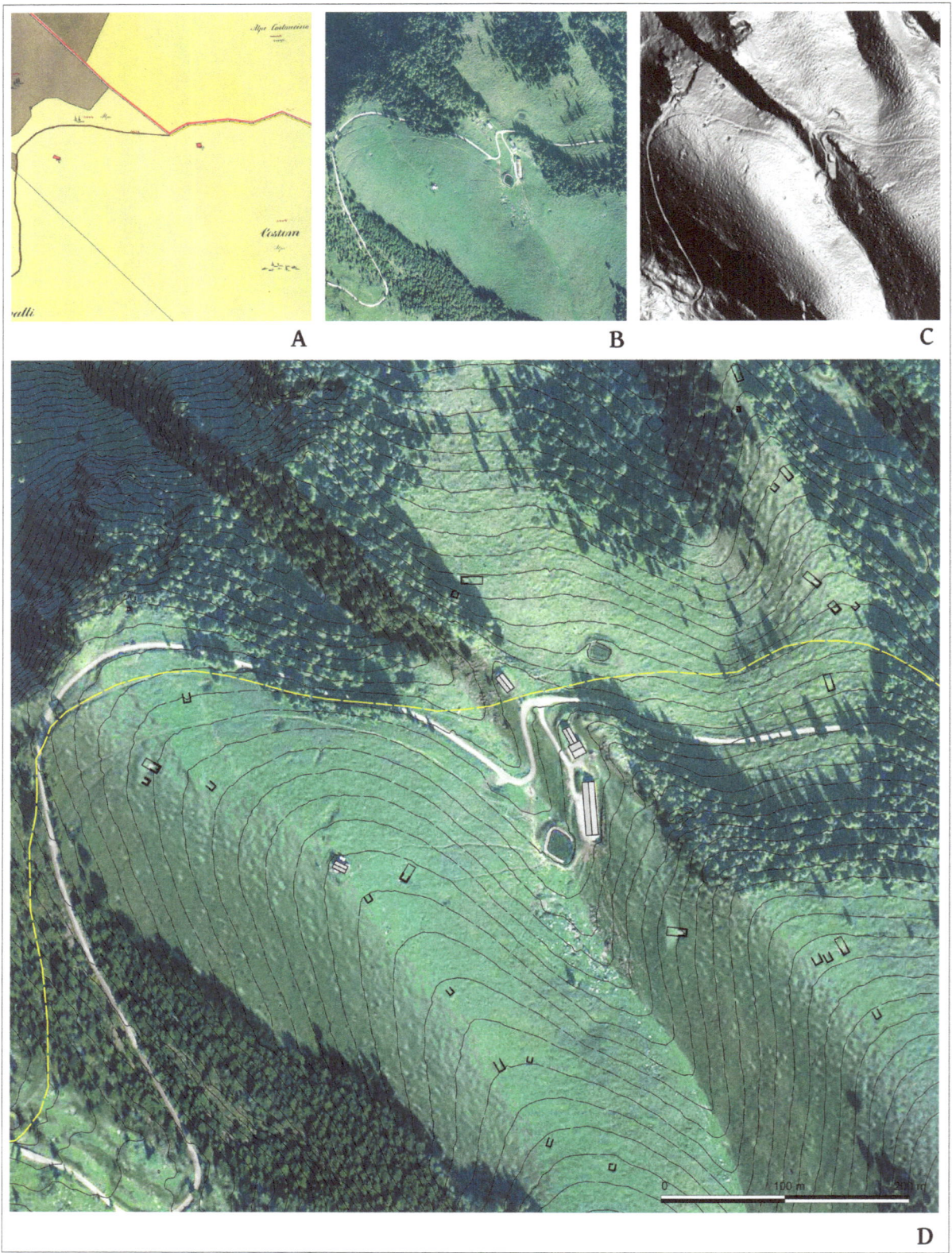

Figure 9.6 Evidence of structures related to dairy production in *Costoni* pasture (3 in Fig. 3.7). A: Habsburgic Cadastre (1859); B: Vertical photo (2006); C: Digital Terrain Model (2008); D: Integrated map of documentary data and field surveys. The black rectangles show the identified structures.

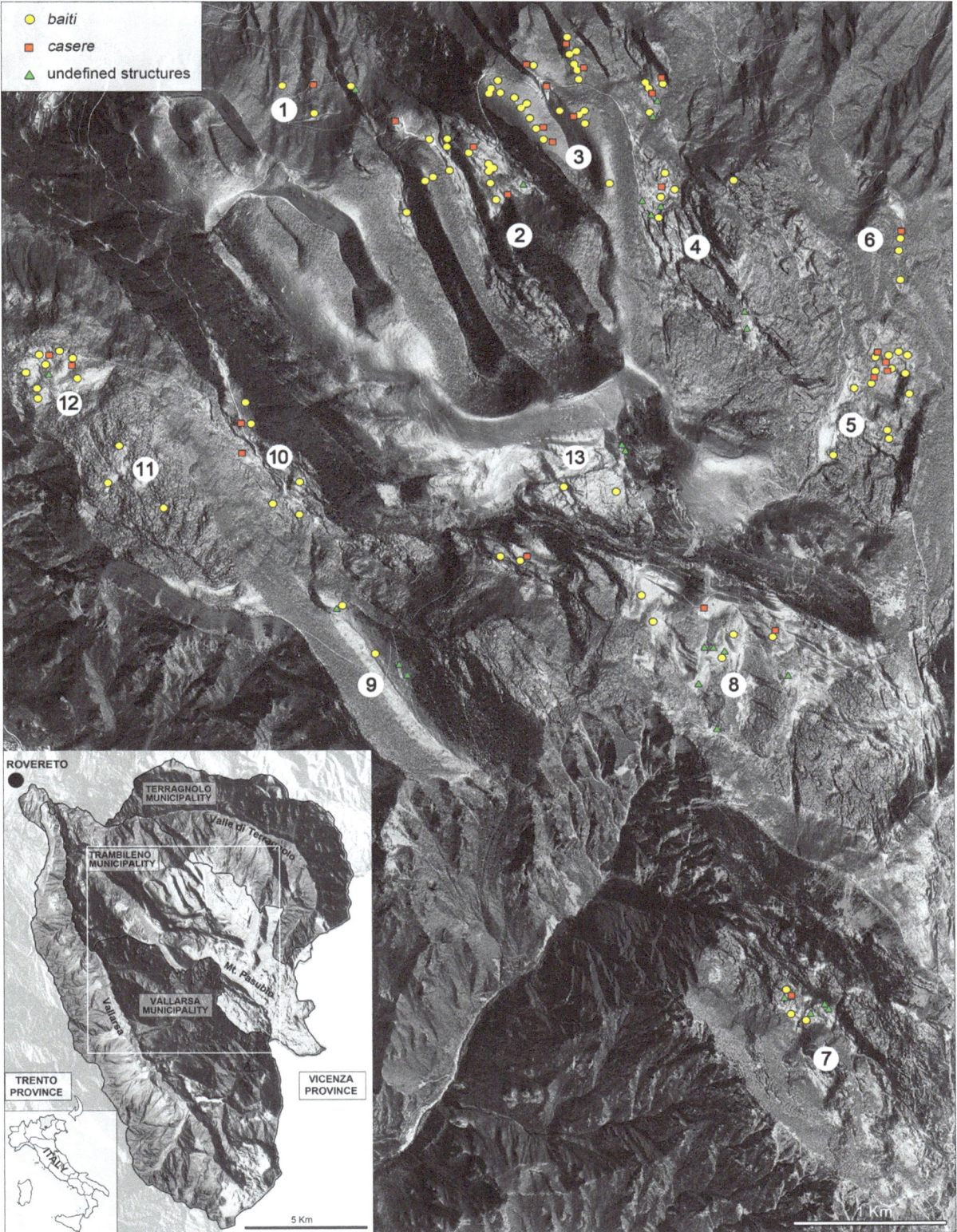

Figure 9.7 The study area in the Pasubio Massif. Pasture locations: 1: Fratielle (Fra); 2: Valli (Val); 3: Costoni (Cos); 4: Corona (Cor); 5: Campobiso (Cmb); 6: Pezzi (Pez); 7: Cosmagnon (Csm); 8: Pozze (Poz); 9: Zocchi (Zoc); 10: Cheserle (Che); 11: Buse (Bus); 12: Monticello (Mon); 13: Lastè (Las).

Figure 9.8 a: Stones alignment of casaria in Campobiso pasture (Cmb15, 16th century); b: Structure without wall remains in Costoni pasture (os7, 17th century); c: 'Baito' remains marked in the Habsburgic Cadastre in Corona pasture (Cor1, 1859); d: 'Baito' remains (d1) and its deposit (d2) in Campobiso pasture (Cmb7 and Cmb8, 18th-19th centuries), at the centre the wide casera built in 1853.

possible to infer a relative dating for each feature. Thanks to a cross reference study with historical documents found in the municipal archives, it was also possible to specify the function of the remains.

a: Depressions with little morphological emphasis and without traces of masonry, mainly quadrangular in shape and divided into two main dimensional classes ranging from 3.5 to 4×4m to 5×6m approximately. These structures were the foundations of shelters of which no trace remains and are the first evidence of the intensive exploitation of the mountain territories. They correspond to the wooden mobile huts (*casaria*) mentioned in documentary sources (Fig. 9.8a,b).

b: These structures show clear evidence of building schemes; they are the remains of rectangular features characterised by a longitudinal axis perpendicular to the contours with a subterranean first half, and the second half represented by an artificial terrace. They were connected to a second shape, smaller and quadrangular, lacking the terrace that encloses the front space towards the valley. To these two shapes correspond the ones that historical documents called *baiti* and *casere* (Fig. 9.8d).

b1: The *baiti* had a width between 5.5 and 6.5m and a length between 12 and 14m. The lack of variety in the construction forms of the structures of the 18th and 19th century demonstrates a lack of morphological evolution. The only variation is represented by the different size of the interior spaces. The milk parlour size increased during the 18th and 19th century to a greater extent than that of the fire room. The latter shows masonry walls that did not exceed one meter and a half in height. They also act as retaining walls where the slope is steep. The stones are superimposed on each other, with no mortar; probably the need to move and rebuild the *baito* at each new leasing, required the walls to be 'removable' and the stones reusable (Fig. 9.8c,d). In some facilities, it was possible to identify the most important structural element, the centre of production and life of the *malga*: the fireplace. This was located in the centre of the fire room, against the up-slope wall and showed a morphological evolution from a single limestone slab (18th century) to semicircular structures which define a combustion chamber of about 80–90cm in diameter (19th century).

b2: The *casere*, the storage buildings, appear as square depressions with an entrance facing the valley. Where the slope is not particularly steep, there are soil supports on either side of the walls. The *casere* served several *baiti*. The oldest (17th–18th century) had internal dimensions of 2.5 to 3×4 m, with a storage area of about 10–12 square metres. In the second half of 18th century, these storage rooms increased in size reaching an internal area of 18 to 20 square metres (Fig. 9.8d). In its most advanced stage (mid-19th century), the *casera* was a solid building (inside area: 4 to 4.5m × 8.10m) which consisted of a single masonry barrel-vaulted room, above which a stone or wooden structure supported the heavy limestone slabs. The walls had a thickness of 100–120cm and consisted of limestone blocks of considerable size (approximately 50×70×100cm) arranged in a regular way; the flooring was also made of limestone slabs. There was only one entrance, generally arc shaped and protected by a sturdy wooden door (Fig. 9.8d).

Elements of material culture

The limited accumulation of soil common to the mountain environment facilitated the discovery of traces of material culture within or in the immediate vicinity of the identified structures. Among them, some coins of small value were found, which aided the identification of different phases in the use of the structures (Fig. 9.9). Agricultural tools, materials connected with animal husbandry, clothes, objects for daily use, and materials related to the collapse and decay of the above-ground structures of the buildings were also found. Iron nails of different sizes and functions represent the most abundant objects; they all belong to light carpentry kits, very likely related to the wooden structures of the milking parlours and of the roofs (Fig. 9.10, 10–29).

Among the items of clothing found were shoe nails (Fig. 9.10, 1–9), and shoe and belt buckles (Fig. 9.11, 1–7), together with knee-buckles in use between the 17th and 19th century. Some waistcoat or trouser buttons (Fig. 9.11, 10–11) date the use of the structures to the 17th century. Some items are related to the daily life of highlanders: the handle of a spoon and scissors from the 18th century and a pocket watch case from the late 19th century (Fig. 9.11, 9, 14, 15).

A few fragments of pottery and a piece of a cauldron hanger for a copper pot (Fig. 9.14, 4-10) are related to the stages of milk processing, among them, a fragment of Kröning pottery dating to the 18th century (Fig. 9.14, 7). An iron cow bell (Fig. 9.14, 1) from the 18th century, some 18th–19th century goat bell clappers (Fig. 9.14, 2–3), and horse harness and horse nails (Fig. 9.14) confirm the presence of cattle, goats and horses.

The discovery of an old 'Roman' stud of the 3rd–4th century (Fig. 9.11, 6), was an exceptional find, perhaps an indication of a more ancient settlement.

Furthermore, some items not directly connected with dairy production were found near the remains of an isolated structure (*Val11*) (knife, a spear butt, girth strap fitting, horse harnesses, and an incised/painted jug), suggesting a human frequentation of the uplands during the medieval period (15th–16th century) (Fig. 9.12, 1–4).

The variation of population density in time and possible interactions with the climate

Documentary data, construction methods, space organisation, lichenometric analyses and elements of material culture allowed the dating of all the studied structures, in order to locate their chronological distribution between the 15th and 20th century.

From these analyses, storage structures (which could serve multiple production buildings and be used for long periods), structures without a clearly recognisable function and buildings for the storage of ice were excluded.

The *baiti* seem to be best suited to interpret the dynamics of population and exploitation of a territory. These are the structures most closely related to the productivity of pastures. It therefore appears reasonable to use their growth, decline and distribution to assess the ease or difficulty of living at high altitude.

The buildings were divided into fifty-year-long age groups and the resulting 'building density' showed fluctuations over time. Analysing these variations throughout the study area (84 *baiti*), a progressive increase in the number of structures, with a clear surge in the 19th century, is demonstrable. The surge data are likely to be distorted though by the fact that these traces are more recent and by the habit of building again in the same areas and / or on the same building sites.

It was therefore decided to focus the analysis on a more limited area (Valli-Costoni-Corona-Campobiso pastures) in which there is less overlap of structures and therefore a greater possibility of recognising each building in its different stages of construction. In this area, after a relative high building density in the first half of the 17th century, an almost complete abandonment of the pastures in the second half followed. During the 18th century the number of buildings increased significantly. In the first half of the 19th century a slowdown in growth in the number of structures occurred, when compared to the previous fifty years. Buildings again increase in number in the second half of the 19th century.

This tendency matches well with the trend of the demographic curve of the same territories. The curve also shows a fall between 1650 and 1700 and a gradual increase that accelerates in the second half of the 19th century. The epidemics of the first half of the 17th century, largely responsible for the fall in population, are

Summer Farms: Seasonal exploitation of the uplands from prehistory to the present

Figure 9.9 Some coins found near structures useful for dating: 1: Mezzo soldo, bezzone, 1619–?, Venezia (Val7); 2: 1 soldo, 12 bagattini, 1709–1722, Venezia, Giovanni Corner Doge, (Cos1b); 3: 1 soldo, 1739, Trento, Karl VI (Cos1b); 4: 1 soldo, 1769, Regno Lombardo Veneto, Joseph II von Habsburg-Lothringen (Cmb8); 5: 1 Kreutzer, 1816, K.K. Oesterreichische, Franz II Erwählter Römischer Kaiser (Las1); 6: 5 centesimi di Lira, 1822, Regno Lombardo Veneto, Franz II Erwählter Römischer Kaiser (Cor3); 7: 1 soldo, 1862, Regno Lombardo Veneto, Franz Joseph I von Österreich (Cmb13); 8: 2 heller, 1897, K.K. Oesterreichische, Franz Joseph I von Österreich (Cor1).

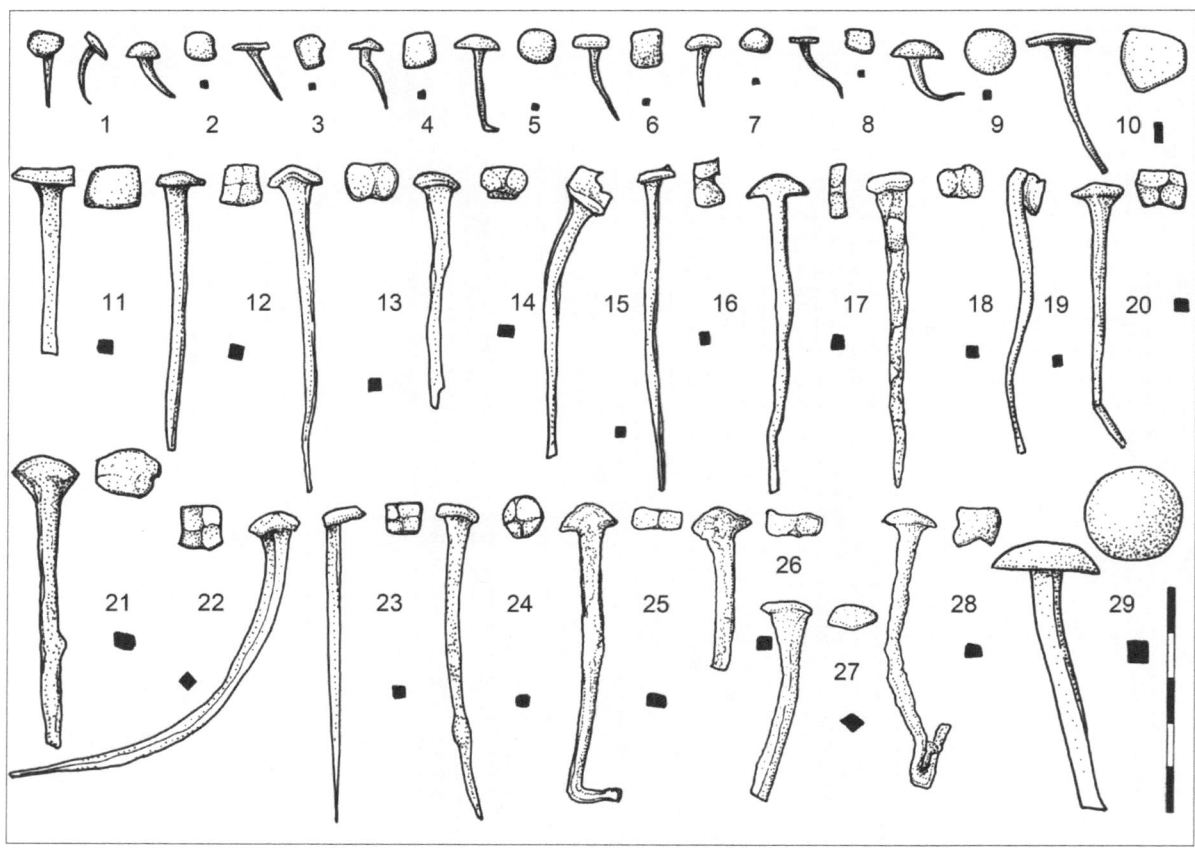

Figure 9.10 Iron nails of different size and functions: 1–9) shoe nails; 10–29) carpenter nails. 1: 18th century (Cos2); 2–3: 17th–18th century (Cmb11); 4–6: 18th–19th century (Cmb7); 7: 18th century (Val9); 8: 19th century (Cos1); 9: 19th century (Cos15); 10: 19th century (Cor5); 11: 18th century (Cos11); 12–13: 18th century (Val9); 14–15: 18th century (Cos35); 16–17: 18th–19th century (Cor3); 18: 19th century (Cos4); 19: 18th century (Cos1b); 20: 19th century (Cos14); 21: 18th century (Cmb4); 22–24: 19th century (Cos1a); 25–28: (Cmb1), 19th century; 29: 19th century (Poz1).

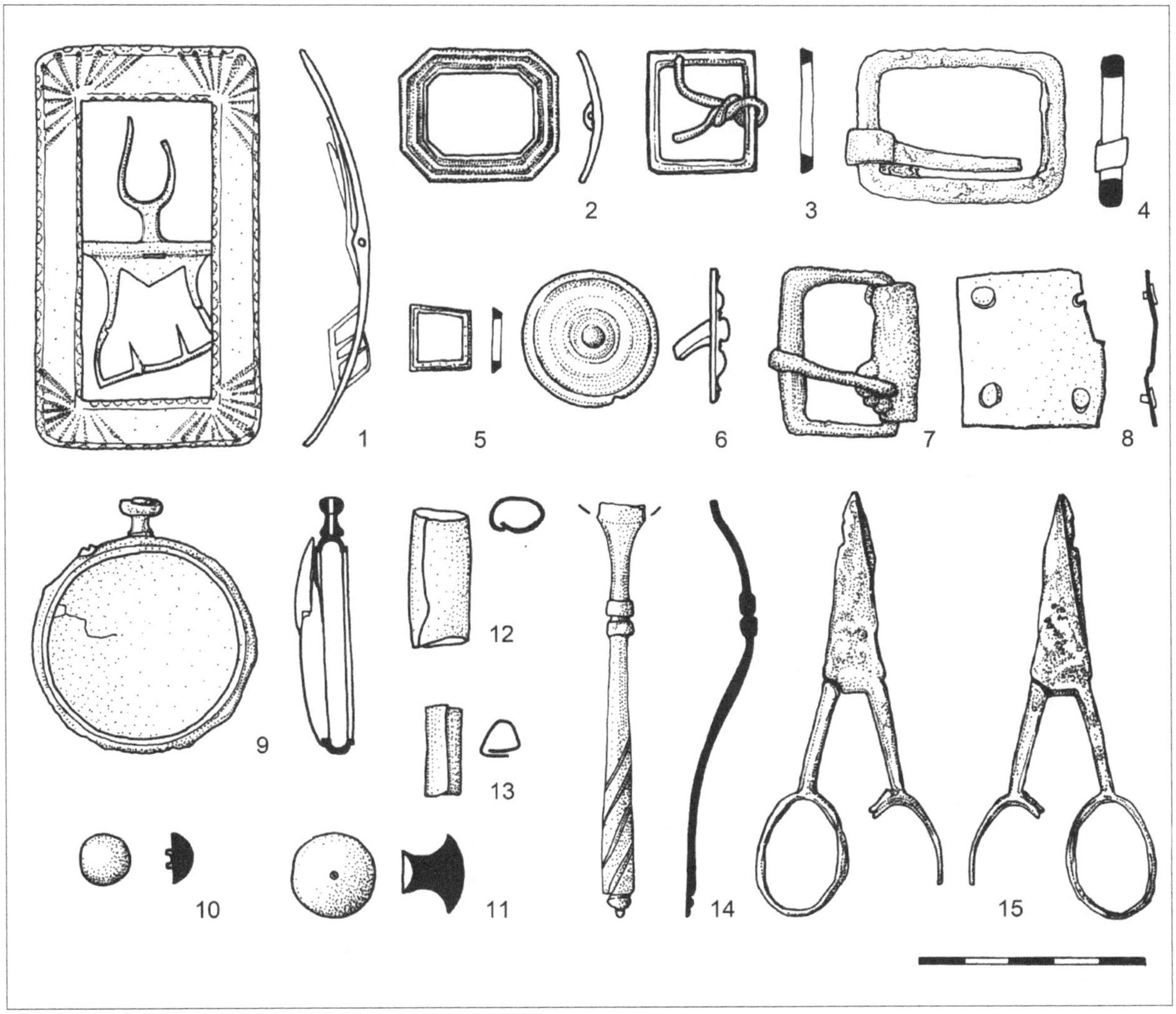

Figure 9.11 Items of personal adornment and daily use objects: 1–5, 7: buckles; 6: boss; 9: pocket watch; 10–11: buttons; 12–13: brass lace cap; 14: spoon handle; 15: scissors. 1: 18th century (Val9); 2: 18th century (Val12); 3: 19th century (Cos9), 4: 19th century (Cmb1); 5: 19th century (Cos16); 6: 3rd–4th century (Val5); 7: 19th century (Val3); 8: 18th–19th century (Cor3); 9: 19th century (Che6); 10: 17th century (Cos7); 11: 19th century (Cmb2); 12: 18th century (Val9); 13: 19th century (Cos4); 14–15: 18th–19th century (Cmb7–Cmb8).

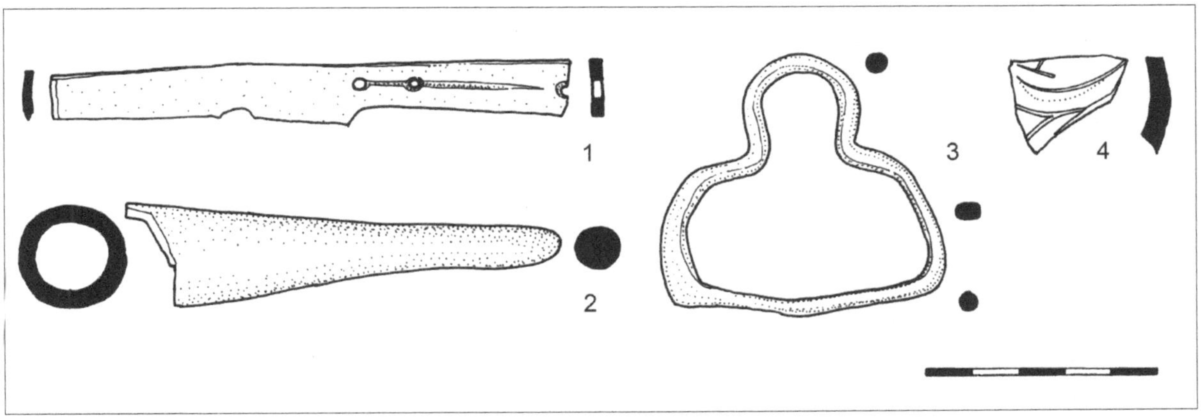

Figure 9.12 Items related to a generic medieval presence: 1: knife; 2: spear butt; 3: girth strap fitting (horse harness); 4: incised/painted jug (Val11).

Summer Farms: Seasonal exploitation of the uplands from prehistory to the present

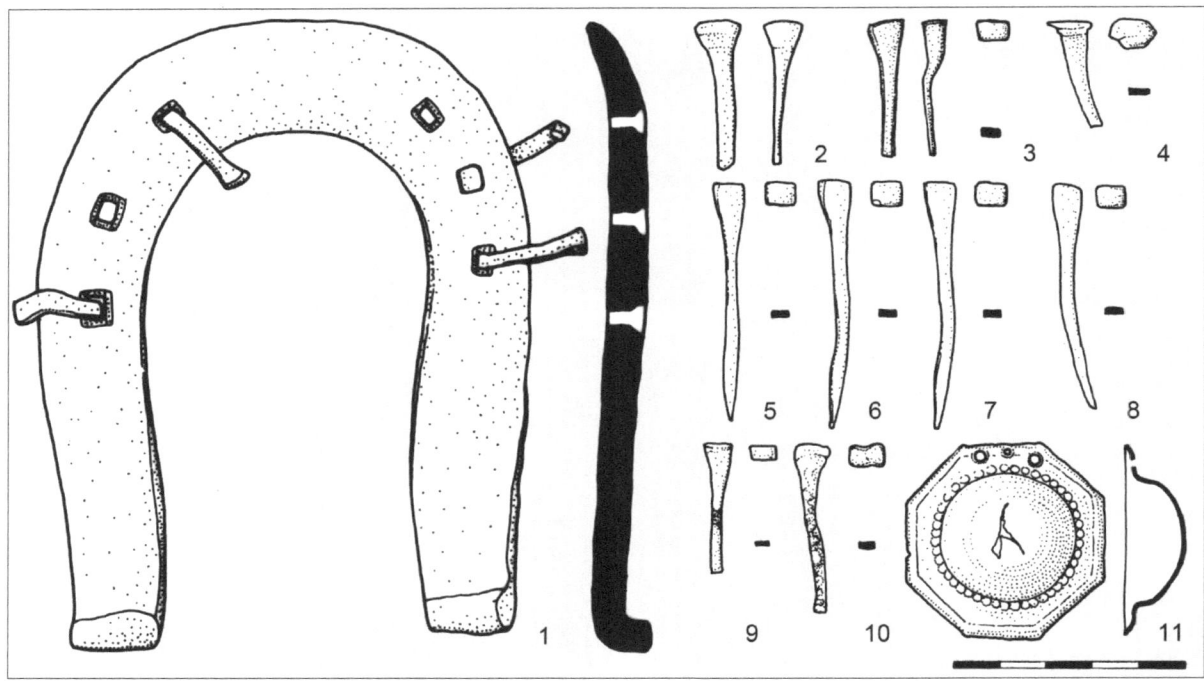

Figure 9.13 Items related to the presence of horses: 1: horseshoe; 2-10: horse nails; 11: stud (horse harness). 1: 19th century (Val4); 2: 18th–19th century (Cmb8); 3: 19th century (Cmb2); 4-10: 19th century (Cmb1); 11: 18th century (Cos28).

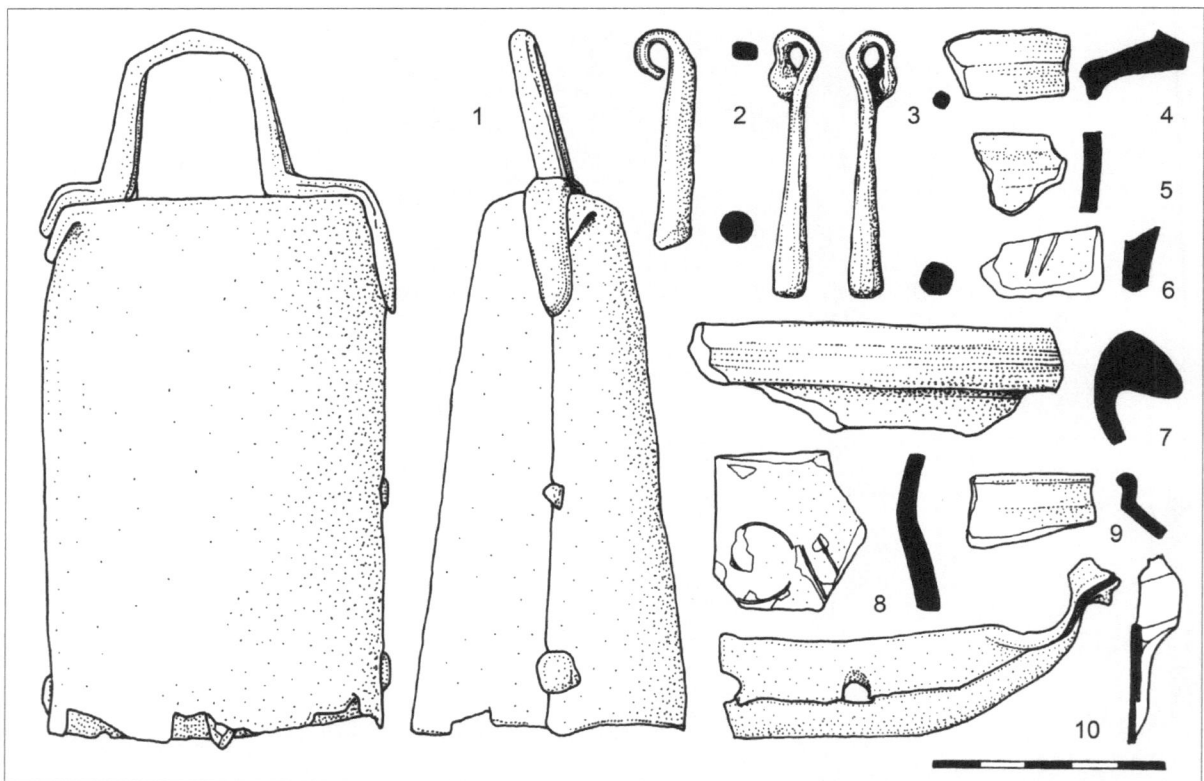

Figure 9.14 Items related to cattle and goat presence: 1: cow bell; 2–3: goat bell clappers; Items related to the stages of milk processing: 4–9: fragments of pottery; 10: piece of a suspension hook from a copper pot. 1: 18th century (Cos8); 2: 18th century: (Cos21); 3: 18th–19th century (Cmb7); 4: 18th–19th century (Cmb7); 5: 19th century (Cos16); 6: 18th century (Cmb4); 7: 18th century (Cos2); 8: 18th century (Cos13); 9: 19th century (Cor5); 10: 19th century (Cmb13).

Demographic curve

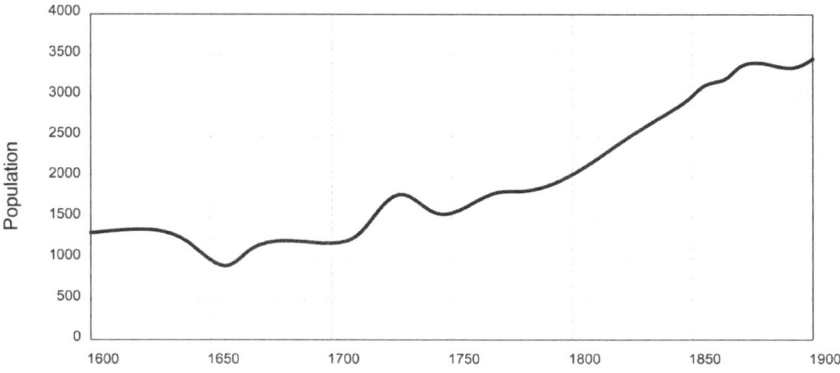

"Building density" curve

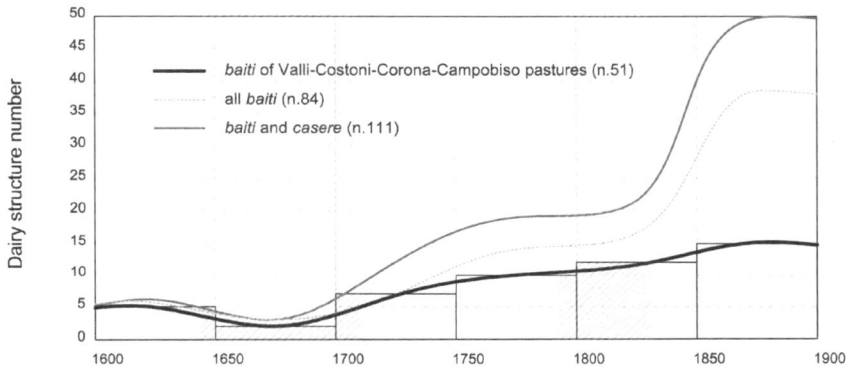

Reconstructed mean annual temperature

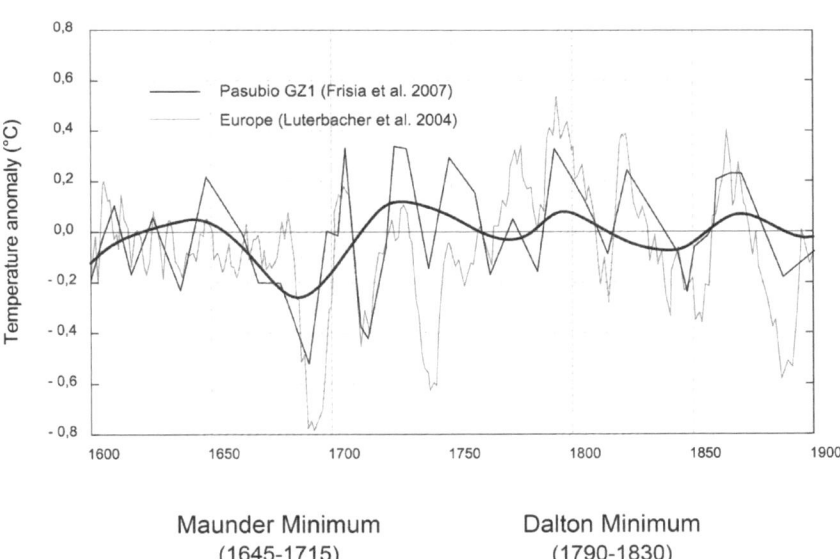

Figure 9.15 Demographic curve, 'building density' curve and temperature oscillations of the study area. The decrease of the number structures between 1650 and 1700 is related to the Maunder Minimum. The number of structures greatly increases after the Dalton Minimum and the revolution in the agricultural economic systems.

Summer Farms: Seasonal exploitation of the uplands from prehistory to the present

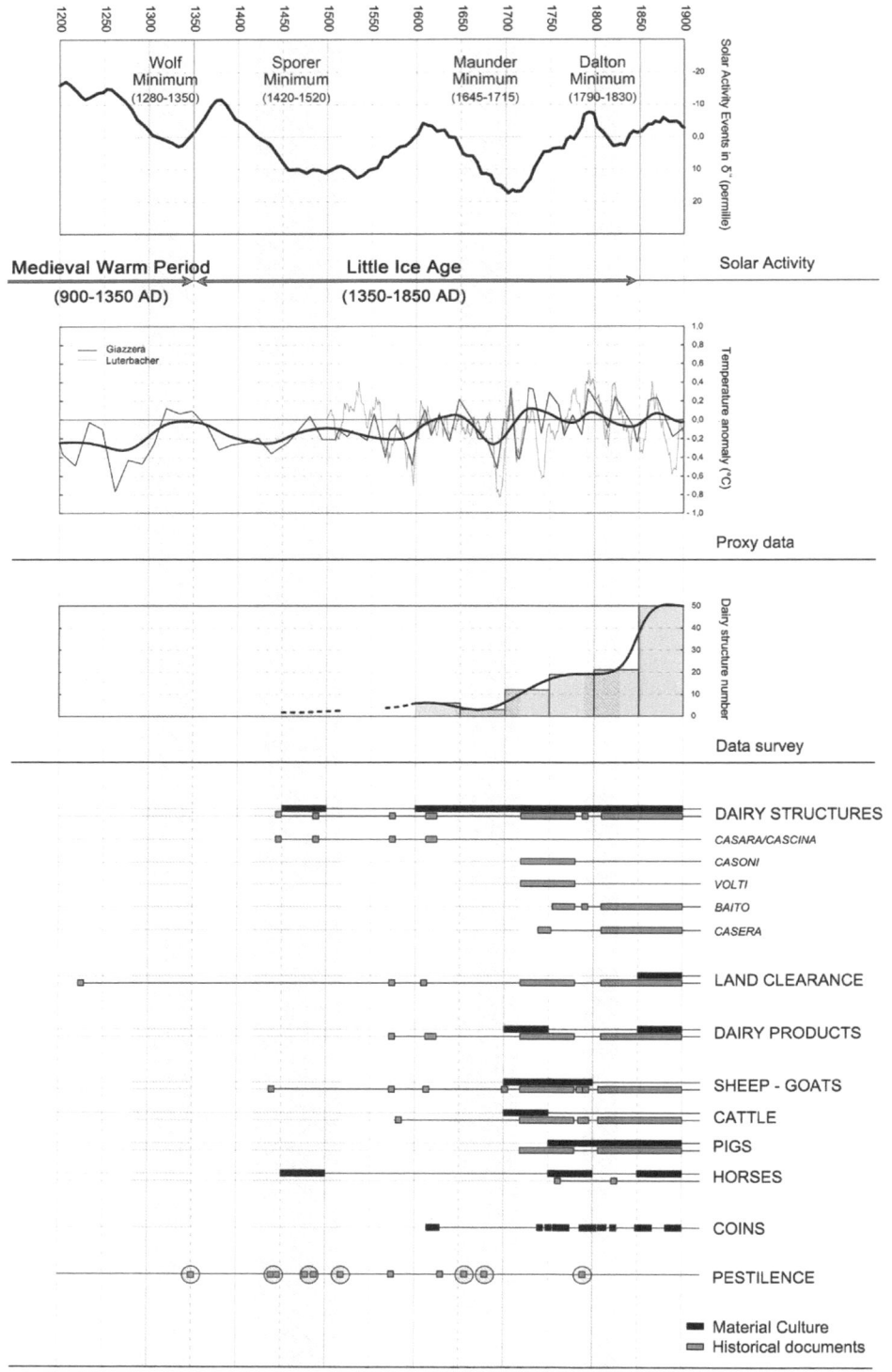

Figure 9.16 A cross reference study of archive records and field data shows strong correlation between population pressure trends and environmental constraints. During prolonged sunspot minimum periods (Sporer, Maunder and Dalton Minimum), a drop in temperatures is recorded. This coincides with a decrease or slowdown in growth in the number of structures and a lack of historical documents indicating a dependence of mountain grazing on climatic oscillations. Correlation between trends is highlighted during the Maunder Minimum, the coldest phase of the Little Ice Age, when epidemics of the first half of the 17th century are associated with the apparent abandonment of grazing at high altitude.

likely to be associated with the apparent abandonment of farming at high altitude.

However, if we cross these two trends (demographic and population at high altitudes) with the curve of temperature oscillations of the same area obtained from proxy data (Fig. 9.15), it can be noted that the decrease in the number of building structures between 1650 and 1700 is also connected with the sharp drop in temperature common across the Alps and related to the 'Maunder minimum' of solar activity.

The subsequent rise in temperature corresponds to an increase in population in the highlands which, although marked by increasing intensification, suffered a slowdown during the first half of the 19th century. The negative temperature oscillation linked to the 'Dalton minimum' (1790–1830) and the subsequent peak of the LIA could be responsible for this slowdown, due to the poor growth of grazing and its difficult exploitation.

The rise of temperature from the beginning of the 20th century, in connection with a revolution in the agricultural economic systems, then led to a strong growth in high altitude structures.

Conclusions

The cross reference study of the archive records related to the exploitation of mountain areas between the 15th and 19th century with field data, showed a discontinuity of land use and the presence of anthropic infrastructures in the same area (Fig. 9.16).

It also seems to demonstrate that the exploitation of the highlands in the post-medieval period in this area of the Alps was strongly affected by environmental constraints, according to different phases that can be summarised as follows:

a. collapse of the agricultural economy and of the use of higher pastures in the second half of the 17th century, connected with a marked negative temperature oscillation (Maunder minimum);

b. a rise of activities at the beginning of the 18th century with subsequent stagnation in the first half of the 19th century, also linked to a negative climatic oscillation (the Dalton minimum);

c. a reduction of the direct relationship between environmental constraint and population during the second half of the 19th century, when an increase in technology and specialisation in the highlands overcomes the close dependency on the climate and natural environment.

Bibliography

Anfodillo, T. 2007. *Cambiamenti climatici e dinamica di popolazione al limite superiore del bosco: importanza delle ricerche di lungo termine*. Forest@ 4:3–5.

Baglioni, M., Macagno, G., Nunes, P. and Travisi, C. 2009. Natura 2000 Network, Agricultural Pressures and Biodiversity Conservation: an Italian outlook. Analysing the potential of the Natura 2000 database in describing the linkages between agricultural pressures and biodiversity. *Report of the EXIOPOL*, Deliverable.II.3.b.1. part 2, Milan, Italy.

Bosello, F., Zhang, J. 2005. *Assessing Climate Change Impacts: agriculture*. FEEM Working Paper 94.

Bozhong, L. 1998. Changes in climate, land, and human efforts; the production of wetfield rice in Jiangnan during the Ming and Qing dynasties. In M. Elvin and L. Ts'ui-jung (eds.) *Sediments of Time; environment and society in Chinese History*. Cambridge: Cambridge University Press.

Büntgen, U., Frank, D.C., Nievergelt, D. and Esper J. 2006. Summer temperature variations in the European Alps, A.D. 755–2004. *Journal of Climate* 19:5606–5623.

Carrer, F., 2012. *Etnoarcheologia dei Paesaggi Pastorali nelle Alpi; Strategie Insediative Stagionali d'Alta Quota in Trentino*. PhD Thesis, Trento University.

Dearing, J.A. 2006. Climate – human – environment interactions: resolving our past. *Climate of the Past* 2:187–203.

Fraser, E.D.G. 2009. *Economic crises, land use vulnerabilities, climate variability, food security and population declines: will history repeat itself or will our society adapt to climate change?* Centre for Climate Change Economics and Policy Working paper No. 1: Leeds and London.

Frisia, S. 2007. Sintesi delle conoscenze sulla variabilità climatica nelle Alpi negli ultimi 1000 anni. *Studi Trentini di Scienze Naturali, Acta Geologica* 82:65–69.

Frisia, S., Borsato, A., Richards, D.A., Miorandi, R. and Davanzo, S. 2007. Variazioni climatiche ed eventi sismici negli ultimi 4500 anni nel Trentino meridionale da una stalagmite della Cogola Grande di Giazzera. *Studi Trentini di Scienze Naturali, Acta Geologica* 82:205–223.

Giacomoni, F. 1998. Comunia et Divisa; l'organizzazione dei prati pascoli e l'ordinamento forestale nella montagna trentina dal XIV al XVIII secolo. In G. Kezich and P.P. Viazzo (eds.) *Equilibri sulle Alpi; saggi in onore di Robert McC. Netting*. Atti di SPEA3 (Seminario Permanente di Etnografia Alpina, 3° ciclo) 1995–1996:97–146. San Michele all'Adige (TN): Annali di San Michele 11.

Giacomoni, F. 2001. La tutela dell'alpeggio nelle carte di regola del Trentino. *Economia alpestre e forme di sfruttamento degli alpeggi*. Pp. 119–150. Bolzano: Athesia.

Gusmeroli, F., Corti, M., Orlandi, D., Pasut, D. and Bassignana, M. 2005. Produzione e prerogative qualitative dei pascoli alpini: riflessi sul comportamento al pascolo e l'ingestione. *Quaderni SoZooAlp* 2:7–27.

Hegerl, G.C., Hoegh-Guldberg, O., Casassa, G., Hoerling, M.P., Kovats, R.S., Parmesan, C., Pierce, D.W. and Stott, P.A. 2010. Good practice guidance paper on detection and attribution related to anthropogenic climate change. In T.F. Stocker, C.B. Field, D. Qin, V. Barros, G.-K. Plattner, M. Tignor, P.M. Midgley, and K.L. Ebi (eds.) *Meeting Report of the Intergovernmental Panel on Climate Change Expert Meeting on Detection and Attribution of Anthropogenic Climate Change*. IPCC Working Group I Technical Support Unit, Bern/Switzerland, University of Bern.

Leonelli, G., Coppola, A., Pelfini, M., Salvatore, M.C., Cremaschi, M. and Baroni, C. 2012. Il segnale climatico e le sue variazioni negli anelli di accrescimento degli alberi da siti estremi al contorno della regione mediterranea. *Rendiconti Online Società Geologica Italiana* 18:24–28.

Luterbacher, J., Dietrich, D., Xoplaki, E., Grosejan, M. and Wanner, H. 2004. European seasonal and annual temperature variability, trends and extremes since 1500. *Science* 3013:1499–1503.

Malanima, P. 2006. Energy crisis and growth 1650–1850: the European deviation in a comparative perspective. *Journal of Global History* 1:101–121.

Mann, M.E. and Jones, P.D. 2003. Global surface temperatures over the past two millennia. *Geophysical Research Letters* 30 (15):1820.

Menzel, A. and Fabian, P. 1999. Growing season extended in Europe. *Nature* 397: 659–661.

Michaelowa, A., 2001. The impact of short-term climate change on British and French agriculture and population in the first half of the 18th century. In P.A. Jones and T. Ogilvie (eds.) *History and climate; memories of the future?* New York: Kluwer.

Migliavacca, M., 2012. Paesaggi pastorali nella montagna veneta: archeologia ed etnoarcheologia. In M.S. Busana and P. Basso (eds.) *La Lana nella Gallia Cisalpina.* Padova: Padova University Press.

Migliavacca, M., Saggioro, F. and Sauro, U. 2013. Ethnoarchaeology of pastoralism: fieldwork in the highlands of the Lessini Plateau (Verona, Italy). In F. Stoppiello, S. Biagetti (eds.) *Ethnoarchaeology: current research and field methods.* Conference Proceedings, Rome, Italy, 13th–14th May 2010. British Archaeological Reports International Series 2472.

Moberg, A., Sonechkin, D. M., Holmgren, K., Datsenko, N. M. and Karlén, W. 2005. Highly variable northern Hemisphere temperatures reconstructed from low-and high-resolution proxy data. *Nature* 433:613–617.

Orlandi, D. and F. Clementel, 2007. *Caratterizzazione produttiva dei pascoli alpini.* From http://www.pianteofficinali.org/main/Alpicoltura/Alpicoltura.htm

Orlandi, D., Clementel, F., Bovolenta, S. and Dovier, S. 2004. Caratterizzazione chimica e nutrizionale delle principali specie pascolive alpine. *Quaderni SoZooAlp* 1:171–180.

Paavola, J., and Fraser, E. 2011. Ecological economics and environmental history. *Ecological Economics* 70:1266–1268.

Pfister, C. and Bradzil, R. 2006. Social vulnerability to climate in the 'Little Ice Age': an example from central Europe in the early 1770s. *Climate of the Past Discussions* 2:123–155.

Rea, R., Rampanelli, G. and Zardi, D. 2003. The temperature series of Trento: 1816–2002. *Proceedings of the 27th International Conference on Alpine Meteorology and MAP-Meeting 2003*, Vol. A. Brig. Pp. 483–486.

Roson, R., 2003. Modelling the economic impact of climate change. *EEE Working Paper No. 9.* International Centre for Theoretical Physics.

Solomon, S., Qin, D., Manning, M., Chen, Z., Marquis, M., Averyt, K.B., Tignor, M. and Miller, H.L. (eds.) 2007. *Climate Change 2007: the physical science basis. Contribution of Working Group I to the Fourth Assessment Report of the Intergovernmental Panel on Climate Change.* Cambridge: Cambridge University Press.

Varanini, G.M., 1991. Una montagna per la città. Alpeggio e allevamento nei Lessini veronesi nel Medioevo (Secoli IX-XV). In P. Berni, U. Sauro and G.M. Varanini (eds.) *Gli alti pascoli dei Lessini veronesi.* Vago di Livagno (VR): La Grafica Editrice. Pp. 13–106.

Xoplaki, E., Fleitmann, D., Diaz, H., Van Gunten, L. and Kiefer, T. 2011. Medieval climate anomaly. *Pages News* 19:1–40.

Ziliotto, U. and Scotton, M. 1993. Metodi di rilevamento della produttività dei pascoli alpini. *Comunicazioni di ricerca ISAFA (TN)* 93/1: 33–42.

Marco Avanzini: MUSE - Museo delle Scienze, Trento, Italy.
Email: marco.avanzini@muse.it

Isabella Salvador: MUSE - Museo delle Scienze, Trento, Italy.
Email: isabella.salvador@mtsn.tn.it

10. Alpine huts, livestock and cheese in the Oberhasli region (Switzerland): Medieval and early modern building remains and their historical context

Brigitte Andres

Oberhasli in the Bernese Alps has produced over 400 new sites, the majority linked to Alpwirtschaft *activities dating from medieval to modern times. Due to its location on the transalpine route over the Brünig, Grimsel and Gries Passes the area profited from the export of livestock and hard cheeses from late medieval times up to the end of the 19th century. The archaeological sites were not excavated which limits the possibilities of dating and interpreting them. The main category of sites was the foundations of buildings which allows an overview of their location, form and method of construction. Written and pictorial sources as well as comparisons with standing buildings and with results from archaeological excavations elsewhere assist in understanding their historical context.*

Ein Prospektionsprojekt in den Alpgebieten der Region Oberhasli im östlichen Berner Oberland brachte rund 400 neue Befunde zu Tage, die mehrheitlich im Zusammenhang mit alpwirtschaftlichen Tätigkeiten stehen und aus dem Mittelalter und der Neuzeit stammen. Dank der Lage an der transalpinen Route über die Pässe Brünig, Grimsel und Gries profitierte die Region vom Spätmittelalter bis zum Ende des 19. Jahrhunderts vom Export von Vieh und Hartkäse nach Norditalien. Die archäologischen Befunde wurden nicht ausgegraben, was die Aussagemöglichkeiten zu Datierung und Interpretation einschränkt. Die wichtigste Befundkategorie bilden die Gebäudegrundrisse, die mit einem Überblick über Standort, Form und Bauart vorgestellt werden. Schrift- und Bildquellen sowie Vergleiche mit bestehenden Alpgebäuden und archäologischen Grabungsresultaten aus anderen Regionen ergänzen ihre historische Einordnung.

Introduction

Archaeological investigations have seldom been carried out in Switzerland's alpine regions. When they are undertaken, they are usually local or regional projects, limited to a single valley or alp. In order to detect changes, such as an intensification or a reduction in utilisation, the focus of surveys is ideally not just on one historical period, but on a settlement area as a whole. This affords the opportunity of finding features from different eras and different areas of activity, such as seasonal alpine pastoral economic activity (*Alpwirtschaft*), hunting, inter- and transalpine communication and exploitation of raw materials.

With Werner Meyer's excavation at Bergeten ob Braunwald in 1971, the investigation of deserted alpine sites in Switzerland blossomed and over the course of the past forty years numerous excavations and prospection projects have been undertaken (Meyer 1998c). The first mention of remains of alpine huts was made by the naturalist Johann Jakob Scheuchzer as far back as the beginning of the 18th century. He noted the existence of what were known to local people as *Heidenhüttchen* (heathens' huts) and reported that they appeared to be ancient, and were built right up against the rocks using a curious masonry technique.[1] A little later, Johann Heinrich Tschudi believed he could explain these puzzling masonry remains as being where a small group of families, having driven their cattle up to the alpine pasture for the summer, set up home for the season.[2] A first excavation took place in the Mühlebach Valley in Canton Glarus in 1846, even before the pile-dwellings were discovered on Lake Zurich. Although no finds came to light, the features were dated to the pre-Roman period. Around 1900, interest in these deserted alpine sites appears to have flickered out again.

The projects from 25 years of research into deserted settlements by Werner Meyer and his colleagues, which were summarised in a book entitled *Heidenhüttli* (Meyer et al. 1998) mainly related to the seasonal alpine pastoral economic activity (*Alpwirtschaft*) of the Middle Ages

---

1. "[…] uralte, nach sonderbarer Bau-Art gemaurte, an den Felsen klebende Hüttlein, welche die Einwohner Heiden-Häusslein heissen", quoted by Meyer 1998c:13.
2. […] dass sonst einige Hauss-Haltungen zu Sommers-Zeit, wann sie ihr Vieh auf die Alpen getrieben, ihre Wohnungen allda aufgeschlagen", quoted by Meyer 1998c:14.

keywords: Alpine Archaeology, survey, alpine economy, building remains, rock shelter constructions, cultural landscape, Middle Ages, modern times, Bernese Alps, Switzerland
Wörter: *Alpine Archäologie, Prospektion, Alpwirtschaft, Gebäudereste, Konstruktionen unter Fels, Kulturlandschaft, Mittelalter, Neuzeit, Berner Oberland, Schweiz*

and the modern era. In order to obtain as much information as possible about the nature of this pre-industrial *Alpwirtschaft*, in addition to undertaking excavations and surveys, written records were investigated and people with local knowledge were interviewed. More recent projects have been undertaken with the same approach in Canton Uri (Sauter 2009), in the Val de Bagnes in Canton Valais (Taramarcaz 2012) and in Oberhasli in Canton Bern (Andres 2012b). The latter project is the subject of the present article. Over the Swiss alpine region as a whole, studies of deserted alpine sites are at various stages of surveying, excavating, analysing and publishing, and there are still a great many gaps. Although large numbers of individual features have been discovered, it has not been possible to do much comparative work because the inventories so far available are often only published in summarised form. It would be helpful to have illustrations of the catalogued features, in order to create an overall picture of regional differences and similarities. Dating the features also poses many problems; even those features which have been fully excavated, or where exploratory trenches have been dug, often cannot be dated. Finds are rare and there are many obstacles to radiocarbon dating, for example if firewood had initially been used as construction timber (Obrecht 2009; Meyer 1998d:364–365).

The Archaeological Service of Canton Bern filled in one of the gaps by undertaking alpine surveys in three valleys in the Oberhasli region of the Bernese Alps in 2003, 2004 and 2006. This involved a systematic search of the terrain above the tree-line for archaeological structures (for procedures and documentation see Ebersbach and Gutscher 2008). The primary aim was to draw up an inventory of archaeological sites, so as to be able to react promptly to planning requests for new building. The high number of almost 400 hitherto unknown individual archaeological features discovered was surprising and encouraging. An evaluation project by the Archaeological Service now followed, to study these structures in more detail.[3]

The surveyed area comprised the two steep valleys of the Gadmen Valley and the Gen Valley, and the Hasliberg basin. This region is crossed by numerous routes across alpine passes and this favourable geographical situation meant that it could participate in the important trade in cattle and cheese with areas to the south. At least since the Middle Ages, *Alpwirtschaft* has been an important part of the local economy, and it is still practised today. At the same time, the region has developed into a popular tourist area and also plays an important role in the generation of hydro-electric power (Fig. 10.1). The result is a cultural landscape in which various different economic activities can be seen operating side by side, each leaving its mark to a greater or lesser extent.

Most of the documented archaeological features are traces of ground-plans, rock shelter constructions, pens and pasture walls, which would have been associated with the *Alpwirtschaft* of the Middle Ages or modern times. Sections of mountain paths, iron ore mines and other industrial facilities were also recorded. The building remains were associated with *Sennerei* or the alpine dairy economy. The structures had served as living quarters, facilities for processing milk, cellars for cooling the milk before skimming off the cream and stores for maturing cheese. Cream, butter and various types of cheese were among the most important products from the alpine pastures. However, due to poor preservation, hardly any archaeological finds can be associated specifically with milk processing. Hard cheese, with its high energy content, was mainly exported, because, due to its consistency, it could withstand being transported across the alpine passes to northern Italy by mule-track.

As the structures were not excavated, the statements which can be made regarding dating and interpretation are limited. Written and pictorial sources, as well as comparisons with existing alpine buildings and the results of archaeological excavations from other regions have therefore been used to assist the study. After an introduction to the area surveyed, the article will discuss the prerequisites and results of the project. Taking the archaeological features as the starting point, it will then review selected types of building, their design, location, construction method and historical context.

The Oberhasli Region

Natural environment and settlement

The Oberhasli region is located at the eastern end of the Bernese Oberland, on the northern slope of the Aare-Gotthard Massif, and encompasses the catchment area of the River Aare from its source in the Grimsel area to the Aare plain, not far from where it flows into Lake Brienz (Fig. 10.2). Apart from the area of the Aare plain in the west, the boundaries of the district, which covers approximately 1229km^2, for the most part follow the natural watersheds. To the north, the Limestone Alps border on the pre-alpine region, while to the south, the gneiss and granite rocks of the Grimsel area give it the appearance of a high-alpine terrain (Veit 2002:21–23). Altitudes range from 600m above sea-level in Meiringen to over 3000m. The Oberhasli climate, thanks to its position on the northern edge of the Alps, is damp and temperate. Average annual precipitation is between 1200 and 2000mm (1971–1990) and is distributed throughout year, with a slight maximum at the height of summer. In winter, the dry, warm air of the *Foehn* can bring pleasant temperatures and those areas which are exposed to the sun quickly become snow-free in spring (Veit 2002:51; Bätzing 2003:34–35, 37). The valleys in Oberhasli are extremely exposed to

---

3. The PhD project at the University of Zurich (May 2010–December 2013) is supervised by PD Dr. Adriano Boschetti and funded by the Archaeological Service of Canton Berne.

Figure 10.1   View looking towards the Grimsel Pass. The landscape is punctuated by dams and electricity pylons (photo: Brigitte Andres).

natural events. As well as avalanches in winter, heavy rainfall in the warmer six months of the year can lead to mudslides. The permanent settlements, which, with the exception of the villages in the Hasliberg basin, are all situated in the valleys, were therefore very carefully chosen, protected from avalanches, mudslides and flooding.

The three areas which were studied – the Gadmen Valley, the Gen Valley and the Hasliberg Basin – have been shaped by rivers and glaciers. The Gadmen Valley runs eastwards from Innertkirchen to where the road over the Susten Pass begins its ascent (Fig. 10.3). Most of the settlements in the valley bottom (850–1250 m.a.s.l.) are small hamlets. The highest of these, Gadmen and Obermad, lie at a good 1200 m.a.s.l. To the northeast, after a rise of almost 400m in altitude, the Wenden Valley continues to the Wenden Glacier, which borders on Mount Titlis to the south. On the southern side of the Gadmen Valley, the valley of the River Triftwasser is the only significant feature to make any inroad into the mountainside. Proceeding up this valley, the Wildbach brings you to the Trift Glacier. The Gen Valley begins northeast of the village of Innertkirchen (625 m.a.s.l.) and leads over the Engstlen Alp, with its idyllic mountain lake, to the Joch Pass, crossing over into Canton Nidwalden (Fig. 10.4). The lower slopes of the Gen Valley, like those of the Gadmen Valley, lie at about 1200 m.a.s.l., but unlike the Gadmen Valley, there are no permanent settlements. The earliest mention of the valley dates from 1323 and from this we know that the valley bottom was used, at least from the 14th century onwards, as summer pasture (Zybach 2008:25). Rocky crags divide the valley walls into several terraces.

The permanently inhabited villages, hamlets and farms in the Hasliberg Basin lie at between 1000 and 1300 m.a.s.l. on a sunny terrace above Meiringen. The alpine pastures lie in a wide natural arena cut through by torrents (Fig. 10.5).

Transport topography

Like every region, the Oberhasli area was shaped by its location, its approach routes and its accessibility. In this case, it is the mountain passes which are of particular importance. They link Oberhasli to all points of the compass and with other parts of Switzerland, so that, although remote from urban centres, the valleys are not cut off from the world. The only route to the Hasli Valley which does not cross a mountain pass leads through the Aare Valley from Brienz to Meiringen. This route links the area with Interlaken, Thun, Berne and the Swiss Plateau, as well as other parts of the Bernese Oberland (Fig. 10.6).

To the north, the Brünig Pass provides a link to Obwalden and Lucerne, along Lake Lucerne to Central Switzerland and the Swiss Plateau. At only 1000 m.a.s.l. it is a relatively easy pass to cross (von Rütte 1992). Parallel to the Brünig Pass, the Joch Pass offers another route to Central Switzerland (von Rütte 1990b).

To the south, the Grimsel Pass and the Gries Pass together formed an alpine transit route from Oberhasli to Domodossola in the Eschen Valley/Val d'Ossola in Italy. Together with the Brünig Pass, the Grimsel-Gries route can also be seen as part of the transport network of Central Switzerland, joining western Central Switzerland with northern Italy. As the route involved crossing two or three passes, it never acquired the economic importance of the more direct routes over the Great St Bernard or the Gotthard Passes. With the opening of the Gotthard Railway in 1882, the demise of pack-animal transport was inevitable and the Grimsel-Gries route lost its importance (von Rütte 1990c).

Within the Bernese Oberland, the route over the Grosse Scheidegg Pass linked Meiringen with Grindelwald. As a mule-track it had great regional importance, especially for the inhabitants of Grindelwald, who not only supplied the markets in Thun and Berne with livestock and cheese but also used this route to channel their wares abroad over the Grimsel Pass. The Grosse Scheidegg Pass was also part of the Oberland Tour, and with the rise of tourism in the 19th century, it became an essential part of any tour of the Alps, not to be missed by visitors to Switzerland (von Rütte 1994).

The only east–west communication route in the Oberhasli region is via the Susten Pass, the stretch from Innertkirchen to Wassen in Canton Uri linking the Aare River with the Reuss Valley, and at the same time linking the great transit routes over the Grimsel and Gotthard Passes with each other. Its historic significance lies in the history of 19th and 20th century road construction rather than with economic activity. It only became important as a trade-route in exceptional circumstances, when the Grimsel route was blocked (von Rütte 1990a).

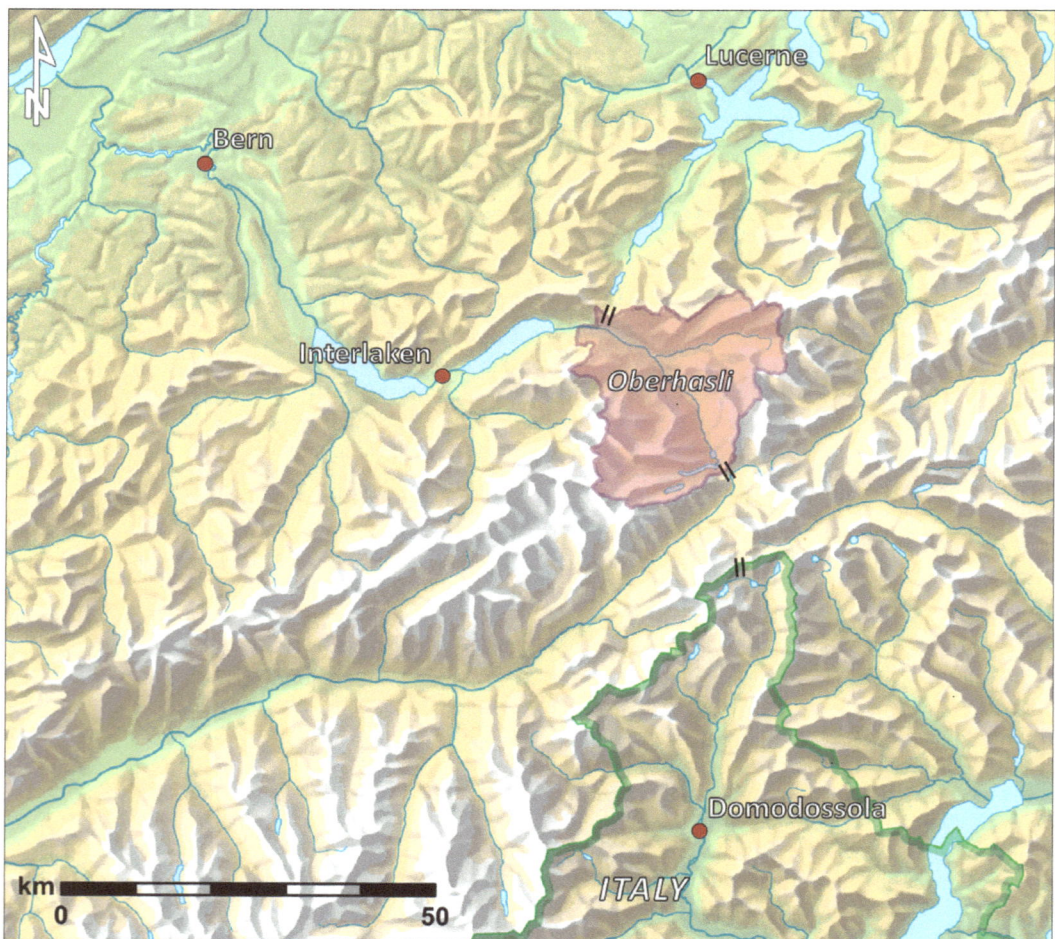

Figure 10.2   The Oberhasli region, situated at the eastern end of the Bernese Oberland (map: Swisstopo, GeoDB Bern, Brigitte Andres).

Brigitte Andres: Alpine huts, livestock and cheese in the Oberhasli region (Switzerland)

Figure 10.3 View of the Upper Gadmental Valley, looking towards the northeast, with the village of Gadmen and Mount Titlis in the centre of the picture (photo: Brigitte Andres).

Figure 10.4 View of the Gen Valley, looking northeast in the direction of the Engstlenalp. The pastures can be seen on the valley floor and on the mountain sides (photo: Andri Spinas).

Summer Farms: Seasonal exploitation of the uplands from prehistory to the present

Figure 10.5　Permanent settlements above Meiringen on the Hasliberg Mountain with the alpine pastures in the background. View to the northeast (photo: Brigitte Andres).

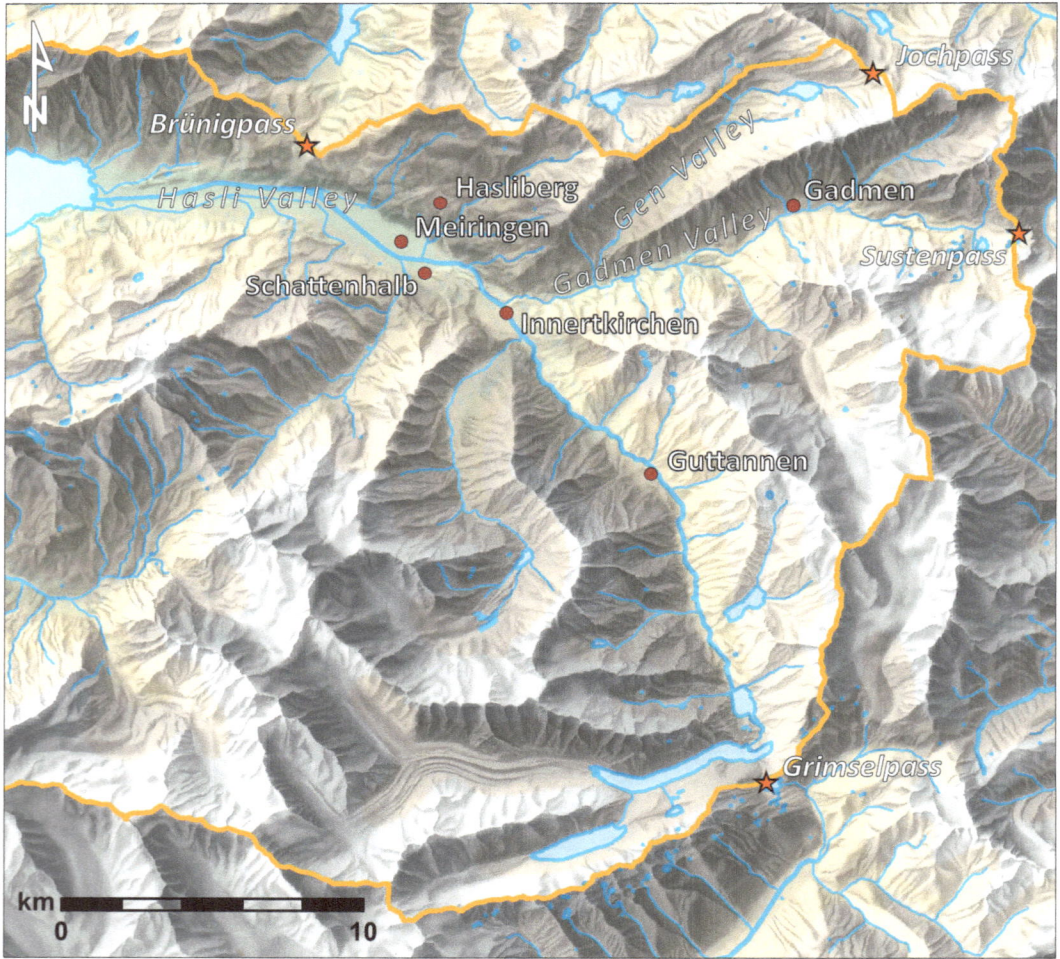

Figure 10.6　Mountain passes link the Oberhasli region in every direction with other areas (map: Swisstopo, Brigitte Andres).

Within the overall framework of the politics of transport in the Middle Ages and the modern era, the Oberhasli region was insignificant. The most important Swiss alpine transport routes were still, as they had been since Roman times, over the Great St Bernard and the Grisons Passes (Bitz 2003:9–10; Roth-Bianchi 2007:8–18). Even the importance of the transport of goods across the Gotthard Pass is often over-estimated for the period before the opening of the railway tunnel in 1882 (Loepfe 2007:12–14). For the population of the northern alpine valleys, from the late Middle Ages onwards, the various local and regional routes across the Alps were important above all for the direct export of livestock and cheese to the markets in northern Italy.

Evidence of prehistoric land use

To date there has been little evidence for prehistoric settlement of the Oberhasli region and known sites have mostly yielded only isolated finds.[4] A Bronze Age flanged axe of the Langquaid type was discovered on the Grindel Alp in the Reichenbach Valley and a Middle Bronze Age palstave and a Roman votive axe were both found in a rock crevice in the area of the Grimsel Pass (Schaer and Martin-Kilcher 2009:264–265). A Neolithic flint arrowhead was also found further afield, on the Brienzer Rothorn mountain (Hafner 2010:82–83). These finds prove that the alpine regions were used at that period. Whether this was permanent or sporadic is unknown.

Further evidence comes from analyses of pollen and macrobotanical remains taken from the sediments of three mountain lakes between Lake Brienz and Grindelwald (Lotter *et al.* 2006; Wick *et al.* 2003; Heiri *et al.* 2003). There is evidence of intrusions on the landscape from as early as the Neolithic and Bronze Age periods. For the moment it remains unclear whether forest clearance at this early period was for the purpose of keeping livestock, or whether the main reason was rather the exploitation of wood and other resources. A significant increase in grass pollen indicates a huge opening-up of the landscape from about AD 800, indicating that areas of forest were being cleared. Natural alpine meadows were able to spread and were kept clear by constant grazing.

Development of the alpine pastures in the Middle Ages

Monasteries often played a crucial role in the development of the alpine pastures, as they pursued economic interests on their extensive land-holdings. In Switzerland, one of the earliest sources for the organisation of alpine pastures, and the laws and taxes to which they were subject, is the Acta Murensia of Muri Abbey, from the period shortly after 1140 or, in fact, a copy from the end of the 14th century (Sablonier 1990; Glauser 1988; Sieber and Bretscher-Gisiger 2012).

4. Archive of the Archaeological Service of Canton Bern.

The earliest written references to alpine pastures and settlements in the Oberhasli region date back to the 14th century. An intensification of the pastoral economy is evident, reflected, amongst other things, in verdicts on boundary disputes and, increasingly, in written regulation of alpine pastures. The regulations governed such matters as the driving of livestock to and from the pastures, the number of animals involved, the clearing of the pastureland, and the winter feed. Alongside these regulations are efforts in order to guard pasturing rights against newcomers and cattle owners from elsewhere and to prevent further alienation of property to them. It may be that the mountains and pastures were originally opened up by nomadic pastoralists, legal pasturing rights only being established when economic use became more intensive, but this is almost impossible to prove (Brülisauer 1984; Sablonier 1990:51–52, 83–85).

Traditional Alpwirtschaft as a multi-level system

In the alpine regions of Switzerland agriculture developed as a multi-level system, in which areas in the valleys, in the mountains and on the alpine pastures were used for different purposes at different times of the year (Stebler 1903:3–5; Weiss 1992:25–29). In traditional *Alpwirtschaft*, the main economic activities were animal husbandry and the manufacture of milk products during the summer season spent on the alpine pastures. In the Oberhasli region, as in Switzerland in general, *Alpwirtschaft* was practised as a three-level system, based on the exploitation of the vegetation at different altitudes (Fig. 10.7).

The first level was in the valley on the home farm. In order to protect the valley meadows and build up hay supplies for the winter, the animals were driven to higher pastures for three to four months in the summer. A series of ascending pastures was used from early summer onwards and then grazed again in descending order in the autumn. A distinction was made between these intermediate pasture levels, used in early summer and autumn, and the high alpine pastures grazed at the height of summer. Within each individual pasture level, there were often several areas, so-called *Stafel*, which were grazed in a particular order (Sprüngli 1760:875). Only essential possessions were transported up to the summer quarters in the high pastures and were brought back to the valley for the winter.

Alpwirtschaft is still part of local agriculture in the Oberhasli region today. Various developments in the alpine pastureland over the past 50 years, however, have resulted in many of the intermediate pastures being abandoned. Many of the *Stafel* at high altitudes can now also be reached by vehicle, so they no longer all have their own infrastructure for milk processing. Often the milk is transported to the dairy in the main pasture area and buildings which were once used elsewhere have been turned into holiday homes or demolished.

Figure 10.7 The three most important levels of the agricultural system exemplified in the Gadmen and Wenden Valleys (photo/drawing: Brigitte Andres).

Export of cheese and livestock

Up until the end of the 19th century, the multi-level system of agriculture including the use of mountain pastures in the summer was the main form of economic activity in the Oberhasli region. Large animal husbandry increased in general from the 14th century in the Alps, and depending on the area, markets for cattle and milk products were found not only in the Swiss Plateau, but also in France, southern Germany and northern Italy. After the bridging of the Schöllenen Gorge around 1200, which opened up the Gotthard route, a growing demand from the cities of Lombardy for livestock and milk products led, especially in the north-alpine valleys, to a move away from an agriculture based on self-sufficiency, with small-animal husbandry and arable farming, to an economy based on livestock and oriented towards trade (Sauerländer 2006; Sablonier 1990; Glauser 1988:158). In cheese production, a transition was made in western central Switzerland and the eastern Bernese Oberland to full-fat, hard cheeses, so-called Sbrinz or Spalen cheeses, which kept well and were firm enough to be exported over the alpine passes. The most important pack-animal route led into the Oberhasli region via the Brünig Pass, from Obwalden to Meiringen, and then on over the Grimsel and Gries Passes to Domodossola (see Fig. 10.2; Roth 1993:1–6; Küchler 2003).

The agricultural changes which had started in the 18th century were already under way when, in 1814, the first village cheese dairy was opened in Canton Bern. Alpine cheese-making, which was only carried on in the summer, quickly declined as the 19th century progressed, and many *Sennen*, the alpine dairy-producers, had to look for other employment (Roth 1977). Although the cheese dairies in the valleys brought about an increase in cheese exports from Canton Bern, the trading route through the Oberhasli region collapsed. Transport by rail replaced the traditional pack-animal convoys, the final straw being the opening of the Gotthard Railway in 1882 (Mösching and von Rütte 1990:6; Küchler 2003:50). At the same time, tourism in the 19th century and hydroelectric power generation from 1900 onwards brought new economic opportunities, which were to leave their mark on the landscape (Zybach 2008; Andres 2012a:282).

Alpine surveys

Procedure and documentation

The catalyst for the survey project by the Archaeological Service of Canton Bern was a planned extension of the 'Hasliberg-Frutt-Titlis' winter sports area (Ebersbach and Gutscher 2008). At the time only a few sites in the Oberhasli region were listed in the Archaeological Service's inventory. These were either isolated finds from the valleys and mountain area or else features such as castles and churches. The primary aim of the surveys was therefore to create an inventory of sites for the alpine pastures which might be affected by the building of new ski-lifts, in other words, the Hasliberg area, together with the Gen Valley and the area around the Engstlen Lake, towards the Joch Pass (Fig. 10.8). Surveys were also carried out in the Gadmen Valley and Wenden Valley and around the Trift Glacier. Comprehensive surveys of all the territory above the tree-line were carried out, and one or two features from lower-lying areas were also included. The main focus was on structures which were still visible, such as masonry remains, and on evidence of medieval *Alpwirtschaft*. Rock shelters and knolls were also examined for traces of (prehistoric) use.

In preparation for the surveys, historical and topographical maps were checked for symbols denoting ruins and for place names, and information was collected from people with local knowledge. The surveys took place in the years 2003, 2004 and 2006, occupying three weeks during August of each year. Groups of two or three cleared vegetation from the structures they discovered in the field and documented each of them with a scale drawing, descriptions and photographs. In 2008, small exploratory excavations were carried out at selected sites where ground-plans had been discovered, in order to recover material for radiocarbon dating. Because the areas excavated were relatively small (around 50×50cm), it was not possible, however, to make any unambiguous identifications of hearths. As photographic documentation had often been limited by impassable terrain littered with scree or boulders, a quadrocopter was used in 2012 to take aerial photographs of some of the larger deserted settlements (Andres and Walser 2013).

For the purposes of the study, separate features, individually recorded and documented as described above, were treated as single units.

Categories of features

As already mentioned, the focus was on features which were still visible in the landscape. Most of the almost 400 newly-documented structures are therefore features with masonry remains (Fig. 10.9). Of these, those categorised as *ground-plans*, *rock shelter constructions*, and *possible ground-plans* are the most frequent. In fact, 200 structures were defined as *ground-plans*, making up over half of all the features. The emphasis on recording evidence of alpine pastoral economy is reflected in the categories *boundary/pasture walls*, *clearance cairn* and *animal pen* or *possible animal pen*. Sections of *constructed path*, which were recorded in various locations, remains of *dry stone walls* which cannot be more closely identified, *iron ore mines*, and other constructions such as old industrial facilities or avalanche wedges, grouped together as *others*, bear witness to the multiple uses of the area. Also recorded in the inventory were potential occupation sites under

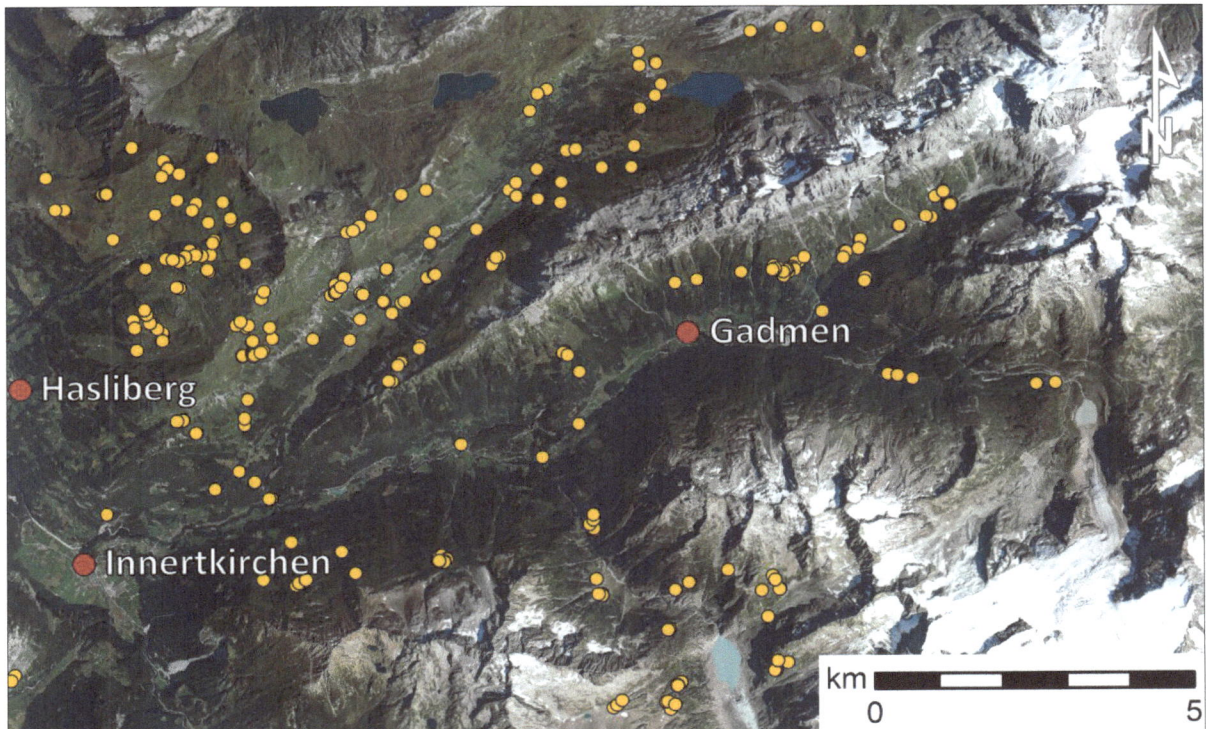

Figure 10.8 Distribution of the features recorded during the surveys (map: GeoDB Bern, Brigitte Andres).

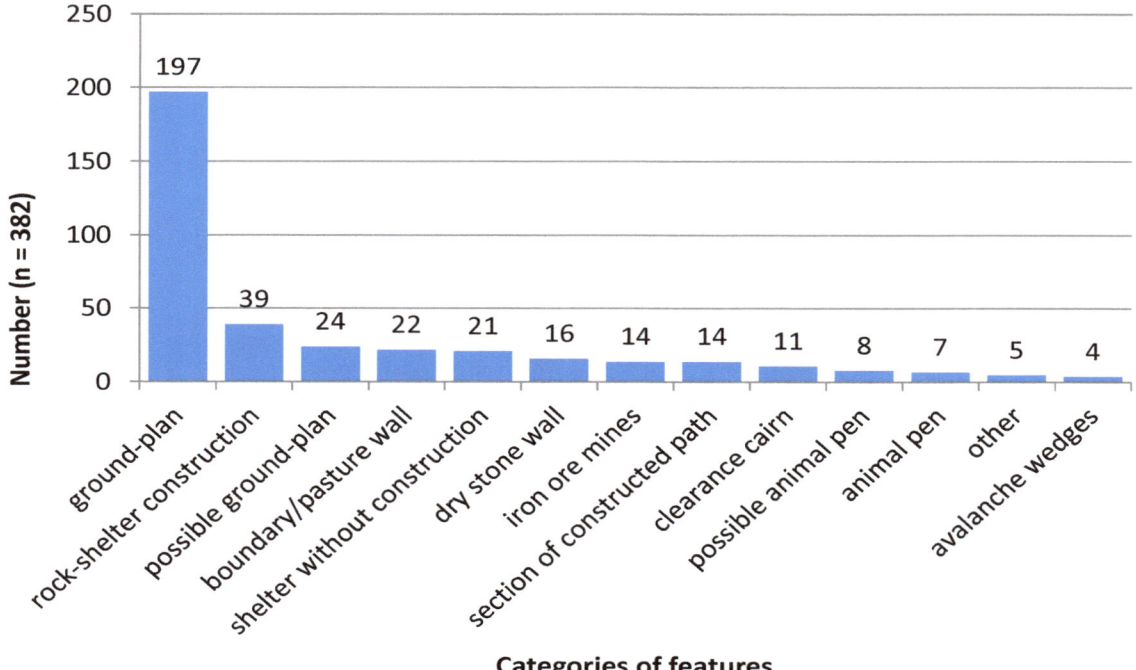

Figure 10.9 Overview of the categories of features. Half of all the features belong to the category of ground-plans (diagram: Brigitte Andres).

rock projections (*shelter without construction*), which did not yield any unambiguous evidence of human use. Occasionally the surveys also included descriptions of older alpine huts still standing or in a state of collapse.

The surveys in the Gen Valley and Gadmen Valley and in the Hasliberg Basin were only systematically carried out in the alpine pasture areas above the tree-line, which lie mostly between 1600 and 1700 m.a.s.l. Most of the features are therefore situated at an altitude of between 1600 and 2000 m.a.s.l. (Fig. 10.10; see also Auf der Maur 1998:318). Nevertheless, it was also possible to record numerous new sites at lower altitudes, especially in the Gen Valley, where the mountain pastures begin as low as 1200 m.a.s.l. At the same time, iron ore mining facilities in the Gen Valley were among the highest-altitude sites found, at between 2200 and 2400 m.a.s.l.

Analysis

Because these were 'only' the results of surveys, rather than excavation findings, there were clear limitations, from a purely archaeological point of view, to their analysis, which was carried out as part of a PhD project. The advantage of comprehensive prospection of a terrain is the large number of features of different categories. Ground-plans and rock shelter constructions were selected for particular analysis, because of their systematic recording. Thanks to their thorough documentation, with scale- or measured- drawings of individual building structures, it is possible to compare features within the region, as one would do with a find assemblage. This makes it easier to see similarities and differences between individual settlement areas, or whether, for example, a particular type of feature is missing or under-represented.

The disadvantage of the absence of excavation results is clear when it comes to interpreting features in terms of function and date. Other sources and methods must be used for dating. It is also impossible to identify multi-stage use or predecessor buildings, let alone the existence of timber structures. It therefore makes sense to broaden the focus and look at the regional history of *Alpwirtschaft* as a whole. Since the alpine pastures which were surveyed are still in use today, it is possible to compare the archaeological remains of past centuries with more recent building types. Together with results from studies of farm houses and excavation results from other regions, written and pictorial sources and oral information from local people with traditional knowledge were also included in the study. The present article will make a detailed presentation of the two selected categories, *ground-plans* and *rock shelter* constructions. Most of the building remains recorded in the surveys were on the high pastures and so buildings in the valleys and on the intermediate pastures will not be discussed here. The other structures recorded, such as animal pens, pasture walls, sections of paths, iron ore mines and other economic facilities, will also not be analysed further here.

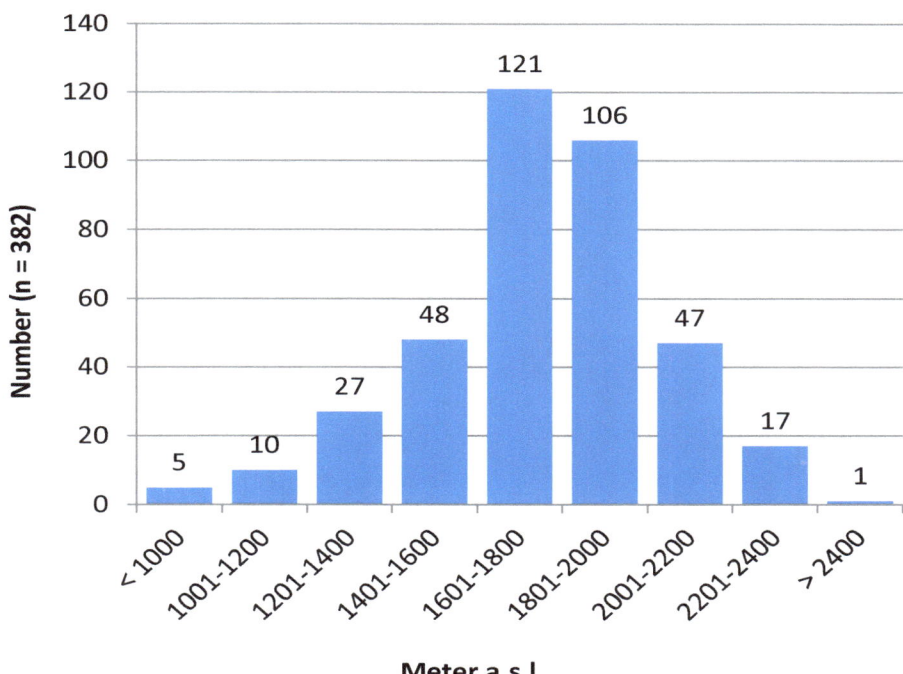

Figure 10.10 Overview of altitude distribution. Most features lie above 1600 m.a.s.l. (diagram: Brigitte Andres).

Archaeologically recorded alpine settlement remains

Ground-plans

Situation

The choice of location for buildings in alpine terrain is very strongly influenced by considerations of available shelter. People have always tried to protect buildings from damage caused by avalanches, mudslides and rockfalls. Although present-day machinery and equipment makes it possible to move larger amounts of earth and stone to build artificial avalanche wedges, the strategies used today to match buildings perfectly to their terrain are still the same as those which can be seen in the deserted settlements. The building remains in the surveyed area are mostly free-standing, though some are in the shelter of a rocky outcrop or ridge. Other buildings were constructed directly against a rock wall or boulder, thus making the best of use of the natural terrain by not only ensuring shelter, but also saving on the building of one wall. Ideally, an overhanging rock-face might also offer protection from the weather by creating a natural roof. If boulders of suitable size are to hand, they are still popular as sites for building today (Fig. 10.11).

Since sites were seldom level, the rear part of the building would often be dug into the slope and the material removed would sometimes be used to make a small terrace at the front. In the case of larger buildings or buildings on steeper sites, there would also be a terrace wall.

Popular sites for alpine settlements in the surveyed area, as in alpine areas generally, are screes and natural terraces. Buildings are easier to construct on level land than on steep slopes. A terrace will provide enough room for several buildings. Depending on the space available, and the prevailing hazards, the buildings will either be well spaced out and incorporated into the terrain, or clustered close together (Fig. 10.12). Many of the new constructions on natural alpine terraces would have been built on the sites of older buildings.

Although screes often occur on steeper terrain, they offer the advantage of supplying sufficient building material on the spot, which would otherwise have to be transported from a distance. It is also easy to find larger boulders and lumps of rock to incorporate into buildings. The distribution of building structures within scree fields is variable in the Oberhasli region. In some deserted settlements the buildings are close together, sometimes even built on to one another, while at Zum See in Innertkirchen, they are scattered around a mountain lake (Fig. 10.13). It is no coincidence that some settlement sites were close to a stream. Despite being heightened, they were close to running water, which was required both for watering the animals and for cheese production.

Surface area and shape

In 171 of the 197 ground-plans surveyed, the inside floor areas could be measured, and vary from 1 to 92m² (Fig. 10.14). Over half (53%) have an inside area of less than 15m² while 75% have an inside area of

Figure 10.11　An alpine building constructed in the shelter of a huge rock in Zum See, Innertkirchen (photo: Brigitte Andres).

Figure 10.12　The buildings of the Baumgarten Alp lie on a terrace on the steep northern slope of the Gental Valley (photo: Brigitte Andres).

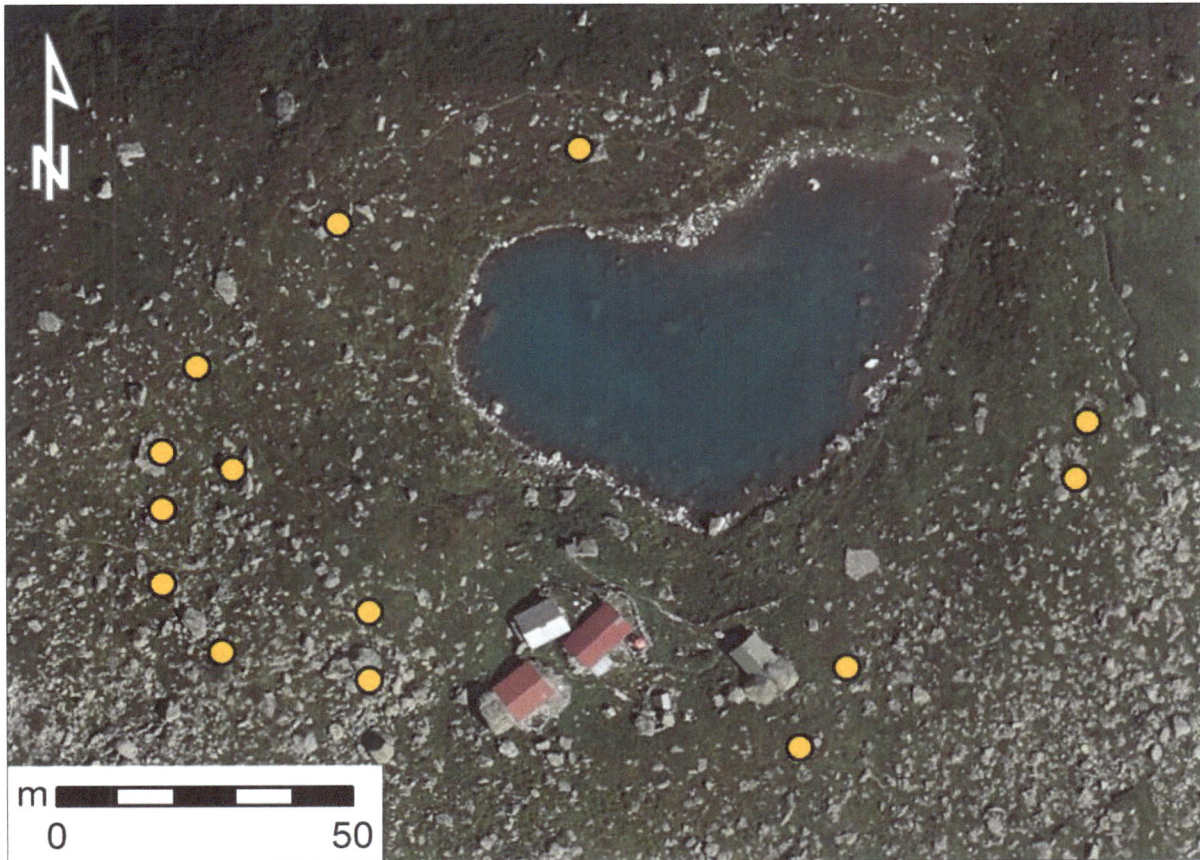

Figure 10.13 The isolated features of Zum See, Innertkirchen, are distributed amongst the scree around the mountain lake (map: GeoDB Bern, Brigitte Andres).

less than 30m². Most of the smaller ground-plans are (approximately) square, or sometimes have a slightly distorted, trapezoid shape. A rectangular shape is less common. Against a boulder or rock wall it is slightly less usual to find ground-plans with an area of more than 15m². Sometimes the inside areas are so small that one wonders what purpose they could have served.

The larger the ground-plans become, the more the (approximately) square shape gives way to a rectangular shape. While small ground-plans mostly consist of only one room, there is more evidence of different types of room division as the inner area increases above 15m². At 10m² and above, inside structures like benches/shelves or hearths begin to appear (Fig. 10.15), while outside, yards and small terraces or lean-to structures like dung heaps become more frequent.

With increased dimensions, there are also more likely to be clues as to the use and function of the buildings. If the part of the building was once used for milk processing, the position of the hearth could sometimes be established, as it was often in an alcove in the back wall. Otherwise, because the ruins were not excavated, it was usually impossible to locate the hearth or other installations. Interpretations as to function often have to be made using analogies with extant alpine buildings, even though most of these only date back to the middle of the 19th century.

Base and superstructure

Almost without exception, the ground-plans consist of dry-stone masonry. The wall remains comprise a variety of building materials. The spectrum ranges from very small stones to large slabs, easy to stack on top of each other. The propensity to build into the slope and therefore straight on to the natural soil lent the walls additional stability. Large blocks of stone, found on-site, were often incorporated directly into the structure (Fig. 10.16). As they are often in an advanced state of collapse, it is so far still an open question as to how high the dry-stone walls originally reached and what the combination of stone- and timber-building techniques looked like. A timber superstructure on top of a dry-stone base can still be seen today in many buildings in the Oberhasli region (Fig. 10.17). The masonry base protects the timber walls from rising damp and helps to level out the uneven terrain. It is also possible that the walls were stone right up to the roof (Kehrli 2008:20, fig. above and 42, fig. below; Stebler 1903:333, fig. 243). A further possibility is that only the walls around

Summer Farms: Seasonal exploitation of the uplands from prehistory to the present

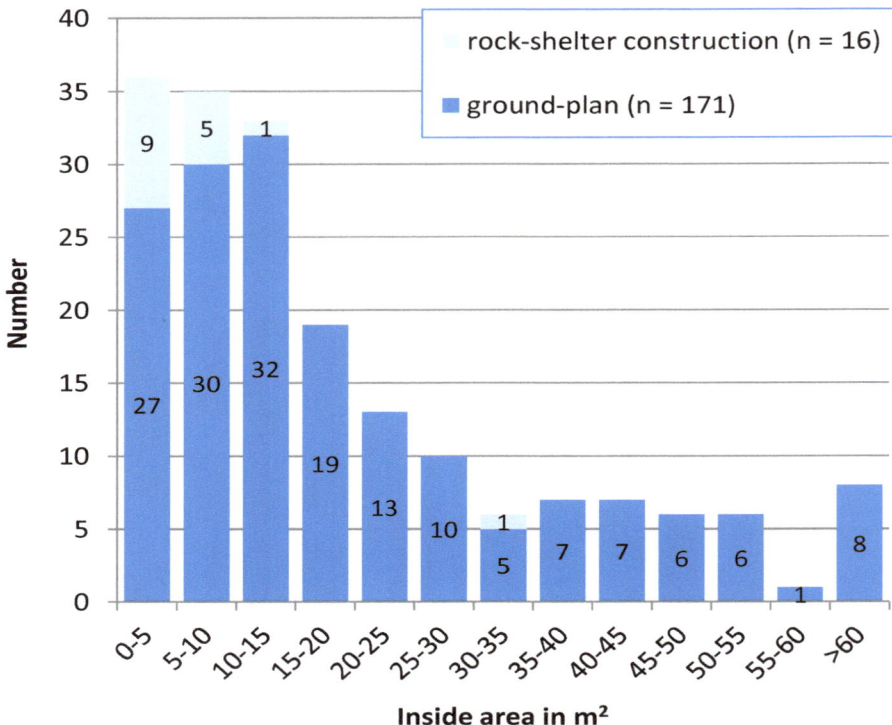

Figure 10.14 Overview of the measurable inside areas. More than half of the ground-plans measure less than 30m² (diagram: Brigitte Andres).

the hearth or the side of the building exposed to the weather were built in stone.

It is almost impossible to draw any conclusions about roof construction from the documented features. From existing alpine buildings it can be seen that freestanding buildings have a saddle roof, while buildings constructed against a rock-face have a single-pitch roof (Fig. 10.18). The situation must have been similar in the deserted settlements (see also Obrecht 1998:118; Meyer 1998a:29). Evidence of a cantilevered domeshaped roof has so far only been noted in one case (Meyer 1998b:58–60; Gschwend 1976:38).[5] The roof could be covered with wooden shingles or, where the local rock type was suitable, flagstones.

Availability must certainly have been an important factor in the choice of building materials. At a period when transport possibilities were limited, timber buildings were probably mainly built in areas close to the forest. From a certain altitude above the treeline, on the other hand, only stone would have been available. One must be cautious about drawing conclusions about the superstructure from the amount of fallen masonry, however, because suitable stones may have been recycled for later buildings.

Rock shelter structures

Rock shelter structures are structures where a rock or a rock-face has been incorporated into the structure and forms at least part of the roof (Fig. 10.19). Elsewhere the terms '*Abri*', '*Balm*', '*Block*' (where rocks are deliberately placed) and, mainly in southern Switzerland, '*Splüi*' are used (Sauter 2009:82; Auf der Maur 1998:318; Meyer 1998d:370; for the term '*Splüi*' see Zappa 2008:40 and Zappa 2005:69–73). The term used here, 'rock shelter structure' includes both rock and rock-face situations. A naturally occurring space was adapted with purposeful interventions and alterations to make it optimally suited to particular requirements, resulting in structures with irregular ground-plans and very varied physical form. The spectrum ranges from a couple of stones stacked on top of each other, to simple dry-stone walls, to elaborate installations (Fig. 10.20). To achieve sufficient head-room under a rock ceiling, it was often necessary to dig soil out or to level off the floor.

Out of the 39 features identified in the study area as rock shelter structures, 14 are built against a rockface and 25 against a rock. Most of them are smaller features with only a few square metres of useable space – often because of the low ceiling under the rock. Rock shelter structures occur at all altitudes, but are most frequently found between 1800 and 2000 m.a.s.l. As well as isolated structures, some also occur within a larger deserted settlement, along with ground-plans and animal pens (Fig. 10.21; see Andres 2011).

Because of the topographical peculiarities of different settlement areas, rock shelter structures form

---

5. Gadmen, Triftttälli unterhalb Gläckblatten 341.025.2004.01.

Brigitte Andres: Alpine huts, livestock and cheese in the Oberhasli region (Switzerland)

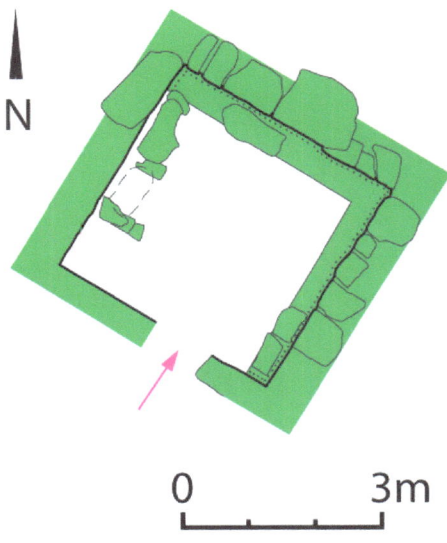

Figure 10.15 Ground-plan in Entlibüöch, Hasliberg, with a stone bench or shelf running along the walls (drawing: Marc Müller).

Figure 10.16 A boulder integrated into the ground-plan at Zum See 1 (photo: Brigitte Andres).

Figure 10.17 Present-day alpine hut with masonry base and timber super-structure in Mägisalp, Hasliberg (photo: Brigitte Andres).

Figure 10.18 The single-pitch roof of this building in Mettlenberg in the Gadmen Valley makes it less susceptible to avalanche damage (photo: Peter Liechti).

a very heterogeneous group. They appear to vary too much in their shape and location to have served any one particular purpose. They were probably used for different things such as rudimentary shelters for people and animals, or cooling cellars for keeping, for example, milk, butter or food provisions (Furrer 1985:411–412). Today the shelters are sometimes used for storage or by sheep seeking shade.

Settlement structure

Alpine settlements can have very different layouts. Some deserted settlement sites consist of the remains of several buildings, some of groups of two or three buildings, and some of single isolated buildings. The arrangement of the buildings, whether clustered together or widely distributed over the terrain, depends very much on the topography.

The deserted settlement of Gadmen, Gries 1 is situated far up the Wenden Valley in a hollow, sheltered from avalanches beneath an outcrop of rock (Fig. 10.22). Four houses stand in a continuous row, built on to each other. Two further buildings of similar size are on the other side of a small, flat, marshy area. The remains of three smaller structures are located nearby. A similar layout is known from the deserted settlement of Bergeten ob Braunwald in central Switzerland, which was excavated in 1971 and dated to the 13th–15th centuries (Geiser 1973; Meyer 1998a). In Bergeten, the continuous row of houses was interpreted as alpine dairies, on account of hearths which were found in them. The smaller buildings scattered close by were probably used as cooling cellars or storage rooms (Geiser 1973:48). Neither settlement had any evidence of animal pens or other enclosing walls. This clustering together of several buildings, furnished with hearths, to form a small alpine village, is an indication of '*Einzelsennerei*' or individual alpine dairy production, meaning that a single unit, for example a family, processes the milk from its own animals into dairy products itself. In the northern Alps, this was usually cow's and goat's milk, while in other areas it might also include sheep's milk.

Other deserted settlements, such as Wendenboden and Wendenläger 1 in Gadmen, are also small alpine

Figure 10.19 The remains of walls built against an overhanging rock-face Hinder Tschuggi 9, Hasliberg (photo: Urs Ryter).

Figure 10.20 A naturally-deposited lump of rock with a man-made masonry entrance and a completely roofed-over chamber in Zum See, Innertkirchen (photo: Brigitte Andres).

Figure 10.21 Purposely enlarged hollow under a naturally-deposited rock in the deserted settlement of Wendenboden in the Gadmen Valley (photo: Nicolas Storck).

villages with multiple structures (Andres 2011). Their location on the edge of screes meant that individual buildings could be built against large blocks of stone (Fig. 10.23).

A quite different situation is seen in deserted villages with large animal enclosure systems. Although these are not found in the study area, they should nevertheless be briefly introduced, using examples from other regions, since they are sometimes used in Swiss settlement archaeology for dating purposes (Meyer 1998d). To the west of the Oberhasli region, in Axalp Chüemad in Brienz, there is an animal enclosure system consisting of several sections. The individual pens are arranged around a drinking trough and a small lean-to structure is built against each compartment (Fig. 10.24). A similar pattern to that in Axalp Chüemad is found in deserted settlements in central Switzerland, such as Elm GL, Ämpächli and Muotatal SZ, Spilblätz, which also have multiple enclosure systems with lean-to structures (Obrecht 1998; Meyer 1998b). These enclosure systems, in which herds were penned in together, suggests small animals being taken to summer pasture, a practise which was widespread in the northern Alps in the High Middle Ages. Since no similar walled enclosure systems have been found so far in the Oberhasli region, the question arises whether there simply were none, or whether they have not survived. As the alpine pastures were often steep and exposed to avalanches, suitable sites were probably hard to find. It is also possible that perishable materials were used. One clue to the former existence of animal pens can be found in place names which retain the concept of an 'enclosure' (*Pferch*, '*Färrich*'). These areas are situated at the altitude of the intermediate pastures, where enclosures, if they once existed, might have been used in the autumn to gather the sheep flock for separating by their respective owners (Weiss 1992:119). Up to now there have been no studies comparing the locations and use of known animal enclosure systems in the Swiss Alps.

Building types associated with Alpwirtschaft and their historical context

Because the ground-plans were not excavated, the masonry features have been supplemented by other sources in the study. As well as written sources, folkloristic research and pictorial sources can also provide evidence regarding the types of building used.

Alpwirtschaft reflected in written sources

In the search for clues for a rough dating of the ground-plans, a review was made of edited and published sources from 1300 to 1900. These sources are mainly legal records, travel journals, topographical descriptions and statistics about the alpine pastures. While very many different aspects of *Alpwirtschaft* are mentioned in these sources, specific descriptions of alpine buildings are rare. The legal records from the Oberhasli

Figure 10.22   The deserted settlement Gries 1 in the Gadmen Valley with a row of structures built on to one another, and some individual buildings (photo: Peter Liechti).

Figure 10.23   The deserted settlement of Wendenläger 1 in the Gadmen Valley with several individual buildings and, beside the largest boulder, what is thought to have been an animal pen (photo: Christoph Walser).

Figure 10.24 The animal enclosure system in the deserted settlement of Axalp Chüemad with the ground-plans of lean-to structures (drawing: Eliane Schranz).

region from 1300 until 1800 contain firstly written references to alpine pastureland, regulations relating to the annual herding of animals to and from the alpine pastures, legislation regulating livestock ownership, pasture boundaries and communal work (Brülisauer 1984). Alpine huts are occasionally described and sometimes even illustrated in early post-medieval travel journals. Like those of Johann Jakob Scheuchzer (1672–1733), the descriptions are often idealisations which do not do justice to regional differences in building practice (Furrer 1994).

In the 18th century there is a sharp increase in written sources. The growing interest in the Alps led on the one hand to study trips by naturalists and appropriate reports about the 'nobility' of the mountain landscape. On the other hand, interest also focussed on the local people, who lived in and in relationship with the Alps. The life of the herdsman was often represented as an idealised, carefree existence in the midst of the unspoilt natural surroundings of the mountains and contrasted with the decadent life in the towns and cities (Gerber-Visser 2012:236–242; Bircher 1979). In the second half of the 18th century, the Economic Society of Bern produced, amongst other things, proposals for improving agriculture. In this context, criticisms were made in the so-called topographical descriptions of the insufficient fertilisation of the alpine pastures. From the suggestions made, it can be indirectly concluded that cow-sheds were not widespread, since they could be used to collect cattle dung so that it could then be spread on the pastures where it was most needed (Sprüngli 1760:875–876).

There are also numerous 19th century travel journals written by travellers in the Alps, who now undertook the popular Oberland Tour from Interlaken to Lauterbrunnen, Grindelwald and over the Great Scheidegg Pass to Meiringen, and who, along with awed descriptions of the mountain landscape, also recorded observations of the lives of mountain dwellers (e.g. Wyss 1816; Kasthofer 1822; von Weissenfluh der Ältere 1910; von Weissenfluh der Jüngere 1910). There is also a noticeable increase in criticism of man's treatment of the natural surroundings. The deterioration of the climate caused by the Little Ice Age had its effect on the agriculture of the Swiss Plateau as elsewhere. The blame for flooding was set at the door of the inhabitants of the Alps, who were held to have ruthlessly overexploited natural resources and thereby played a part in these catastrophes (Gerber-Visser and Stuber 2009:77).

The descriptions of alpine buildings were not as detailed as had been hoped. First written references or descriptions of alpine regulations in the legal sources are useful, in that they offer evidence pertaining to the types of husbandry being followed. The numerous early post-medieval travel descriptions, although they seldom describe activities relating to *Alpwirtschaft* nevertheless provide information about the state of *Alpwirtschaft* as well as the travellers' view of alpine life. The change in their point of view from the 18th to the 19th century reflects both economic progress, in that missed economic opportunities were noticed, and the anxieties that the deteriorating climate produced.

Development of building types

One-roomed buildings

Present-day research sees a development from the small, irregular, trapezoidal or square single-roomed buildings to larger, rectangular buildings. This means that the small, irregular ruins, often found at the edges of screes, are taken to be of an earlier date. These buildings served as shelters and in the northern Alps, as facilities for producing *Fettziger*, a type of sour milk cheese made from cow's or goat's milk. After the transition to rennet cheese, greater quantities of cow's milk were available and the smaller buildings were replaced by a more suitable infrastructure of larger buildings (Bitterli-Waldvogel 1998:410–412). Excavations and the study of farmhouses have shown an increase in multi-roomed buildings from the 16th century in several regions such as central Switzerland, the Valais, the Grisons and the western Bernese Oberland (Bitterli-Waldvogel 1998:397). No such early post-medieval development leading to multi-roomed structures could be deduced from the features in the Oberhasli region. Along with land tenure, methods of husbandry and topography, the type of cheese production can also have an influence on the infrastructure and types of buildings.

In three-roomed alpine buildings, the rear half of the building was divided into a space for cooling the milk and a space for storing cheese. For butter production, the milk was put to cool, until the cream rose to the surface and could be skimmed off. The front half of the house served as a working and living space, where the milk was processed into half-fat cheese and the cream already collected was made into butter (Weiss 1992:101–102).

In the eastern Bernese Oberland, on the other hand, the production of full-fat cheese was widespread. Because it was firm and kept well, it was suited to export to northern Italy, which was obviously more lucrative than the sale of butter. This resulted in a lack of butter which was often bemoaned from the Late Middle Ages on, as demand increased in the Bernese towns (Bircher 1979:100). From the 17th century onwards there are also reports that in the Bernese Alps the increase in cattle rearing was leading to a lack of butter, because the milk was left for the young animals (Bircher 1979:104–105). The change in emphasis to full-fat cheese could also have contributed to the fact that one-roomed buildings continued to dominate in the Oberhasli region well into modern times, whereas two- or three-roomed buildings became widespread elsewhere.

Johann Rudolf Wyss noticed on his Oberland Tour in 1817 that alpine buildings in the western Bernese Oberland had integrated cooling rooms, which were absent in the eastern Bernese Alps. According to Wyss, the milk was kept in wide, shallow wooden vessels, called '*Gebsen*', stored on milk benches in the alpine

Figure 10.25 This ground-plan at Stäfelti 4 in the Gental Valley was dug into the mountainside and may have been used as a cooling cellar. The inside area measures 1.4×2.5m and the dug-out space is up to 1.5m high at the back. The back wall is formed by existing rock (photo: Nicolas Storck).

Figure 10.26　This double cheese store in Mägisalp, Hasliberg, is raised above the ground on a timber frame (photo: Brigitte Andres).

huts. Also, not every hut had a butter churn (Wyss 1817:556–559). Outside the houses, the most likely structures to have served as cooling rooms were the small constructions under rocks or dug deep into the mountain slopes (Fig. 10.25). These were not necessarily used for cooling milk, however, but during the season on the mountain pastures, could also have been used for keeping provisions.

In the eastern Bernese Oberland, the use of storehouses for keeping alpine cheese is attested to by inscriptions from the 16th century onwards (Affolter, Känel and Egli 1990:171). These buildings, situated close to the alpine dairies, usually have stone paving beneath a timber framework which supports the log construction above, keeping it elevated from the ground (Fig. 10.26). On the one hand, this keeps out vermin, and on the other hand it allows for the circulation of air below the floor of the storehouse and helps to keep the atmosphere dry inside. Although this building type remained almost unaltered over centuries, it is hard to identify archaeologically because of the absence of masonry.

Melkhütten (milking huts) in the eastern Bernese Oberland

Some types of building, such as the *Melkhütten*, can be shown from the sources to have had a long history. The milking hut is a type of building which so far has only been found in the eastern Bernese Oberland (Fig. 10.27). It is a one-roomed log structure with a dry-stone plinth and no windows, in which were located a hearth and the infrastructure for processing milk. Part of the building projected, resting on supports, and served as sleeping quarters. Below it was the so-called *Melkgang*, a sheltered space for milking in bad weather. The roof must originally have been covered with wooden shingles weighted down with stones. On the Axalp near Brienz, to the west of the Oberhasli region, two milking huts were recorded by the Archaeological Service, as it was planned to transfer one of them to the nearby open-air museum at Ballenberg. Dendrochronological analysis of the log-built structures gave dates of AD 1501 and 1519 respectively (Gutscher 2002).

Inscriptions on milking huts giving dates in Grindelwald, southwest of the Oberhasli region, are evidence that this type of structure was still being built in the 18th and 19th centuries (Wetli 2010). Also, milking huts still occur in 19th century illustrations (Fig. 10.28). This seems, therefore, to have been a typical form of alpine building over a period of several centuries. J.R. Wyss left a very detailed written description of the appearance and the interior fittings of a milking hut in his travel journal (Wyss 1817:551–561).

The advent of cow-sheds around 1800

The absence of cow-sheds was already deduced from the written sources. The first cow-sheds were probably erected towards the end of the 18th century, under the influence of the Economic Society of Bern. However, the criticism is still being made in the Alpine Pasture Inspection Report of 1902 that there are far too few cow-sheds on the Oberhasli mountain pastures (Bernisches statistisches Bureau 1902). As

Figure 10.27   Milking hut in Axalp Litschentellti in Brienz, with its typical sheltered milking space (photo: Brigitte Andres).

Figure 10.28   Picture of a milking hut in Gsteigwiler, Breitlaunen, drawn in 1822 by G. Lory jr. (from Roth 1993).

Figure 10.29 Existing alpine building in Hinder Tschuggi, Hasliberg. The alpine dairy in the rear section is recognisable by its chimney. A small animal pen has been built on to the outside of the cattle shed. Over to one side is a small pig sty (photo: Brigitte Andres).

Figure 10.30 Outline of an alpine dairy with a cow-shed and a lean-to dung heap in Gadmen, Mettlenberg (drawing: Marc Müller).

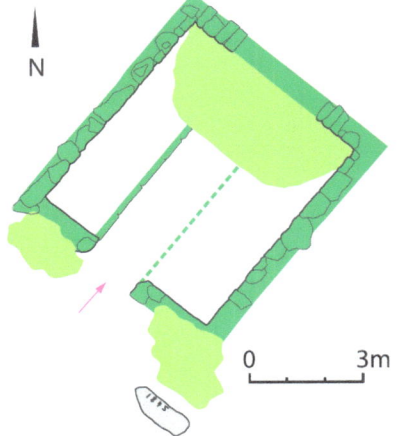

Figure 10.31 Ground-plan of a cow-shed in Hasliberg, Hääggen with a central passageway and cattle stalls to either side (drawing: Marc Müller).

a first stage in their development, a shed was often built on to an existing building. There are examples of buildings where the milking hut can still be identified as the oldest part of an extended structure (Affolter, Känel and Egli 1990:160). Other buildings were newly erected as multi-purpose structures. Today there are still alpine buildings in existence in the Oberhasli region which combine a milk-processing area with an area for housing cattle (Fig. 10.29).

In the archaeological context the larger buildings identified as 'milk-processing facility with cattle shed' (Fig. 10.30) are also easily identifiable by the way the space is divided. The smaller milk-processing area is usually on the side of the building nearest to the mountain. It is recognisable by the hearth, which in the Oberhasli region is often constructed as an alcove in the back wall. The larger cattle shed area is usually accessed from the side in existing buildings, for reasons of topography, and has a central passageway with cow stalls on either side. The small buildings like dung heaps (above ground heaps revetted by dry-stone walls) or small animal sheds are also evidence of the housing of cattle. There are also buildings which were only used as cattle-sheds. These rectangular ground-plans also have a central passage with cow stalls on either side (Fig. 10.31).

Conclusions

Analysis of the ground-plans newly recorded by the surveys on the alpine pastures of the Oberhasli region shows that one-roomed buildings are in the majority. They seem to have been the predominant building type in the region for centuries. Written evidence for the use of the alpine pastures goes back to around 1300. However, it has so far been impossible to date any structures

Figure 10.32 New tourist transport facilities pose a threat to archaeological sites (photo: Brigitte Andres).

with certainty to the Middle Ages. From comparisons with the features examined in archaeological excavations in central Switzerland it seems probable that some of the deserted settlements situated at the edge of screes, with several small and often irregular groundplans, may be of a medieval date. At the same time, there are no remains of the large animal-enclosure systems which were used elsewhere for sheep husbandry.

From existing buildings in the eastern Bernese Oberland, as well as from written and pictorial sources, it has been shown that milking huts, so far known only in this region, were in use from 1500 to the 19th century. With its roofed milking area and the short distance between the cow's udder and the milk-processing facilities, this early alpine building type was already well adapted to an intensified large-animal husbandry and dairy economy. Working, living and sleeping space were accommodated in a single room equipped with a hearth. There must also have been one-roomed alpine buildings without a milking area, especially on alpine pastures where milk processing did not take place.

Without precise dates, it is impossible at present to establish a chronological sequence, especially with regard to the one-roomed structures.[6]

From the 19th century onwards the infrastructure in the Oberhasli region became increasingly focused on tourism. Nowadays the landscape is shaped not only by alpine buildings but also by hotels, cable cars and ski lifts (Fig. 10.32). The largest intrusions into the landscape were made by energy companies with the building of the huge dams in the Grimsel area. Meanwhile the region is criss-crossed by a complex system of tunnels and reservoirs. Archaeological features, as a result, are continually being threatened by new construction, which is why inventories of archaeological sites, even in mountainous areas, are essential, and their creation must be expedited.

Big changes in agriculture and with it in *Alpwirtschaft* have often led to the abandonment of buildings. Whereas in the past they were left to fall to pieces, nowadays they are removed without trace or converted into holiday homes. As late as the 1960s, countless alpine buildings from earlier centuries disappeared, owing to the restructuring of *Alpwirtschaft*, and were replaced by new buildings with modern infrastructure. Unfortunately, no attempt was made over the past 50 years to systematically record the stock of old alpine buildings. In the study of farm houses, the focus was mainly on buildings in the valley and on the intermediate pastures. For research into deserted settlements, therefore, regional studies of the archaeological features are urgently needed. For although Swiss *Alpwirtschaft* displays a similar form in many places, there are typical regional economic traditions and peculiarities which can only be identified through thorough observation of local conditions. Innovations like the transition to rennet cheese-making or the introduction of the use of cow-sheds are almost always subject to influences from outside. Historical and archaeological research can help to define more exactly the historic timeframe of these developments.

Summary

The surveys carried out in 2003, 2004 and 2006 in three valleys in the Oberhasli region resulted in the recording of nearly 400 new archaeological structures, which were for the most part related to *Alpwirtschaft* in the Middle Ages and modern period. At these new sites, which mostly lay at altitudes of between 1600 and 2000 m.a.s.l., various different categories of fea-

---

6. When it is finally published, the analysis will include a catalogue of drawings and descriptions of all the groundplans and rock shelter structures, which will form a basis for further comparisons both within the Bernese Alps and beyond.

tures were recorded, including ground-plans, rock shelter structures, sections of mountain paths, iron ore mines, pasture walls and animal pens.

The most numerous categories of features, the ground-plans and the rock shelter structures, are presented in more detail. The buildings against a rock or rock face described as rock shelter structures, with their various shapes and irregular usable spaces, formed a very heterogeneous find category with certain distribution concentrations.

In many cases, the choice of site for the ground-plans was seen to have been determined by considerations of shelter. The majority of measurable ground-plans proved to have an interior area of less than 15m^2. The bigger the structures became, the more likely they were to be rectangular in shape rather than square. Room divisions and interior fittings such as benches and hearths or lean-to structures such as dung heaps also became more frequent.

Because the results were obtained from surveys rather than excavations, non-archaeological sources play an important role in the interpretation and dating of the structures. Written and pictorial sources as well as comparisons with existing alpine buildings and the results of archaeological excavations from other regions supplement the analysis. One-roomed buildings predominated in the Oberhasli region over a period of centuries. Only after 1800 were they gradually replaced by multi-purpose buildings, consisting of a milk-processing area and an area used as a cow-shed.

Even in the Oberhasli region, the greatest economic significance of *Alpwirtschaft* for the area lies within the context of well-known medieval developments north and south of the Alps. The growing demand from the northern Italian towns for large animals and dairy products led, from the 14th century onwards, to an intensification of the pastoral economy especially in the valleys of the northern Alps. Energy-rich hard cheese was used chiefly for export, as it was firm enough to withstand transport on mule tracks over the Grimsel and Gries Passes to the south.

Acknowledgement

I would like to thank Sandy Haemmerle for the excellent translation of my German text.

Bibliography

Affolter, H.C., von Känel, A. and Egli, H.-R. 1990. *Die Bauernhäuser des Kantons Bern: das Berner Oberland*. Basel, Schweizerische Gesellschaft für Volkskunde (Die Bauernhäuser der Schweiz 27).

Andres, B. 2011. Gadmen, Wendenboden: eine alpine Siedlungs-wüstung im Oberhasli. *Archäologie Bern. Jahrbuch des Archäologischen Dienstes des Kantons Bern*, pp. 48–53.

Andres, B. 2012a. Alpine summer farms: upland animal husbandry and land use strategies in the Bernese Alps (Switzerland). In Wiebke Bebermeier *et al.* (ed.) *Landscape Archaeology. Proceedings of the International Conference Held in Berlin, 6th–8th June 2012*:279–283. eTopoi. Journal for Ancient Studies (Special Volume 3). URL: http://journal.topoi.org/index.php/etopoi/article/view/113 [2013-04-04].

Andres, B. 2012b. Hanglage mit Gletscherblick. Alpine Wüstungen im Oberhasli. *Archäologie Bern. Jahrbuch des Archäologischen Dienstes des Kantons Bern* pp. 220–236.

Andres, B. and Walser, C. 2013. Drohnen in der alpinen Archäo-logie: Luftbildaufnahmen von Alpwüstungen im Oberhasli. *Archäologie Bern. Jahrbuch des Archäologischen Dienstes des Kantons Bern*, in press.

Auf der Maur, F., 1998. Alpine Wüstungen im Kanton Schwyz. In Werner Meyer *et al.* (ed.), *«Heidenhüttli»: 25 Jahre archäologische Wüstungsforschung im schweizerischen Alpenraum*), pp. 315–327. Basel, Schweizerischer Burgen-verein (Schweizer Beiträge zur Kulturgeschichte und Archäologie des Mittelalters 23/24).

Bätzing, W. 2003. *Die Alpen: Geschichte und Zukunft einer europäischen Kulturlandschaft*. München, Beck.

Bernisches Statistisches Bureau 1902. *Ergebnisse der Alpstatistik im Kanton Bern pro 1891–1902*. Bern, K.J. Wyss.

Bircher, R. 1979. *Wirtschaft und Lebenshaltung im schweizerischen «Hirtenland» bis Ende des 18. Jahrhunderts*. Bern, K.J. Wyss.

Bitterli-Waldvogel, T., 1998. Spätmittelalter und Neuzeit. In Werner Meyer *et al.* (ed.), *«Heidenhüttli»; 25 Jahre archäologische Wüstungsforschung im schweizerischen Alpenraum*, pp. 394–418. Basel, Schweizerischer Burgenverein (Schweizer Beiträge zur Kulturgeschichte und Archäologie des Mittelalters 23/24).

Bitz, V., 2003. Strassen haben eine Geschichte; vom internationalen Verkehr zum sanften Tourismus. In Bundesamt für Strassen ASTRA (ed.) *Historische Verkehrswege im Kanton Wallis*, pp. 8–21. Bern: Inventar historischer Verkehrswege der Schweiz.

Brülisauer, J., 1984. *SSRQ BE II 7. Das Recht des Amtes Oberhasli*. Aarau, Sauerländer.

Ebersbach, R. and Gutscher, D. 2008. Alpine Prospektion im Oberhasli. Vorbericht 2003–2006. *Archäologie Bern. Jahrbuch des Archäologischen Dienstes des Kantons Bern* pp. 189–196.

Furrer, B., 1985. *Die Bauernhäuser des Kantons Uri*. Basel, Schweizerische Gesellschaft für Volkskunde (Die Bauern-häuser der Schweiz 12).

Furrer, B., 1994. *Die Bauernhäuser der Kantone Schwyz und Zug*. Basel, Schweizerische Gesellschaft für Volkskunde (Die Bauernhäuser der Schweiz 21).

Geiser, W. (ed.) 1973. *Bergeten ob Braunwald; ein archäologischer Beitrag zur Geschichte des alpinen Hirtentums*. Basel, Werner Geiser.

Gerber-Visser, G. 2012. *Die Ressourcen des Landes; der ökonomisch-patriotische Blick in den Topographischen Beschreibungen der Oekonomischen Gesellschaft Bern (1759–1855)*. Baden, Hier + Jetzt (Archiv des Historischen Vereins des Kantons Bern 89).

Gerber-Visser, G. and Stuber M. 2009. Brachliegende Ressourcen in Arkadien; das Berner Oberland aus der Sicht Albrecht von Hallers und der Oekonomischen Gesellschaft Bern. *Mitteilungen der Naturforschenden Gesellschaft in Bern* 66:61–83.

Glauser, F. 1988. Von alpiner Landwirtschaft beidseits des St. Gotthards 1000–1350; Aspekte der mittelalterlichen

Gross- und Kleinviehhaltung sowie des Ackerbaus der Alpenregionen Innerschweiz, Glarus, Blenio und Leventina. *Der Geschichtsfreund* 141:5–173.

Gschwend, M. 1976. *Die Bauernhäuser des Kantons Tessin, Band 1; der Hausbau*. Basel, Schweizerische Gesellschaft für Volkskunde (Die Bauernhäuser der Schweiz 4).

Gutscher, D. 2002. Axalp, Litschentellti. *Jahrbuch der Schweiz-erischen Gesellschaft für Ur- und Frühgeschichte* 85:341.

Hafner, A. 2010. Brienz, Brienzer Rothorn, Ober Stafel/ Schonegg; eine prähistorische Silex-Pfeilspitze aus alpinem Gebiet. *Archäologie Bern, Jahrbuch des Archäologischen Dienstes des Kantons Bern*, pp. 82–83.

Heiri, O. *et al.* 2003. Holocene tree immigration and the chironomid fauna of a small Swiss subalpine lake (Hinterburgsee, 1515 m.a.s.l.). *Palaeogeography, Palaeoclimatology, Palaeoecology* 189 (1–2):35–53.

Kasthofer, K. 1822. *Bemerkungen auf einer Alpen-Reise über den Susten, Gotthard, Bernardin, und über die Oberalp, Furka und Grimsel; mit Erfahrungen über die Kultur der Alpen und einer Vergleichung des wirthschaftlichen Ertrags der Bündenschen und Bernischen Alpen; nebst Betrachtungen über die Veränderungen in dem Klima des Bernischen Hochgebirgs*. Aarau, Heinrich Remigius Sauerländer.

Kehrli, O. 2008. *Alte Ansichten vom Oberhasli; Bilder von den Gemeinden Gadmen, Guttannen, Hasliberg, Innertkirchen, Meiringen und Schattenhalb aus den Jahren 1869–1960*. Visp, Otto Kehrli.

Küchler, R. 2003. *Obwaldens Weg nach Süden durch Oberhasli, Goms und Eschental*. Sarnen, Historischer Verein Obwalden (Obwaldner Geschichtsblätter 24).

Loepfe, A. 2007. Historische Verkehrswege im Kanton Uri; kurzer Einblick in die Verkehrsgeschichte. In Bundesamt für Strassen ASTRA (ed.) *Historische Verkehrswege im Kanton Uri*, pp. 8–25. Bern, Inventar historischer Verkehrswege der Schweiz.

Lotter, A.F. *et al.* 2006. Holocene timber-line dynamics at Bachalpsee, a lake at 2265 m.a.s.l. in the northern Swiss Alps. *Vegetation History and Archaeobotany* 15 (4):295–307.

Meyer, W. 1998a. Die Ausgrabungen auf Bergeten ob Braunwald GL 1971. In Werner Meyer *et al.* (ed.) *«Heidenhüttli»; 25 Jahre archäologische Wüstungsforschung im schweizerischen Alpenraum*, pp. 24–36. Basel, Schweizerischer Burgenverein (Schweizer Beiträge zur Kulturgeschichte und Archäologie des Mittelalters 23/24).

Meyer, W. 1998b. Die Wüstung «Spilblätz» auf der Charetalp SZ 1981. In W. Meyer *et al.* (ed.), *«Heidenhüttli»; 25 Jahre archäologische Wüstungsforschung im schweizerischen Alpenraum*, pp. 48–70. Basel, Schweizerischer Burgenverein (Schweizer Beiträge zur Kulturgeschichte und Archäologie des Mittelalters 23/24).

Meyer, W. 1998c. Einleitung; ein forschungsgeschichtlicher Rückblick. In Werner Meyer *et al.* (ed.) *«Heidenhüttli»; 25 Jahre archäologische Wüstungsforschung im schweizerischen Alpenraum*, pp. 13–17. Basel: Schweizerischer Burgenverein (Schweizer Beiträge zur Kulturgeschichte und Archäologie des Mittelalters 23/24).

Meyer, W. 1998d. Früh- und Hochmittelalter bis 1300. In Werner Meyer *et al.* (ed.), *«Heidenhüttli»; 25 Jahre archäologische Wüstungsforschung im schweizerischen Alpenraum*, pp. 364–393. Basel, Schweizerischer Burgenverein (Schweizer Beiträge zur Kulturgeschichte und Archäologie des Mittelalters 23/24).

Meyer, W. *et al.* (ed.) 1998. *«Heidenhüttli»; 25 Jahre archäo-logische Wüstungsforschung im schweizerischen Alpenraum*. Basel, Schweizerischer Burgenverein (Schweizer Beiträge zur Kulturgeschichte und Archäologie des Mittelalters 23/24).

Mösching, H. and von Rütte, H. 1990. IVS Dok. BE 17.5: Meiringen–Obergestelen, Grimselpass. Fahrstrasse 19. Jahrhundert. In Inventar historischer Verkehrswege der Schweiz (ed.), *IVS Dokumentation*, 1–2. URL: http://cw-ivs2b.bgdi.admin.ch/beschr/de/BE00170500.pdf [2013-01-03].

Obrecht, J. 1998. «Ämpächli», Elm GL 1984. Archäologische Untersuchung einer hochmittelalterlichen Alpsiedlung. In Werner Meyer *et al.* (ed.), *«Heidenhüttli»; 25 Jahre archäologische Wüstungsforschung im schweizerischen Alpenraum*, pp. 105–123. Basel, Schweizerischer Burgen-verein (Schweizer Beiträge zur Kulturgeschichte und Archäologie des Mittelalters 23/24).

Obrecht, J. 2009. Datierung von Gebäuderesten längst aufgelassener Schwyzer Alpstafel. *Mitteilungen des Historischen Vereins des Kantons Schwyz* 101:11–15.

Roth, A.G. 1977. *Talkäsereien; zur Aufnahme des Betriebes in der Schweiz*. Burgdorf, G. Roth.

Roth, A.G. 1993. *Der Sbrinz und die verwandten Bergkäse der Schweiz*. Burgdorf, Schweizerische Käseunion.

Roth-Bianchi, W. 2007. Die Geschichte des Kantons Graubünden ist die Geschichte seiner Alpenpässe und Verkehrswege. In Bundesamt für Strassen ASTRA (ed.) *Historische Verkehrswege im Kanton Graubünden*, pp. 8–21. Bern, Inventar historischer Verkehrswege der Schweiz.

Sablonier, R. 1990. Innerschweizer Gesellschaft im 14. Jahrhundert; Sozialstruktur und Wirtschaft. In Historischer Verein der Fünf Orte (ed.) *Innerschweiz und frühe Eidgenossenschaft. Jubiläumsschrift 700 Jahre Eidgenossenschaft; Band 2: Gesellschaft, Alltag, Geschichtsbild*, pp. 11–233. Olten, Walter-Verlag.

Sauerländer, D. 2006. *Viehwirtschaft, Kap. 2: Hochmittelalter bis frühe Neuzeit*. URL: http://www.hls-dhs-dss.ch/textes/d/D26236.php [2012-11-22].

Sauter, M., 2009. *Wüstungsforschung im Kanton Uri: Ergebnisse der hochalpinen Prospektion im Brunni- und im Schächental, auf Haldi und dem Surenenpass durch Studenten der Hochschule Luzern (Technik und Architektur) begleitet von Walter Imhof und Marion Sauter*. Altdorf: Gamma.

Schaer, A. and Martin-Kilcher, S. 2009. Das Heiligtum und sein Umland. In Stefanie Martin-Kilcher and Regula Schatzmann (ed.) *Das römische Heiligtum von Thun-Allmendingen, die Regio Lindensis und die Alpen*, pp. 257–283. Bern, Bernisches Historisches Museum (Schriften des Bernischen Historischen Museums 9).

Sieber, C. and Bretscher-Gisiger, C. 2012. *Acta Murensia; die Akten des Klosters Muri mit der Genealogie der frühen Habsburger*. Basel, Schwabe.

Sprüngli, J. 1760. *Beschreibung des Hassle-Lands im Canton Bern*. Zürich, Heidegger (Der Schweizerischen Gesellschaft in Bern Sammlungen von landwirtschaftlichen Dingen 4).

Stebler, F.G. 1903. *Alp- und Weidewirtschaft; ein Handbuch für Viehzüchter und Alpwirte*. Berlin, Parey.

Taramarcaz, C. 2012. *Economie alpestre dans le Val de Bagnes; «îtres» et bâtiments d'alpage*. Mémoires, Université de Neuchâtel.

Veit, H., 2002. *Die Alpen; Geoökologie und Landschaftsentwicklung*. Stuttgart, Ulmer.
von Rütte, H. 1990a. IVS Dok. BE 15: Meiringen–Wassen, Sustenpass. In Inventar historischer Verkehrswege der Schweiz (ed.), *IVS Dokumentation*, pp. 1–7. URL: http://cw-ivs2b.bgdi.admin.ch/beschr/de/BE00150000.pdf [2013-04-04].
von Rütte, H. 1990b. IVS Dok. BE 16: Meiringen–Engelberg, Jochpass.In Inventar historischer Verkehrswege der Schweiz (ed.), *IVS Dokumentation*, pp. 1–5. URL: http://cw-ivs2b.bgdi.admin.ch/beschr/de/BE00160000.pdf [2013-04-04].
von Rütte, H. 1990c. IVS Dok. BE 17: Meiringen–Obergesteln; Grimselpass. In Inventar historischer Verkehrswege der Schweiz (ed.), *IVS Dokumentation*, pp. 1–6. URL: http://cw-ivs2b.bgdi.admin.ch/beschr/de/BE00170000.pdf [2013-04-04].
von Rütte, H. 1992. IVS Dok. BE 13: Brienz–Brünigpass. In Inventar historischer Verkehrswege der Schweiz (ed.), *IVS Dokumentation*, pp. 1–8. URL: http://cw-ivs2b.bgdi.admin.ch/beschr/de/BE00130000.pdf [2013-04-04].
von Rütte, H. 1994. IVS Dok. BE 100: Lauterbrunnen–Meiringen, Oberland-Tour. In Inventar historischer Verkehrswege der Schweiz (ed.), *IVS Dokumentation*, pp. 1–17. URL: http://cw-ivs2b.bgdi.admin.ch/beschr/de/BE01000000.pdf [2013-01-07].
von Weissenfluh der Ältere, J. 1910. Chronik 1792–1821. In Andreas Fischer (ed.), *Aufzeichnungen zweier Haslitaler*. Bern, A. Francke.
von Weissenfluh der Jüngere, J. 1910. Alpenreisen 1850–1851. In Andreas Fischer (ed.), *Aufzeichnungen zweier Haslitaler*. Bern, A. Francke.

Weiss, R. 1992. *Das Alpwesen Graubündens; Wirtschaft, Sachkultur, Recht, Älplerarbeit und Älplerleben*. Reprint der Originalausgabe von 1941. Chur, Octopus.
Wetli, E. 2010. *Bauinventar der Gemeinde Grindelwald; nur Teilbereich gemäss Perimeter*. Bern, Denkmalpflege des Kantons Bern.
Wick, L. et al. 2003. Holocene vegetation development in the catchment of Sägistalsee (1935 m.a.s.l.), a small lake in the Swiss Alps. *Journal of Paleolimnology* 30 (3):261–272.
Wyss, J.R. 1816. *Reise in das Berner Oberland, Band 1*. Bern: Burgdorfer.
Wyss, J.R. 1817. *Reise in das Berner Oberland, Band 2*. Bern: Burgdorfer.
Zappa, F. 2005. Genutzte Felshöhlen im Muotatal; Milchbalm-Höhle, Eiskeller und Siten-Balm; ein Vergleich mit den «splüi» südseits der Alpen. In Franz Auf der Maur, Walter Imhof and Jakob Obrecht, *Alpine Wüstungsforschung, Archäozoologie und Speläologie auf den Alpen Saum bis Silberen, Muotatal SZ*. Mitteilungen des Historischen Vereins des Kantons Schwyz 97:69–73.
Zappa, F. 2008. *I segni visibili e invisibili del paesaggio rurale; Stein e Bétti, due alpi walser*. Aosta, Assoziazione culturale Augusta.
Zybach, A. 2008. *«Im indren Grund»; Chronik von Innertkirchen*. Münsingen, Fischer.

Brigitte Andres, Viktoriastrasse 105, 3084 Wabern, Switzerland.
Archaeological Service of the Canton Bern, Postfach 5233, 5001 Bern, Switzerland.
email: brigitte.andres@erz.be.ch

11. Driving forces and variability in the exploitation of a high-altitude landscape from the Neolithic to Medieval Periods in the southern French Alps

Kevin Walsh and Florence Mocci

With contributions from: *M. Court-Picon, J.-L. de Beaulieu, F. Guiter, S. Richer and B. Talon*

The aim of this paper is to assess the development of summer activities in the high-altitude zone of the southern French Alps between the Neolithic and the Middle Ages. During these periods, there was enormous variety in the nature of high-altitude activity in these valleys. The Bronze Age witnessed the establishment of the first stone-built pastoral structures at 2200m and above. This marked an important change in the engagement with this landscape, with high-altitude summer pasturing emerging as a new activity. The Iron Age and Roman era are characterised by a dearth of archaeological structures, but continued palaeoecological signals for pastoral (and possibly mining) activity. The medieval periods saw a substantial increase in activity; a combination of pastoralism and mining, with some large high altitude settlements created which imply the wholesale summer movement of communities from valley-bottom to the high altitude zones.

Introduction

The aim of this paper is to consider the evolution of summer-activities in the high-altitude zone of the southern French Alps between the Neolithic and the Middle Ages. During these periods, there were variations in the nature of high-altitude activity in these Alpine valleys. This contribution considers the role of ecological models, or principles that underpin many assessments of movement in and out of high altitude areas. More specifically, we consider the range of motivations for the development of pastoralism in the southern French Alps.

Problematic

More often than not, ecological models inform the assessment of the development of upland activities, where vertical ecological zonation is considered to be a fundamental control on the potential range of activities that can take place in a mountainous landscape. Alpine landscape terminology is important here, and vertical zonation in any mountain range influences vegetation and geomorphic processes, and clearly affects what can and cannot be done by people. The sub-alpine (or pseudo-alpine) zone is situated between c. 1600/1700 and 2200/2400m. Today, permanent human settlement rarely goes beyond 1650m. The alpine zone comprises the altitudes between 2200/2400m and 3000m, with the nival zone beyond this altitude possessing little or no vegetation. Simply put, temperature drops with altitude and has an effect on soil development and vegetation growth, thereby influencing the range of agricultural activities that can take place across each of these zones. The reduction in temperature of one degree centigrade for every 300m climbed is one of the key defining environmental variables in mountainous areas. Vertical ecologies underpin the framework and models for much research in mountains (Fig. 11.1). Consequently, biogeographical models have influenced the ways in which we interrogate and interpret human-environment engagements in mountains.

One important notion in mountain ecology is the idea that the high altitudes are 'buffered' by lowland forests, grasslands or deserts in the same way that islands are buffered by seas; moreover, the fact that high altitude zones tend to be cold adds another form of buffer (Billings 1979:97). We must, however, avoid the uncritical application of such models in the assessment of human colonisation and life in high altitude zones. There are, however, rudimentary biogeographical rules; pedogenesis decreases with altitude, and therefore the number of different species of plants and animals also decreases. These characteristics are directly influenced by levels of exposure to sunlight, wind, slope, geology etc. (Billings 1979:105). The exposure to sunlight via the orientations of mountain slopes is *the* defining characteristic in many mountainous zones, and leads to specific designations for north- and south-facing slopes, such as *ubac* and *adret* in French. Slope orientation is an important control on vegetation

keywords: French Alps, Ecrins National Park, Neolithic, Bronze Age, Medieval, high altitude pastures, vertical zoning, transhumance, secondary products revolution, radiocarbon dates

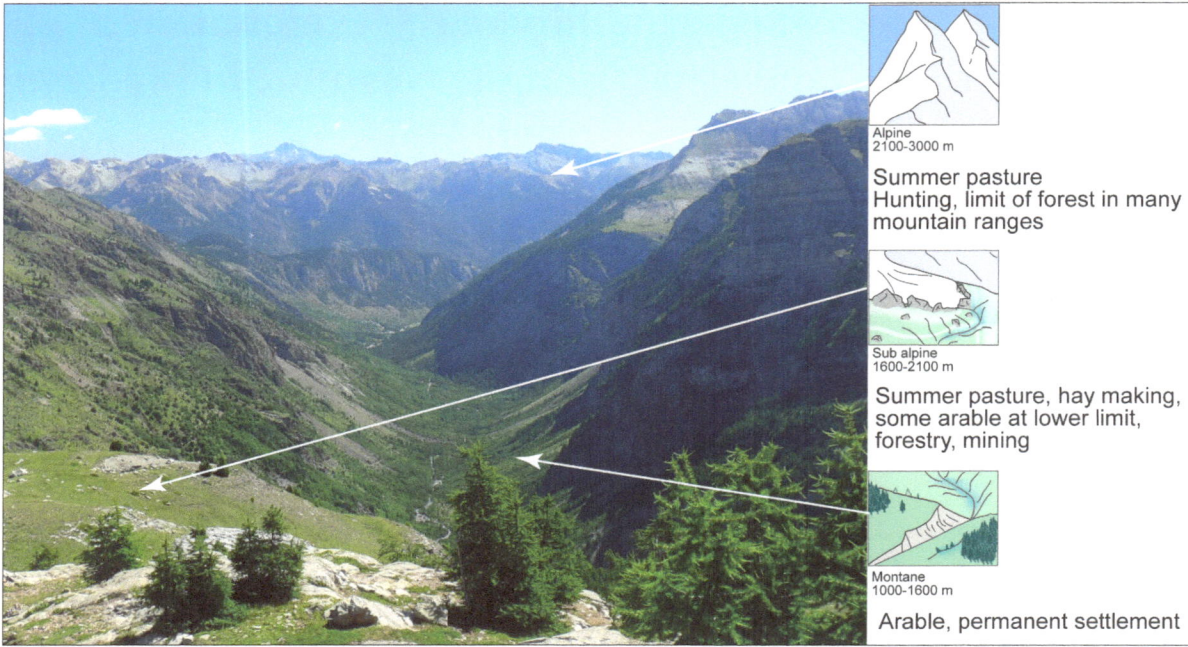

Figure 11.1   Vertical ecological zonation and typical activities within those zones.

growth, and the upper limit of the forest. The natural limit of the forest is the timberline (the limit of the closed canopy); the tree line is the limit of tree growth. These limits advance and retreat, and are influenced by changes in climate and human exploitation of the forest. Whilst the primary productivity of an alpine zone will influence what people can and cannot do at high altitude in terms of farming and hunting, there are certain resources, especially minerals and ores, that attract people to live and work in these zones no matter what the primary ecological productivity. However, humans are usually absent from high altitude zones during the periods when plants are dormant, a situation that is not true of most other environments.

Vertical zones and annual cycles

There are many publications (in the fields of history and geography) that present studies of recent and contemporary uses of mountains by people, as well as attempts by archaeologists to identify the origins of long-distance transhumance. These questions have been dominant themes in mountain research in the Mediterranean. However, as Halstead noted (1987), there are problems with extrapolating contemporary, traditional forms of Mediterranean agriculture back into the past. Therefore, we have to develop locally based research that considers specific regional processes, such as Cherry's work in Greece (1988).

To date, many explanations for the development of high altitude farming are founded on the notion that ecological necessity 'forces' the development of upland summer farming – i.e. stresses on niche capacity due to demographic pressure, or more specifically, when assessing the development of transhumance, the pressure imposed by the increase in animals requiring pasture. Consequently, the combination of ecological premises, and the desire to extrapolate contemporary and traditional practices back into the past, places an emphasis on identifying periods of increased population and economic activity across the Mediterranean where people were 'forced' into exploiting notionally marginal land.

Another issue is the premise that environmentally marginal landscapes are settled and exploited by those groups on the periphery of society is another element in the framework that influences our assessment of high altitude activities: "Too often anthropologists have rushed to explain the economies of recent pastoral peoples by placing them on a continuum of putative ecological adaptation" (Koster and Chang 1994:14). This is one element in human ecological models that are founded on the assessment of the reasons why people move in and out of the high-altitude vertical zones. We also have to consider whether we are looking at the development of nomadic pastoralists, or pastoralists who were in fact integrated with other economic structures (Salzman 2004). Nomadic pastoralists today are considered to operate beyond the dominant, orthodox economic and social structures.

Research in the Southern French Alps

The Southern French Alps Landscape project (based in the Ecrins National Park, southern France) has radically changed our understanding of the use of the high altitude zones (above 2000m) in this region.

In 1998, a landscape archaeology project was established with the aim of investigating the development of human activity in the high altitude zone (above 2000m) across a series of Alpine valleys. Although projects in the Italian and Swiss Alps have produced evidence for specific periods of activity (Della Casa 1999; Curdy and Praz 2002; Fedele 1999), no other project has produced detailed evidence for radical fluctuations and variation in the intensity and nature of human activity and environmental change over the entire Holocene. The archaeological and palaeo-ecological results have been extraordinary: our earliest evidence includes an Upper Palaeolithic tool, but abundant evidence for early human activity at altitude comprises Mesolithic and Neolithic flint scatters, Bronze Age animal enclosures, and then structures running through to the Post-Medieval Period.

The identification of the origins and characteristics of high altitude pastoralism in the Alps is one of the key aims of the Southern French Alps Landscape Project. Prior to this research, many archaeologists considered Roman economic expansion and intensification as the drivers for transhumant pastoralism and expansion into the higher altitudes. Our work has gone some way to testing this notion, and demonstrates that the antiquity of upland farming (in particular, pastoralism) goes back to the Late Neolithic. The possibility that this development should be considered part of the secondary products revolution is a question discussed in this paper.

The study area

The Ecrins are situated just to the west of the upper Durance River in the Provence-Alpes-Côte d'Azur region in France, in the administrative department of the Hautes-Alpes. Our study area embraces the valleys of the Champsaur to the west, and the Freissinières, Fournel and Onde Valleys to the east (Fig. 11.2). The Parc National des Ecrins is a complex Alpine zone centred on a series of high mountains up to 4000m (Claudin and Miellet 2000). The core of this sector comprises glaciers, and a series of long, deep valleys running off in all directions from this core. The geological formations are dominated by limestone, flysch, and sandstones, with areas of gneiss. Whilst there is no autochthonous flint, there are sources of quartz and minerals, including lead, silver and copper, and even some gold, minerals that have attracted people to these areas in the past.

Human-environment interactions from the Neolithic to the Medieval Period

Below, a synthesis of the key archaeological and palaeo-environmental evidence is presented with a view to assessing the development of high altitude pastoral activities in our study area.

Figure 11.2   Location map of the study area.

Prior to the Neolithic, there is little evidence for human impact on the vegetation in any of the altitudinal zones. Probable human activity during the Neolithic contributed to the contraction of the low-mid altitude forest across parts of the French Alps (Ali *et al.* 2005a). The palynological research in our valleys suggests Early Neolithic agro-pastoral activity in valley bottoms, with little evidence for the expansion of these activities up into the higher altitudes. This early impact on the lower forest might be seen as the start of 'creeping' pressure on vegetation, gradually forcing the lower tree-limit upwards over time. The pressure on vegetation on the upper edge of the forest (pushing the tree-line downwards), is the other process that we need to consider in any assessment of the development of alpine pastoral landscapes. This pressure (from the 'top down') was initially caused by climate change (normally climatic cooling), but with time, once people started to exploit these upper altitudes, human pressure contributed to this process.

In the southern French Alps the tree line (dominated by *Pinus cembra*) was at *c.* 2400m during the

Neolithic, descended to 2200m during the Bronze Age, and then returned to 2400m during the Iron Age (Ali *et al.* 2005b). It is often difficult to differentiate between human and climatic forcing, however for much of the Holocene, vegetation change at the upper edge of the forest was largely influenced by climate, with people manipulating relatively small areas within or at the edge of the forest. Whilst Neolithic activity in the valley bottoms, comprising arable and pastoral agriculture, undoubtedly put some pressure on the local vegetation, there is no evidence for extensive human activity and concomitant pressure on the vegetation at high altitude. The only archaeological evidence attesting to Middle Neolithic activity in the Alpine zone of the Ecrins is lithic material. This material demonstrates how hunting remained an important activity in the high altitude zone. For the moment, the palaeo-ecological evidence only allows us to suggest that pastoral activity took place at the lower and mid-range altitudes (probably up to *c.* 1500m). There is no convincing evidence for summer transhumance in the high altitude pastures (above 2000m) during the Early to Middle Neolithic (Nicod 2008).

The late third and second millennia BC

It is the late third and second millennia that saw a radical change in the use of the high altitude areas (above 2000m). Our palaeo-ecological and archaeological evidence demonstrates the development of high altitude pastoralism during this period. The details of the archaeological and palaeo-ecological evidence have been published elsewhere (Mocci *et al.* 2009; Walsh 2005; Walsh *et al.* in press), therefore, a synthesis of that evidence is presented prior to the discussion below of the possible reasons for the development of high altitude pastoral activity during this period.

Synthesis of key evidence

A series of ten sites in the Ecrins, in the Freissinières Valley, the Haute Fournel and Champsaur (Fig. 11.3), have been excavated and radiocarbon dated to the period covering the later third and second millennia BC (Mocci and Tzortzis *et al.* 2006; Palet-Martinez *et al.* 2003; Walsh and Mocci *et al.* 2007). About twenty other similar structures have been located and recorded, but have not been excavated: There is little doubt that many of these are contemporary with the

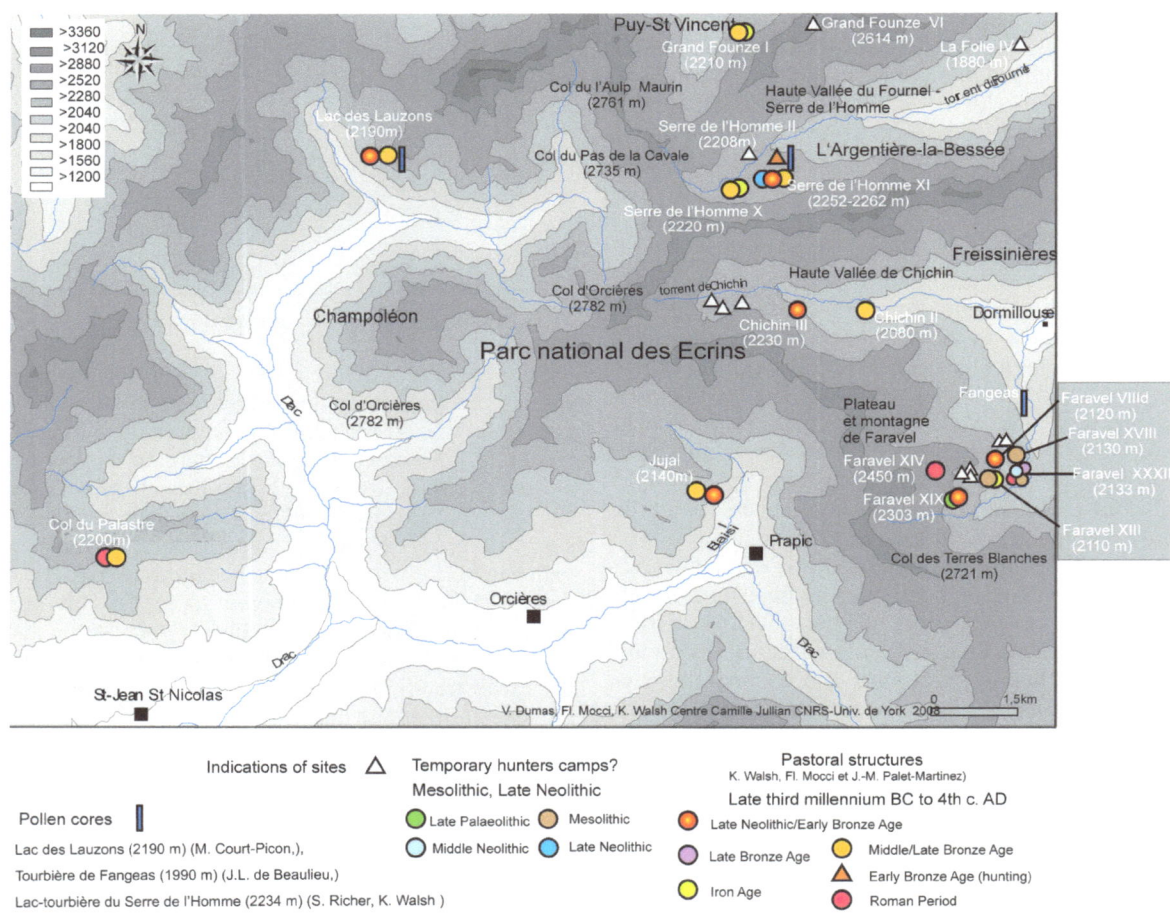

Figure 11.3 Key pre- and proto-historic archaeological and palaeoecological sites in the Ecrins (plan – V. Dumas, F. Mocci, K. Walsh).

sites described below. Many of the sites share similar characteristics; they often comprise a stone-built structure made from large rocks or boulders (normally between 20 to 70cm in diameter) arranged into broadly ovoid, or sub-circular enclosures, some of which comprise sub-divided spaces. The dry-stone walls (or foundations) normally survive with two or three 'courses' of superimposed rocks. Eight of these structures have been excavated (for a full list of dates, calibrations and lab' references, see Table 11.1).

Despite the similarities in construction technique, there is variation in the size and specific shapes of structures across this set of sites. For example, the two sites located adjacent to one another in the Chichin Valley (see Fig. 11.3 for location) are different in form despite similar radiocarbon dates: Situated at 2200m, Chichin IIIb is a small, roughly circular structure of about 8m^2 (radiocarbon dated to 2580-2400 BC); traditionally seen as the Neolithic/Early Bronze Age transition. Three metres to the southeast of this structure, Chichin IIIa (2460-2200 BC) is a relatively large structure of about 60 m^2. As with all of these sites, it comprises a zone of large blocks (from 20cm up to almost 1m in diameter) that delimit a roughly ovoid zone. This 'wall' measures between 1m and 3m in thickness. At Faravel (3.5km to the south), two enclosures (one at 2200m and another at 2300m) have yielded radiocarbon dates that are exactly the same, both dated to 2150–1920 BC (Fig. 11.4). Although the general form of the structures is similar, one covers

Figure 11.4 Bronze Age 'enclosure sites' excavated in the Ecrins. Faravel XIX (alt. 2303 m; date = 2150–1920 BC) and Faravel VIIId (alt. 2170 m date = 2150 – 1920 BC). N.B. Two enclosures at two different altitudes with same 14C date (photos K. Walsh).

an area of 100m², and the other 20m². The fact that these structures comprise large open spaces with little material culture, leads us to interpret these as animal enclosures. The acidic soils mean that animal bones are rare to absent. We are therefore reliant upon inferences from pollen indicator species and coprophilous spores from our peat and lake cores in the assessment of activities associated with these structures.

Serre de l'Homme: a case study in the long-term development of an upland pastoral site

Serre de l'Homme comprises one of, if not *the* most complex series of Bronze Age high altitude archaeological structures anywhere in the Alps. Many of these sites are linked to one another by common walls, and are centred on the site of Serre de l'Homme XI (Fig. 11.5). Radiocarbon dates from the excava-

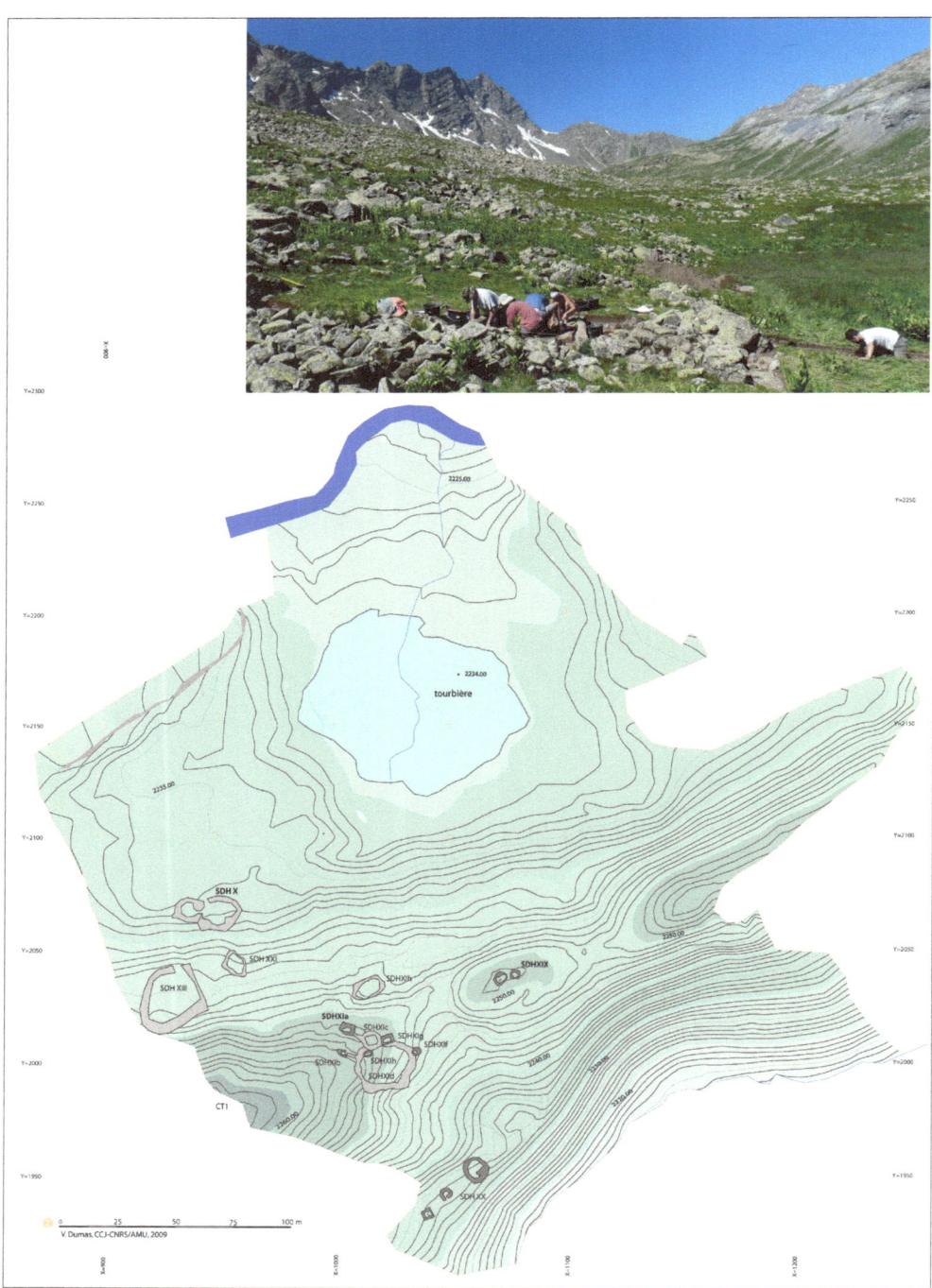

Figure 11.5   The settlement complex at Serre de l'Homme – here a sequence of structures spread along this moraine span the entire Late Neolithic through to the Late Bronze Age (plan – V. Dumas, F. Mocci, K. Walsh).

tions of different enclosures in this sequence demonstrate how new enclosures were added from the Late Neolithic/Chalcolithic and throughout the Bronze Age.

The first, somewhat ephemeral structure, Serre de l'Homme XIX, is dated to 3090–2900 BC. Lower on the moraine over which most of these sites were built, other structures saw several phases of activity. The first phase of activity at the central structure, Serre de l'Homme XI, is dated to 2480–2280 BC, with a second phase dated to 1750–1600 BC. This site is one structure amongst a series of enclosures spread across c. 400 m². Two radiocarbon dates (1510–1400 BC and 1500–1310 BC) represent a later phase of activity, and the spread of charcoal around the sites suggests anthropogenic burning of the local vegetation. This suggests deliberate pasture creation, or enhancement. Lower down the moraine, abutting the flat, seasonally wet zone, Serre de l'Homme X (2236m) is a substantial ovoid enclosure with an occupation phase dated to 1460–1310 BC. This sequence of dates reveals a pattern of enclosure expansion along and down the flank of the moraine from the Late Neolithic through to the Middle Bronze Age.

The pollen diagram from the lake at Serre de l'Homme comprises the following important characteristics. The very base of the diagram shows a pollen spectrum that is consistent with a warm and dry climate. High levels of tree pollen, dominated by *Pinus* sp. show that the treeline would have been close to the lake which is adjacent to the moraine, whilst low levels of *Sparganium*-type suggest that the area may have been somewhat drier. This phase is likely to represent the warmer period of the Early Bronze Age, probably around 1800–1600 BC. This warm phase appears to be followed by a period of climatic cooling in the second half of the Bronze Age (1400–1100 BC), represented by a decline in tree pollen and the first appearance of *Sparganium*-type. This corresponds with higher lake levels just to the north in Savoie (Magny *et al.* 2009) and other evidence from just to the south of the Alps. The presence of some nitrophilous plants and coprophilous spores indicates pasturing animals in this area during the Bronze Age (see Walsh *et al.* in press for a detailed review of the palaeo-ecological evidence). In summary, the Bronze Age provides the earliest evidence for direct manipulation of vegetation, plus the first constructions within the high altitude alpine landscape. What we have is evidence for some groups moving beyond localised (probably, valley-bottom) husbandry to pastoralism.

Regional and local processes have to be considered if we are to explain these developments. Perhaps one of the most striking characteristics of the series of dates from the sites in the Ecrins is the group of dates at the top of Table 11.1, between 2580 to c. 2200 BC. Despite the steep step in the calibration curve between 2500 and 2450 cal BC (Voruz 1996), there is little doubt that there is consistency in the radiocarbon estimates for these first structures in the high altitude zone. The onset of high altitude pastoralism appears to coincide with the emergence of so-called Beaker (*Campaniform*) cultures in the region. This period saw an increase in nomadic pastoralism across parts of Europe (Lichardus *et al.* 1985:359). The possibility that some populations moved across Europe at this time (associated with Beaker pottery) is a process that must be given credence: Regional and sub-regional mobility during the late 3rd and 2nd millennia mobility seems commonplace (Evans *et al.* 2006). This period might have seen influx of new people and we should consider the possibility that accessible, useful, open land might have been in demand. Local environmental characteristics (dense low-altitude forest, short supply of pasture) would have rendered the higher altitudes more attractive, as the exploitation and/or creation of pasture would have been easier as tree density decreased with altitude. At the same time, an influx of people with different attitudes towards mobility and movement (perhaps characterised by a willingness to explore and exploit the unknown) might also explain new interactions with the high altitude zones. There is little doubt that this period witnessed the development of new networks of economic, social, and cultural processes (perhaps related to the development of groups who used Beakers) that led to a very different construction of attitudes towards the high altitude zones. The complex interplay of the ideologies associated with Beakers and possible influences of incoming nomadic pastoralists, such as the Yamnaya, should also be considered (Harrison and Heyd 2007). There is isotopic evidence that suggests migrations or movements of peoples associated with Bell Beakers in south-central Europe (Price *et al.* 2004).

Iron Age and Roman era

During the Iron Age, the lower altitudes (valley floors and abutting slopes) witnessed intensive human activity; there are numerous, isolated portable antiquities found across the southern Alps, and a great deal of funerary evidence (Willaume 1991; Garcia 2008). However, few settlement sites have been found in our valleys; this is probably due to the extreme taphonomic processes that have either masked or destroyed any such site evidence. In the Champsaur, two tombs are known from Orcières, whilst a silver torque was discovered some years ago in the Freissinières Valley. Despite this low altitude evidence, the archaeological record for high altitude activity is limited. The one site dated to this period, Faravel XIII (2150m), comprises postholes dated to the early 8th century BC. Consequently, it seems likely that the later protohistoric periods saw a modified use of the sub-alpine environment. Arable agriculture and pastoralism certainly took place in the lower altitude areas, and the palynological evidence suggests the continuation of pastoralism at higher altitudes, but there is no suggestion that this

Summer Farms: Seasonal exploitation of the uplands from prehistory to the present

COMMUNE	SITE	ALT	Exc date.	TAXON	Lab ref	DATE BP	Error	CAL. BP 95	CAL. BC/AD	CHRONOLOGY	INTERPRETATION
ARGENTIEROIS-VALLOUISE											
FREISSINIERES	Abri Faravel	2133	2011	*Pinus Cembra*	Poz-43854	1725	30	1708-1557 BP	243-393 AD	Late gallo-roman - Late Antiquity	Burnt post-hole (TP1)
Freissinières	Abri Faravel	2133	2011	*Pinus Cembra*	Poz-45433	1245	30	682-870 BP	682 - 870 AD	Early Medieval	Localised burning/fire (structure against shelter?)
Freissinières	Abri Faravel	2133	2011	*Pinus Cembra*	Poz-43853	3530	35	3892-3706 BP	1949-1751 BC	Early - Middle Bronze Age	Thick layer of localised burning on edge of shelter
Freissinières	Abri Faravel	2133	2011	*Pinus Cembra*	Poz-45434	940	30	926-790 BP	1025-1160 AD	Medieval	Ancient post hole
Freissinières	Chichin IIa	2230	2003	*Larix Picea*	Poz-5603	3220	35	3554-3370 BP	1540-1410 BC	Middle Bronze Age	Bonfire in structure (cabin) with lithic material
Freissinières	Chichin IIb	2074	2003	*Rhododendron*	Poz-5499	685	35	683-561 BP	1264-1391 AD	Medieval	Colluvial layer beneath wall (cabin)
Freissinières	Chichin IIIa	2230	2003	*Betula*	Poz-5498	3845	35	4407-4153 BP	2460-2200 BC	Neolithic - Bronze Age Transition	Fire place/bonfire in ovoid structure
Freissinières	Chichin IIIb	2230	2003	*Pinus Cembra*	Poz-5500	3970	35	4524-4299 BP	2580-2400 BC	Neolithic - Bronze Age Transition	Buried soil within enclosure
Freissinières	Fangeas VIIa	2003	2002		Pa 2235	1180	80	1276-957 BP	674-993 AD	Early Medieval	Localised burning/fire in structure
Freissinières	Fangeas VIIb	2003	1998		Pa 1800	500	40	631-496 BP	1390-1460 BC	Medieval	Localised burning/fire in structure
Freissinières	Faravel VIIIb	2170	1998		Pa 1728	857	40	905-688 BP	1045-1263 AD	Medieval	Fire place/bonfire in structure (cabin)
Freissinières	Faravel VIIId	2170	1999		Pa 1841	3665	40	4143-3878 BP	2150-1920 BC	Early Bronze Age	Localised burning/fire in structure (with lithic material)
Freissinières	Faravel XIIa	2175	2000			680	30	680-561 BP	1270-1390 AD	Medieval	Mixed with wall structural elements
Freissinières	Faravel XIIb	2175	1999		Pa 1843	650	40	673-552 BP	1281-1400 AD	Medieval	Localised burning/fire
Freissinières	Faravel XIIIb	2150	2001		Pa 2113	2460	50	2711-2360 BP	770-400 BC	Late Bronze Age/ Early Iron Age Transition	Post-hole fill
Freissinières	Faravel XIV	2450	2001		Pa 2097	1985	50	2060-1823 BP	50 BC - 80 AD	Early gallo-roman	Fire place/bonfire in structure (cabin)
Freissinières	Faravel XIX	2303	2002		Pa 2209	3670	45	4146-3876 BP	2150-1920 BC	Early Bronze Age	Fire place/bonfire in structure (cabin)

L'ARGENTIERE-LA-BESSEE									
L'Argentière-la-Bessée	Serre de l'Homme IIb	2208	2005	Pinus Cembra	Poz 13919	3965	35	4524-4296 BP	2580-2370 BC	Neolithic - Bronze Age Transition	Fire place/bonfire within structure (cabin)
L'Argentière-la-Bessée	Serre de l'Homme IIb	2208	2005	Pinus Cembra	Poz 14893	3960	35	4522-4295 BP	2580-2370 BC	Neolithic - Bronze Age Transition	Fire place/bonfire within structure (cabin)
L'Argentière-la-Bessée	Serre de l'Homme IIc	2208	2005		Pa 2364	3665	45	4145-3869 BP	2200-1910 BC	Early Bronze Age	Localised burning/fire
L'Argentière-la-Bessée	Serre de l'Homme IIc	2208	2005		Pa 2363	3890	30	4418-4239 BP	2470-2280 BC	Neolithic - Bronze Age Transition	Fill of TP2
L'Argentière-la-Bessée	Serre de l'Homme X	2236	2009	Larix Picea	Poz-32463	3125	35	3442-3260 BP	1460 - 1310 BC	Middle Bronze Age	Charcoal on occupation layer in enclosure SDH X (Phase 3)
L'Argentière-la-Bessée	Serre de l'Homme X	2236	2009	Larix Picea	Poz-32465	2435	35	2701-2353 BP	760-400 BC	Late Bronze Age/Early Iron Age Transition	Localised burning/fire (logs) internal to enclosure SDH X (Phase 4)
L'Argentière-la-Bessée	Serre de l'Homme X	2236	2009	Larix Picea	Poz-32466	2395	35	2688-2342 BP	740-390 BC	Late Bronze Age/Early Iron Age Transition	Localised burning/fire (logs) internal to enclosure SDH X (Phase 4)
L'Argentière-la-Bessée	Serre de l'Homme X	2236	2009	Larix Picea	Poz-32467	3075	30	3365-3216 BP	1420-1260 BC	Middle Bronze Age	Localised burning/fire upslope et ante structure SDH X
L'Argentière-la-Bessée	Serre de l'Homme XIa	2250	2007	Larix Picea	Pa 2463	3375	30	3696-3490 BP	1750-1600 BC	Early Bronze Age	Localised burning/fire structure SDH XIa (Phase 3)
L'Argentière-la-Bessée	Serre de l'Homme XIa	2250	2007	8 frag de Larix Picea et 2 frag Pinus Cembra	Pa 2462	3895	35	4423-4192 BP	2480-2280 BC	Neolithic - Bronze Age Transition	'Combustion'
L'Argentière-la-Bessée	Serre de l'Homme XIa	2250	2008		Poz-32472	4475	35	5290-4976 BP	3350-3020 BC	Middle Neolithic	Localised burning/fire under structure SHD XIa (Phase 1a)
L'Argentière-la-Bessée	Serre de l'Homme XIa	2250	2008		Poz-32470	3175	30	3453-3353 BP	1510 - 1400 BC	Middle Bronze Age	Localised burning/fire in structure SDH XIa (Phase 7)
L'Argentière-la-Bessée	Serre de l'Homme XIa	2250	2008		Poz-32471	3140	30	3444-3269 BP	1500 - 1310 BC	Middle Bronze Age	Fire place/bonfire in structure SDH XIa (Phase 5)

Table 11.1 Radiocarbon dates for all excavated sites/structures from the Ecrins National Park (continued over).

Summer Farms: Seasonal exploitation of the uplands from prehistory to the present

COMMUNE	SITE	ALT	Exc. date	TAXON	Lab ref	DATE BP	Error	CAL. BP 95	CAL. BC/AD	CHRONOLOGY	INTERPRETATION
L'Argentière-la-Bessée	Serre de l'Homme XI CPA	2250	2008		Poz-32469	4105	35	4815-4452 BP	2870 - 2500 BC	Middle Neolithic/ Late Neolithic	Localised burning/fire - colluviation - ante structure SDH Xa (Etat 1b)
L'Argentière-la-Bessée	Serre de l'Homme XIX	2250	2008		Poz-32473	4365	35	4815-4452 BP	3090 - 2900 BC	Middle Neolithic	Localised burning/fire contemporary SDH XIX
L'Argentière-la-Bessée	Serre de l'Homme SPA	2242	2009	Larix Picea	Poz-32481	3210	35	3554-3361 BP	1610 - 1410 BC	Late-Middle Bronze Age	Localised burning/fire zone upslope from SDH X
L'Argentière-la-Bessée	Serre de l'Homme SPA	2242	2009		Lyon-8553	2990	30	3320-3075 BP	1367-1126 BC	Middle Bronze Age	Localised burning/fire zone upslope from SDH X
L'Argentière-la-Bessée	Serre de l'Homme SPA	2242	2009	Betula	Lyon-8554	3030	30	3345-3084 BP	1389-1211 BC	Middle Bronze Age	Localised burning/fire zone upslope from SDH X
L'Argentière-la-Bessée	Serre de l'Homme SPA	2242	2009		Lyon-8555	3055	30	3360-3170 BP	1408-1260 BC	Middle Bronze Age	Localised burning/fire zone upslope from SDH X
L'Argentière-la-Bessée	Serre de l'Homme SPA	2242	2009		Lyon-8556	3535	30	3899-3708 BP	1943-1771 BC	Early Bronze Age	Localised burning/fire zone upslope from SDH X
VALLOUISE	Grand Founze IA	2210	2007	4 frag Larix Picea et 1 frag Pinus cembra	Pa 2461	2275	40	2352-2156 BP	410-200 BC	Early Iron Age	Buried soil in animal enclosure
Vallouise	Grand Founze IB	2210	2007	Larix Picea	Pa 2460	1095	30	1060-936 BP	890-1020 AD	End of Early Med / High Mid Ages transition	Fire place/bonfire in rock shelter
Vallouise	Grand Founze IC	2210	2007	Larix Picea	Poz-22633	2745	30	2924-2767 BP	980-810 BC	Late Bronze Age	Residual charcoal from fire place/bonfire in small structure
CHAMPOLEON	Cheval de Bois	2360	2001		Pa 2138	1550	40	1528-1355 BP	420-610 AD	Late Antiquity/ Early Medieval	Fire place/bonfire ?
Champoléon	Cheval de Bois II	2360	2001		Pa 2142	895	35	911-735 BP	1030-1220 AD	Medieval	Localised burning/fire in structure (cabin)
Champoléon	Jas des Provençaux	1980	2003		Pa 2296+2309	660	50	680-549 BP	1270-1410 AD	Medieval	Fire place/bonfire dans structure (cabin)
Champoléon	Jas des Provençaux	1980	2003		Pa 2303	700	50	729-556 BP	1220-1400 AD	Medieval	Occupation layer within structure
Champoléon	Jas du Cros	2260	2002		Pa 2238	955	70	1050-726 BP	960-1250 AD	Medieval	Occupation layer within enclosure
Champoléon	Lac des Lauzons II	2190	2000		Pa 1971	3180	60	3559-3265 BP	1610-1310 BC	Middle - Late Bronze Age	Fire place/bonfire dans structure (cabin)

Site	Location			Lab code			BP	Date range	Period	Context
Champoléon	Lac des Lauzons II	2190	2000	Pa 1973	3470	100	3984-3474 BP	2050-1500 BC	Early Bronze Age	Occupation layer within structure
Champoléon	Vallon de la Vallette	2180	2003						Gallo-roman	Charcoal with medieval ceramics
Champoléon	Vallon de la Vallette	2180	2003	Pa 2304	1900	30	1923-1737 BP	20-220 AD	Gallo-roman	Charcoal with medieval ceramics
ORCIERES	Cabanne de la Barre	2200	2001	Pa 2299	1210	60	1275-985 BP	680-980 AD	Early Medieval	Occupation layer within enclosure
Orcières	Chapeau Roux	2340	2003	Pa 2139	1105	35	1080-931 BP	880-1020 AD	Medieval	Fire place/bonfire within structure (cabin)
Orcières	Chapeau Roux	2340	2003	Pa 2315	1715	60	1816-1447 BP	130-440 AD	Gallo-roman	Occupation layer within structure
Orcières	Chapeau Roux	2340	2003	Pa 2302	555	40	647-515 BP	1300-1440 AD	Medieval	Occupation layer within structure
Orcières	Chapeau Roux	2340	2003	Pa 2300	995	40	970-795 BP	970-1160 AD	Medieval	Occupation layer within structure
Orcières	Chapeau Roux	2340	2003	Pa 2295	200	30	305-4 BP	1640-1950 AD	Modern	Abandonment layer within structure (cabin)
Orcières	Chapeau Roux	2340	2003	Pa 2297	275	45	478-3 BP	1480-1680 AD	Modern	Occupation layer within structure
Orcières	Jujal I	2140	2001	Pa 2145	2945	35	3238-2975 BP	1270-1010 BC	Late Bronze Age	Localised burning/fire within enclosure
Orcières	Jujal I	2140	2001	Pa 2141	3145	60	3548-3212 BP	1530-1250 BC	Middle - Late Bronze Age	Localised burning/fire within enclosure
Orcières	Jujal III	2140	2001	Pa 2140	3275	40	3611-3400 BP	1690-1440 BC	Middle - Late Bronze Age	Localised burning/fire within enclosure
ST JEAN ST NICOLAS	Clot Lamiande	2140	2002	Pa 2240	1620	55	1691-1386 BP	320-570 AD	Late Antiquity	Occupation layer within structure
St-Jean-St-Nicolas	Clot Lamiande	2140	2002	Pa 2207	1230	30	1261-1069 BP	680-890 AD	Early Medieval	Pre-construction layer beneath structure (cabin)
St-Jean-St-Nicolas	Clot Lamiande	2140	2002	Pa 2205	960	35	934-790 BP	1000-1170 AD	Medieval	Occupation layer within structure
St-Jean-St-Nicolas	Clot Lamiande II	2140	2002	Pa 2206	1145	50	1220-938 BP	770-1000 AD	Early Medieval	Occupation layer within structure
St-Jean-St-Nicolas	Col du Palastre	2200	2002	Pa 2208	240	30	424-4 BP	1630-1810 AD	Modern	Fire place/bonfire within structure (cabin)
St-Jean-St-Nicolas	Col du Palastre	2200	2002	Pa 2239	1915	80	2059-1628 BP	110BC-260 AD	Late Iron Age - gallo-roman	Erosion layer within structure (cabin)
St-Jean-St-Nicolas	Col du Palastre II	2200	2002	Pa 2236	2770	95	3161-2740 BP	1220-790 BC	Late Bronze Age	Occupation layer within enclosure

Table 11.1 Radiocarbon dates for all excavated sites/structures from the Ecrins National Park (continued over).

activity was intensive or extensive. Notwithstanding evidence for increased activity at lower altitudes, and regional evidence for complex settlement hierarchies (Garcia 2002, 2005), the notion that such demographic pressure 'pushes' activities into the higher altitudes does not necessarily hold true. This seemingly counter-intuitive phenomenon also characterises the Roman period.

The Roman period saw the expansion of agricultural activities and the development and diversification of cultivated plants in the valley bottoms (Court-Picon 2007; Richer 2009): *Juglans* (walnut), with some *Castanea sativa* (chestnut) emerged as ubiquitous cultivated trees here. The emergence of densely settled areas comprised of complex agricultural systems exploiting most niches in the lower altitudes and slopes of the valleys characterised this period. *Brigantium* (modern-day Briançon) was the key town just to the northeast of our valleys and *Vapincum* (modern-day Gap) was situated in the plain immediately to the south of the Champsaur.

In some valley bottoms, evidence for activity only comes from isolated finds such as those from Saint-Laurent du Cros in the Champsaur where fragments of sculpture have been found, along with an altar dedicated to Mars and a stamped bronze pot at Forest-Saint-Julien (Roman 1888; Reynier 1923 (re-edited 1992); Simon 1990; Ganet 1995; Ricou 2002; Segard 2009). The Durance Valley saw the development of the main communications route from southern Provence into the Alps. This included the construction of a *mutatio* (road station) at Rama on a terrace of the Durance River, close to the entrance of the Freissinières Valley. Excavations here revealed substantial archaeological remains (Mocci *et al.* 2010). Conversely, there is little archaeological evidence for intensive pastoralism; the palaeo-ecological evidence does imply continued high-altitude pastoral activity, but not the increase that one might expect during this period. In fact, a contraction in activity is implied in some areas. There are few built structures from this period in the high altitude zone. One site, Faravel XIV (2450m, Freissinières, 50 BC– AD 80), is constituted by an ephemeral circular foundation that encloses an area of about four square-metres. Meanwhile, two areas within multi-period structures from the Champsaur have also yielded radiocarbon dates that correspond with the Roman period.

Whilst the palynological and archaeological evidence demonstrates that there was an increasingly intensive use of lower altitudes for arable agriculture and arboriculture, there is little evidence for valley bottom pastoral activity. Even though there is little doubt that pastoral activity did occur in these areas, we should ask if more intensive pastoralism took place in other parts of the region. It is possible that there was no need for extensive/intensive high altitude pasture if lowlands were cleared and integrated. We have to consider the possibility that the Roman economy was organised in such a way that intensive pastoralism took place in areas beyond the valleys that we have studied. For example, the well-known enclosures on the Crau (to the east of Camargue in the south of France) constitute the evidence for what we assume is winter pasturing of sheep. There is however a possibility that the Crau was an element within a system of reverse transhumance, this area being used during the summer (Leveau and Segard 2004:101–102), or a system of local transhumance with flocks moving between the Crau, the Camargue and the hilly areas to the north (the Alpilles). Some enclosures on the Crau have been dated to the 2nd and 3rd centuries AD, and are considered part of an intensive wool production system (Leguilloux 2003). Despite this extensive evidence for Roman pastoralism at low altitude, this still leaves us with the problem of limited structural evidence for high altitude summer pasturing.

Medieval Periods

Moving into the historic periods, we see the intensification of activity across all altitudinal zones. This process is one that continues despite climatic deterioration during the Post-Medieval Period. Regional and supra-regional economic processes became increasingly important, with mineral extraction (especially silver) developing as a strong economic force in our region. Whereas mining enticed some, other factors, including political and religious motivations, attracted people to the southern French Alps. Religious dissention, comprising the movement of peoples attempting to escape the Catholic Church in some urban centres, led to significant influxes of people during the 15th century (Audisio 1998). The other important economic activity was of course the development of long-distance transhumance. All of these economic and political processes intersected to influence the nature and intensity of summer upland activities.

From as early as the 6th/7th centuries we see the establishment of settlements and evidence for human activity across most of the attitudinal zones. Changes in vegetation composition (as inferred from our pollen diagrams) indicate the development of intensified agricultural activities in the valley bottoms of the Lower Champsaur and the Durance Valley. As with the preceding periods, we present here a brief overview of the nature of evidence for activity at high altitude.

The palynological evidence demonstrates increases in anthropogenic herbaceous indicators, as well as the intensification of pastoral activity. Some consider that early medieval (6th–10th centuries) long-distance transhumance was unlikely (Coste 1994:66); therefore, it is possible that the pastoral signals seen in our study areas reflect local, short-distance transhumance. The most impressive expansion of pastoral activity occurred between the 10th and 11th centuries. This sudden increase of pastoral activity might represent the origins of the great summer transhumance of sheep from the winter pastures of Provence up into to the southern Alps.

The high altitude archaeological evidence for this period is constituted by numerous structures that date to both the Medieval and Post-Medieval Periods (Fig. 11.6). We have found and plotted roughly 200 structures across our study area, including one group of structures that could be characterised as a high altitude hamlet at Serre de l'Homme (2200m).

In the Upper Champsaur Valley, four archaeological sites have been dated to the period, AD 1000–1300, all located at altitudes between 1950m and 2357m. Although two of these sites appear to have been abandoned at some point during the 10th or 11th centuries, other sites attest to activity during the 11th–12th centuries (Palet-Martinez et al. 2003; Mocci et al. 2006).

In the Freissinières Valley, the Medieval Period is represented by four excavated sites, plus many other structures that share similar morphological characteristics. Two phases of medieval activity have been identified at Faravel: the Early Medieval Period and a later phase (mid-11th century and late 13th century) attested to at other structures on this plateau. Two types of structure have been identified: huts, and corrals or animal pens. Animal pens have also been identified in the Upper Champsaur; these can be oval, circular, or polygonal, and they vary in size (Fig. 11.7). They always comprise dry-stone walls that were often 'adapted' to the terrain in that they exploit existing natural ditches or rock falls. They are usually accompanied by one or more rectangular or oval cabins (10–20m^2).

It is important to reiterate the fact that there are numerous, unexcavated, typologically similar structures across our study area. There is no doubt that the Medieval Period saw periods of extensive and intensive activities across all of the altitudinal zones.

Whilst we have emphasised the importance of pastoral activity, we should not lose sight of the other key economic activities that would have taken place in our valleys, in particular, mining. There are a number of silver mines developed in this area, many centred on the appositely named town of Argentières-la-Bessée (Ancel 1998, 1999; Py 2007; V. Py 2009). There are at least three or four mines in the immediate area around Faravel and Fangeas. These mines would have attracted a good number of people into the high altitude zones, and we have to consider the probability that many of the groups of structures were in fact seasonal shelters for groups of miners, or entire families. This period therefore witnessed the development of a dual economy, the combination of pastoralism and mining.

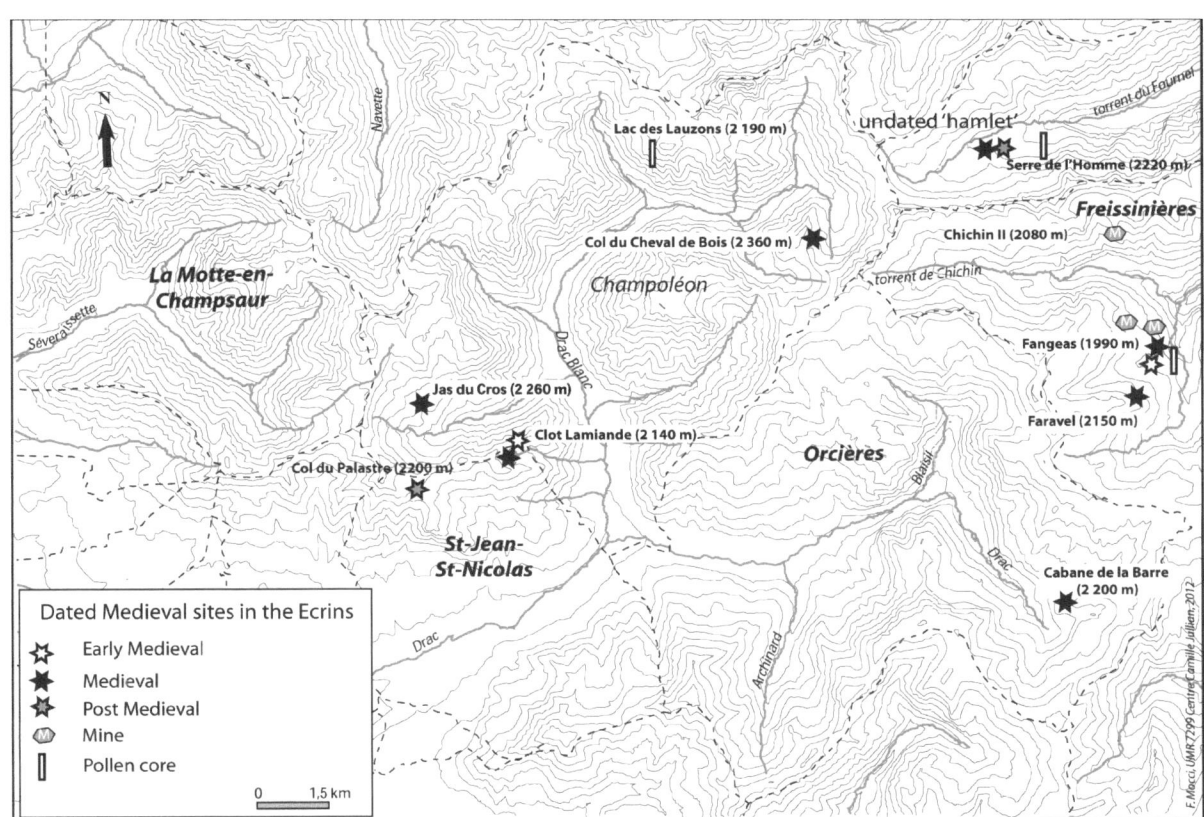

Figure 11.6   Key (excavated) high altitude medieval and post medieval sites in the Ecrins (plan - V. Dumas, F. Mocci, K. Walsh, J.-M. Palet Martinez).

Figure 11.7　The Roman and post-medieval enclosure of Col du Palastre in the Upper Champsaur (photo K. Walsh).

In fact, one may have benefited the other. Mining requires wood and timber for two processes: fire-setting, and then structural supports for shafts (even in shallow mines). Consequently, large areas of forest would have been cleared. These clearances would have facilitated pasture creation.

Discussion

Whilst there is no doubt that medieval and post-medieval activity levels were higher than those in the Bronze Age, there is no evidence for incremental or continuous increases in human activity between these two periods. What we do see is diachronic variations in the extent and intensity of human activity in the high altitude zone. These variations in the levels of activity do not appear to be directly influenced by changes in climate; the development of a 'busy' high altitude landscape during the Little Ice Age attests to this as much as evidence from any other period.

It is apparent that climatic variation does not really affect niche capacity in the southern French Alps. Whereas some have argued for climatic deterioration (lower mean temperatures and increased precipitation) as a constraint on human activity (Bocquet 1997), we have to consider how the opposite climatic trend is problematic; dry summers would have posed a greater threat to pasture resilience. If rain does not fall towards the end of the summer, then pasture productivity is lower in the following year (Mosimann *et al.* 2004). It is not climate that affects niche capacity, but intra-annual weather patterns.

As well as considering the influence of climate on the nature and levels of activity in upland areas, we also need to assess the notion that high altitude zones act as a demographic 'buffer', or pressure release. More specifically, this notion is often situated within the broader conception of pastoral activity as an enterprise that developed as agricultural activity reached its ecological limits in low altitude areas.

The development of pastoralism and the concomitant exploitation of hitherto unexploited niches *can* be considered part of the secondary products revolution. There is no doubt that many parts of southern Europe saw an intensification in the exploitation of domestic animals during the later Neolithic and into the Bronze Age (Sherratt 1981; Greenfield 2010). Research in the Alps has emphasised the importance of not only identifying overall increases in animal exploitation, but also gauging how this change in agricultural practice manifested itself in new engagements with the landscape, including new forms of mobility and exploitation of certain niches that had not been intensively used prior to this period. A related issue is that we do not know if our sites in the French Alps were nodes within a system of nomadic or transhumant pastoralism, and whether we can really differentiate between Halstead's 'pastoralism or household herding' (1996).

In the lower altitudes of central and southern Provence (and the Alpes-de-Haute-Provence), there are four times as many sites dated to the Late Neolithic than the preceding Middle Neolithic. Faunal assemblages from Late Neolithic sites suggest the intensification of the exploitation of both sheep and goats (Blaise 2005:192-5). Whereas a high proportion of animals were killed when they were young (for their tender meat), many animals were kept alive, presumably for their wool and/or milk. During the Couronnien and Rhône-Ouvèze cultural periods, there were equal numbers of low-lying sites and sites on higher (hilly) zones. During the early Beaker phase, the proportion of sites on higher ground was greater, with these proportions reversing during the later Beaker period (Blaise 2005:207). Consequently, it seems likely that the lower areas of Provence witnessed the development of local transhumance, or quite possibly, household, or village herding, the pattern and organisation of which fluctuated with time.

Although these landscapes in lower Provence are not as mountainous as the alpine areas just to the north, similar issues would have been common to both zones. For local transhumant systems, available pasture might be limited by geomorphology and vegetation, and low-lying land would either have been allocated to arable production, or covered with relatively dense forest. Moreover, in the alpine zones, narrow valley bottoms would have been susceptible to extreme flooding and erosion events that would have put pressure on valley bottom pasture. It is important to note that there are very few known riverside settlements across the region, probably due to the fact the Alpine and Mediterranean rivers do experience regular flash-flooding, thus rendering settlement and agricultural activity problematic. Consequently, there may well have been environmental processes that did encourage the development of high altitude pastoralism, in particular, the presence of dense low-altitude forest that would have restricted the clearance of land for pasture – any open land was probably exploited for arable agriculture, as is suggested by our pollen diagrams. With low-altitude land at a premium, movement into unwooded or easily manipulated woodland at high altitude might have made sense; the upper forest edge is easier to manipulate and clear via burning (as attested by our charcoal assemblages). It is likely that the secondary products revolution, or at least that part concerned with increased pasturing, took place in this part of the Alps during the mid-to-late 3rd millennium. In other areas of southern Europe, this period saw the expansion and intensification of the exploitation of these secondary products (Greenfield 2010). We should not forget that evidence for enhanced wool and milk production in central Europe dates to the Late Neolithic (Greenfield *et al.* 1988).

Whilst the general principles of Sherratt's model (Sherratt 1981, 1983) are still relevant, some authors have demonstrated the importance of assessing regional variability and, where possible, looking carefully at the faunal evidence (Marciniak 2011). In the Balkans, evidence for transhumant pastoralism (as inferred from the zoo-archaeological record) suggests a possible Eneolithic–Bronze Age origin. However, as the authors admit, the zoo-archaeological evidence (age profiles of culled animals across sites from different altitudes) does not convincingly support the contention that transhumance was taking place (Arnold and Greenfield 2004).

We should of course question purely utilitarian models, and consider the possibility of changes in ritual activities with newly emerging complex societies placing more emphasis on ritual feasting (Keswani 1994). There is little doubt that many aspects of human lifeways, including ritual activity, did change during the late 3rd and 2nd millennia (Kristiansen and Larsson 2005).

Evidence for intensified domestic and 'ritual' consumption of animals is attested across a number of sites in France from the Late Bronze Age into the Iron Age (Meniel 2005). There is no doubt that the Iron Age in particular saw an increase in ritual feasting in Gaul. This was even true in high altitude zones. The burnt mound of Les Sagnes, situated at 1900m in the Alpes-de-Haute-Provence, is one enigmatic example where large quantities of burnt bone and pottery constitute a substantial mound (Garcia *et al.* 2007). However, despite this evidence for feasting at altitude, few data suggest an increase in high altitude pastoral activity. Pollen evidence suggests maintenance of pre-existing open upland areas, and most surprisingly, there are no convincing structures dated to this period in the high altitude zones. We have to consider the possibility that a site such as Les Sagnes represents a very localised feasting event within the context of southern French Alps, or that pastured animals were maintained at lower altitudes, with some brought up to the high altitude zone for fattening and slaughter. A mixture of sheep, cattle and pig occur in equal proportions in some of the contexts. Deer also constituted a sizeable proportion of the total assemblage (Columeau 2003).

Regional evidence from the south of France suggests substantial variations in the proportions of sheep/goat and cattle across major sites; what is apparent is that these two species were more popular than pig. The Late Bronze Age and Early Iron Age did see the development of pastoralism across the region. Ovicaprids dominate the total number of animals in many assemblages, but unsurprisingly the contribution of their meat is less important than other animals: 8–13% came from sheep/goat, and –65% from cattle, 15–25% from pig. Ovicaprids did of course provide a range of secondary products. The presence of many spindle whorls and loom weights across a number of sites attest to the importance of textile production (M. Py 2012:79–80), along with the first examples of sheep-shears, which appear during the Late Iron Age (M. Py 2012:258–60). What is perhaps more interesting is the reduction in ovicaprid numbers (along with cattle) from the 2nd century BC into the Roman period. Py suggests the development of a more 'sedentary' form of animal husbandry in the region (2012:155) during the later Iron Age, despite the reduction in the quantities of meat from game (notably from *c.* 375 BC onwards), ovicaprid meat consumption fell back, whilst pork consumption increased, a trend that continued into the Roman period.

In many ways, the Roman Period witnessed the maintenance of pre-Roman 'indigenous' levels of activity in the high altitude areas. Despite the extensive evidence for increased activity towards the lower altitudes, there is little convincing evidence for demographic pressure 'pushing' people up into the higher altitudes. In fact, we might even argue that new urban areas, concomitant route-ways, and stations (*mutationes*) actually attracted people down towards these lower elevations, as these were the poles of economic, cultural, and political activity.

Across the wider region (southeastern France) Columeau's synthesis of late Roman faunal remains from a range of different site types (towns, villas, farms) demonstrates the importance of sheep and goat, although, by volume of meat, cattle still provided a substantial proportion of the total meat consumed. It was from the 7th or 8th centuries AD that the rearing of cattle started to diminish (Columeau 2000:355). In some areas, such as the Var, there were substantial variations in the proportions of cattle and ovicaprids across different villa sites. In some instances cattle were more common, with pig also comprising a reasonable proportion of the assemblages on some villas (Leguilloux 1989), whilst in Marseille, the faunal assemblages from late Roman sites around the city are dominated by sheep/goat (Leguilloux 1995). It seems likely that the assemblages from Roman and early medieval sites with relatively high numbers of ovicaprids probably participated in systems of local transhumance within southern Provence, not long-distance transhumance between this area and the Alps (Columeau 2001:134).

As we move into the Medieval Period, we see the development of extensive and intensive forms of upland activities, with the possibility that some groups operated two transhumant systems in parallel. As Carrer's ethno-archaeological research in the Italian Alps has demonstrated, it is quite feasible for two forms of summer high altitude pastoralism to co-exist; for example, short-term migrations, between a valley-bottom alpine and nearby high-altitude pastures (*alpeggio* (alpine pastoralism)), whilst long-distance *transhumanza* constitutes the movement of sheep from the plains up in to the mountains (Carrer 2013:2).

Chang's and Tourtellotte's ethno-archaeological research demonstrated for the Grevena area in Greece that certain regions support a wide range of mobile pastoral strategies "...ranging from seasonal pastoral transhumance to year-round, village-based agro-pastoralism" (1993:261). We might consider that the large, multi-structure sites from the Medieval Period were the product of something akin to nomadic pastoralism, with flocks accompanied by the residential group (Salzman 2004), although there is little doubt that the families concerned came from the same valley, or an adjacent one.

The Post-Medieval Period comprises the Little Ice Age. Whilst traditional, climatically determined models of human activity in high altitude zones might retrodict declines in high altitude activity, this period actually witnessed the zenith of high altitude activity, including mining, summer hay-making, and pastoralism. In our study area, we have three zones that comprise groups of medieval structures, in one instance a set of structures and probable enclosures that constitute a settlement that could easily be described a hamlet. This settlement is located at above 2200m in the Fournel Valley (Serre de l'Homme) and must represent the movement of one or more entire families, or groups of shepherds and miners into this area each summer.

Demographic pressure does not appear to be a factor before the Post-Medieval Period. The fact that the low altitude areas might have been susceptible to enhanced environmental hazards (such as flooding) might have rendered the high altitudes more attractive. Moreover, certain forms of climatic determinism can in fact be inverted; for example, enhanced precipitation levels during the Little Ice Age in a supra-Mediterranean mountain range will probably enhance grass growth (pasture richness) and thereby increase slope stability (we should note the richness of pasture in the northern Alps, where cows can in fact be pastured for the production of large quantities of milk during the summer.

Conclusions

In the southern French Alps, the development of high altitude farming activities, in particular, pastoralism, can be traced back to the end of the Neolithic, with the Bronze Age witnessing the establishment of many built structures above 2000m. Traditional ecological and demographic models have been reassessed, and their shortcomings considered. It is unlikely that demographic pressure 'pushed' pastoralists into the high altitude zones during the late 3rd and 2nd millennia. Moreover, there is little evidence to suggest that people were responding to long-term trends in climate and vegetation change (e.g. the movement of the tree line). A complex network of cultural and economic processes controlled the variation in the extent and intensity of high altitude activity. The late 3rd and 2nd millennia saw the intensification of secondary products exploitation (including wool and milk (by-products). This, combined with the possible influence of new 'Beaker' populations, or cultural ideas, engendered the development of different attitudes towards high altitude zones. One scenario, where population pressure might have encouraged exploitation of high altitude areas, is that where dense, valley-bottom forest actually rendered the low altitude expansion of arable and pastoral production difficult, and therefore, the high altitude areas, within the treeline, or just beyond, became attractive. Although it is logical that pastoral transhumance would not necessarily have existed in subsistence economies, and that lowland market forces triggered the development of more intensive pastoralism, it is quite possible that local subsistence based transhumance did develop in certain landscapes where movement into the high altitude zone is not necessarily considered risky and where valley bottom land was at a premium.

After the Bronze Age, our palaeo-ecological evidence suggests similar levels of high altitude activity running through the Iron Age and into the Roman Period. However, there are very few built structures dating to these periods. Despite undisputed increases in settlement and population levels in the valley bottoms, there is no evidence for demographic pressure inciting movements towards the higher altitudes. Furthermore, variations in climate (climatic deterioration during the Iron Age, and notional improvement during the Roman Period) seem to have had little influence over the nature and intensity of activity above 2000m. Although there is no doubt that the treeline descended over time, there is no evidence that suggests periods of reduction in pasture biodiversity, something that might conceivably explain the waning of activity at any point in the past.

The Medieval and Post-Medieval Periods witnessed substantial increases in high altitude activity, partly a consequence of mining of precious metals, but also due to the development of long-distance transhumance of large flocks of sheep. These activities continued through the Little Ice Age, once again demonstrating that climate has never actually structured the nature and extent of most alpine activities.

Acknowledgments

The case studies employed in this paper are the fruits of collaborative work. In particular, we would like to thank Professor Philippe Leveau for his support over the last 13 years. We should also like to thank the colleagues who work with us on this collaborative project, in particular, Celine Bressy, Aline Durand, Vanessa Py, Stéphane Renault, Philippe Ponel. We have also benefited from the support and collaboration (on and off site) of Vincent Dumas (Centre Camille Jullian, CNRS), as well as numerous friends and students who have worked with us over the years. The following organisation should also be thanked for financial support and/or support in kind: The Parc national des Ecrins, the Communauté des Communes du Pays des Ecrins, Ministère de la Culture, Service Régional de l'Archéologie (Provence-Alpes-Côte-d'Azur), Conseil Régional Provence-Alpes-Côte-d'Azur, Conseil Général des Hautes-Alpes, and the Arts and Humanities Research Council for Richer's research studentship.

Bibliography

Ali, A. A., Carcaillet, C., Talon, B., Roiron, P. and Terral, J.F. 2005a. *Pinus cembra* L. (arolla pine), a common tree in the inner French Alps since the early Holocene and above the present tree line: a synthesis based on charcoal data from soils and travertines. *Journal of Biogeography* 32/9:1659–69.

Ali, A.A, Roiron P., Guendon J.-J., Poirier, P. and Terral J.-F., 2005b. Holocene vegetation responses to fire events in the inner French Alps (Queyras Massif): data from charcoal and geomorphological analysis of travertine sequences. *The Holocene* 15/1:149–55.

Ancel, B. 1998. La mine du Fournel (L'Argentière-la-Bessée, Hautes-Alpes, France): l'exploitation rationnelle aux Xième-XIVième siècles d'un filon de plomb argentifère. In L. Brigo and M. Tizzoni (eds.) *Mount Caliso and the Silver Deposits in the Alps from Ancient Times till the XVIII Century*. Civezzano: Fornace, pp. 161–93.

Ancel, B. 1999. La mine d'argent du Fournel. In G. Boetsch (ed.) *Histoire et anthropologie des populations de l'arc alpin: université d'été 1998*. Gap, Editions du Fournel, pp. 57–63.

Arnold, E. and Greenfield, H. 2004. A zoo-archaeological perspective on the origins of vertical transhumant pastoralism and the colonization of marginal habitats in temperate southeastern Europe. In M. Mondini, M. Munoz and S. Wickler (eds.) *Colonisation, Migration and Marginal Areas: zooarchaeological approach*. Oxford: Association for Environmental Archaeology, pp. 96–117.

Audisio, G. 1998. *Les Vaudois*. Paris, Fayard.

Billings, W.D. 1979. High Mountain ecosystems. In P.J. Webber (ed.) *High-Altitude Geoecology*. Boulder: Westview Press, pp. 97–125.

Blaise, É. 2005. L'élevage au Néolithique final dans le sud-est de la France; éléments de réflexion sur la gestion des troupeaux. *Anthropozoologica* 40/1:191–216.

Bocquet, A. 1997. Archéologie et peuplement des Alpes françaises du Nord. *L'Anthropologie* 101/2:291–393.

Carrer, F. 2013. An ethnoarchaeological inductive model for predicting archaeological site location: a case-study of pastoral settlement patterns in the Val di Fiemme and Val di Sole (Trentino, Italian Alps). *Journal of Anthropological Archaeology* 32/1:54–62.

Chang, C. and Tourtellotte, P.A. 1993. Ethnoarchaeological survey of pastoral transhumance sites in the Grevena Region, Greece. *Journal of Field Archaeology* 20/3:249–64.

Cherry, J.F. 1988. Pastoralism and the role of animals. In C.R. Whittaker (ed.) *Pastoral Economies in Classical Antiquity*. Cambridge: Cambridge Philological Society, pp. 6–34.

Claudin, J. and Miellet, P. 2000. *Atlas du Parc National des Ecrins*. Gap: Parc National des Ecrins.

Columeau, P. 2000. Consommation de viande et élevage dans la vallée des Baux de l'Age du Fer au Moyen Age d'après les vestiges osseux. *Travaux du Centre Camille Jullian* 26:347–57.

Columeau, P. 2001. Nouveau regard sur la chasse et l'élevage dans le sud et le sud-est de la Gaule, aux IVe et Ve s. ap. J.-C. et l'exemple de Constantine (B.-du-Rh.). *Revue archéologique de Narbonnaise* 34/1:123–37.

Columeau, P. 2003. Rapport Intermediaire: Projet Collectif de Recherche, *Histoire d'une vallée alpine*. D. Garcia and F. Mocci, (eds.) *L'Ubaye des âges des Métaux aux Temps modernes*. Unpublished report held at the Service Régionale de l'Archéologie, Aix-en-Provence.

Coste, P. and Coulet, N. 1994. Que sait-on des origines de la transhumance en Provence? In J.-C. Duclos and A. Pitte (eds.) *L'homme et le mouton dans l'espace de transhumance*. Grenoble: Glenat, pp. 65–70.

Court-Picon, M. 2007. *Mise en Place du Paysage dans un Milieu de Moyenne et Haute Montagne du Tardiglaciaire à l'époque actuelle*. PhD, Sciences et Techniques de l'Université de Franche Comté.

Curdy, P. and Praz, J.C. 2002. *Premiers hommes dans les Alpes de 50,000 à 5,000 avant Jésus-Christ*. Sion, Catalogue d'exposition, Musées cantonaux du Valais.

Della Casa, P. 1999. *Prehistoric Alpine Environment, Society and Economy*. Bonn: Rudolf Habelt GmbH.

Evans, J., Stoodley, N. and Chenery, C. 2006. A strontium and oxygen isotope assessment of a possible fourth century immigrant population in a Hampshire cemetery, southern England. *Journal of Archaeological Science* 33/2:265–72.

Fedele, F.G. 1999. Peuplement et circulation des matériaux dans les Alpes occidentales du Mésolithique a l'Age du Bronze. In A. Beeching (ed.) *Circulations et identités culturelles alpines* Full title? *la fin de la préhistoire*. Valence: Centre d'Archéologie préhistorique, pp. 331–58.

Ganet, I. 1995. *Les Hautes-Alpes*. Paris: Foundation Maison des Sciences de l'Homme.

Garcia, D. 2002. Dynamique territoriale en Gaule Méridionale durant l'âge du fer. In D. Garcia and F. Verdin (eds.) *Territoires Celtiques: espaces ethniques et territoires des agglomérations protohistoriques d'Europe occidentale*. Paris: Editions Errance, pp. 88–102.

Garcia, D. 2005. Urbanization and spatial organization in southern France and north-eastern Spain during the Iron Age. In B. Cunliffe and R. Osborne (eds.) *Mediterranean Urbanization 800–600 BC*. Oxford, Oxford University Press/British Academy, pp. 169–86.

Garcia, D. 2008. Objets exogènes et faciès culturels dans le sud des Alpes occidentales. In H. Richard and D. Garcia (eds.) *Le peuplement de l'arc alpin*. Paris: CTHS (Le comité des travaux historiques et scientifiques), pp. 259–74.

Garcia, D., Mocci, F. and Walsh K. 2007. Archéologie de la vallée de l'Ubaye (Alpes-de-Haute-Provence, France) : présentation des premiers résultats d'un Projet Collectif de Recherche. *Preistoria Alpina* 42:23–48.

Greenfield, H.J. 2010. The Secondary Products Revolution: the past, the present and the future. *World Archaeology* 42/1:29–54.

Greenfield, H.J., Chapman, J., Clason, A., Gilbert, A.S., Hesse, B. and Milisauskas, S. 1988. The origins of milk and wool production in the Old World: a zooarchaeological perspective from the central Balkans [and comments]. *Current Anthropology* 29/4:573–93.

Halstead, P. 1987. Traditional and ancient rural economy in Mediterranean Europe: plus ça change? *The Journal of Hellenic Studies* 107:77–87.

Halstead, P. 1996. Pastoralism or household herding? Problems of scale and specialization in early Greek animal husbandry. *World Archaeology* 28/1:20–42.

Harrison, R. and Heyd, R. 2007. The transformation of Europe in the third millennium BC: the example of Le Petit-Chasseur I + III (Sion, Valais, Switzerland). *Praehistorische Zeitschrift* 82/2:129–214.

Keswani, P.S. 1994. The social context of early animal husbandry in early agricultural societies: ethnographic insights and an archaeological example from Cyprus. *Journal of Anthropological Archaeology* 13:255–77.

Koster, H.A. and Chang, C. 1994. Introduction. In H.A. Koster and C. Chang (eds.) *Pastoralists at the Periphery: Herders in a Capitalist World*. Tucson: University of Arizona Press pp. 1–15.

Kristiansen, K. and Larsson, T.B. 2005. *The Rise of Bronze Age Society: travels, transmissions and transformations*. Cambridge: Cambridge University Press.

Leguilloux, M. 1989. La faune des villae gallo-romaines dans le Var: aspects économiques et sociaux. *Revue archéologique de Narbonnaise* 22/1:311–22.

Leguilloux, M. 1995. Alimentation et élevage à Marseille au Ve siècle après J.-C. d'après les études de faunes. *Méditerranée* 82/3–4:85–92.

Leguilloux, M. 2003. Les bergeries de la Crau: production et commerce de la laine. *Revue archéologique de Picardie* 1/1–2:339–46.

Leveau, P. and Segard, M. 2004. Le pastoralisme en Gaule du sud entre plaine et montagne: de la Crau aux Alpes du sud. *Pallas* 64:99–113.

Lichardus, J., Lichardus-Itten, M., Bailloud, G. and Cauvin, J. 1985. *La Protohistoire de l'Europe: le Néolithique et le Chalcolithique*. Paris, Presses Universitaires de France.

Magny, M., Peyron, O., Gauthier, E., Rouèche, Y., Bordon, A., Billaud, Y., Chapron, E., Marguet, A., Pétrequin, P. and Vannière, B. 2009. Quantitative reconstruction of climatic variations during the Bronze and Early Iron Ages based on pollen and lake-level data in the NW Alps, France. *Quaternary International*, 200/1–2:102–10.

Marciniak, A. 2011. The Secondary Products Revolution: empirical evidence and its current zooarchaeological critique. *Journal of World Prehistory* 24/2:117–30.

Meniel, P. 2005. Sur les traces du mouton en Gaule. *Revue de Paleobiologie* 10:33–45.

Mocci, F., Segard, M., Walsh, K., Golosetti, R., Dumas, V., Cenzon-Salvayre, C. and Talon, B.2010. données récentes sur l'occupation humaine dans les Alpes méridionales durant l'Antiquité – recent research on human settlement in the southern Alps during Antiquity. In S. Tzortzis and X. Delestre (eds.) *Archéologie de la montagne européenne: actes de la table ronde internationale de Gap, 29 septembre–1er octobre 2008 Paris./ Aix-en-Provence*. Errance: Centre Camille Jullian, pp. 308–23.

Mocci, F., Tzortzis, S., Palet-Martinez, J.M., Segard, M. and Walsh, K. 2006. Peuplement, pastoralisme et modes d'exploitation de la moyenne et haute montagne depuis la Préhistoire dans le Parc national des Ecrins (vallées du haut Champsaur et de Freissinières, Hautes-Alpes). In A. Bouet and F. Verdin (eds.) *Territoire et Paysages de l'âge du Fer au Moyen-Age: mélanges offerts à Philippe Leveau*. Bourdeaux: Université de Bourdeaux, pp. 197–212.

Mocci, F., Walsh, K., Richer, S., Court-Picon, M., Talon, B., Tzortzis, S., Palet-Martinez, J.M. and Bressy, C. 2009. Archéologie et paléoenvironnement dans les Alpes méridionales françaises: hauts massifs de l'Argentièrois, du Champsaur et de l'Ubaye (Hautes-Alpes et Alpes de Haute Provence) (Néolithique final – début de l'Antiquité). *Cahiers de Paléoenvironnement* 6:235–54.

Mosimann, E., Stevenin, L., Girard, D., Lúscher, A., Jeangros, B., Kessler, W., Huguenin, O., Lobsiger, M., Millar, N. and Suter, D. (2004). Consequences of drought on grazing management in Switzerland. In A. Lúscher, B. Jeangros, W. Kessler, O. Huguenin, M. Lobsiger, N. Millar and D. Suter (eds.) *Land Use Systems in Grassland Dominated Regions: proceedings of the 20th General Meeting of the European Grassland Federation, Luzern, Switzerland, 21–24 June 2004.* Verlag der Fachvereine an den schweizerischen Hochschulen und Techniken, Zurich, pp. 665–7.

Nicod, P.Y. 2008. Les premières sociétés agropastorales dans les Alpes Occidentales. In J.P. Jospin and T. Favrie (eds.) *Premiers Bergers des Alpes*. Gollion: Infolio, Musée Dauphinois, pp. 45–51.

Palet-Martinez, J.M., Ricou, F. and Segard, M. 2003. Prospections et sondages sur les sites d'altitude en Champsaur (Alpes du sud). *Archeologie du Midi Medieval* 21:199–210.

Price, T.D., Knipper, C., Grupe, G. and Smrcka, V. 2004. Strontium isotopes and prehistoric human migration: the Bell Beaker period in central Europe. *European Journal of Archaeology* 7/1:9–40.

Py, M. 2012. *Les gaulois du midi: de la fin de l'âge du Bronze à la conquête romaine*. Paris, Editions Errance.

Py, V. 2009. Mine, bois et forêt dans les Alpes du Sud au Moyen Âge: approches archéologique, bioarchéologique et historique. Unpublished PhD, *Laboratoire d'archéologie médiévale méditerranéenne (LAMM)*. Université de Provence: Aix-Marseille I.

Py, V. and Ancel, B. 2007. Exploitation des mines métalliques de la vallée de Freissinières (Hautes-Alpes, France): contribution à l'étude de l'économie sud-alpine aux IXe–XIIIe siècles. In P. Della-Casa, and K. Walsh (eds.) *Interpretation of Sites and Material Culture from Mid-High Altitude Mountain Environments*. Special Issue of *Preistoria Alpina*, Trentino, pp. 83–93.

Reynier, J.A. 1923 (re-ed 1992). *Notice sur Faudon et les deux Ancelles? (Hautes-Alpes).* Gap: Librairie des Hautes-Alpes.

Ricou, P. 2002. *Structuration d'un paysage de montagne: clapiers et bocage, l'exemple du Champsaur.* Unpublished Masters Dissertation, Université de Provence, Aix-Marseille.

Richer, S. 2009. *From Pollen to People: the interaction between people and their environment in the mid- to high- altitudes of the southern French Alps.* Unpublished PhD. University of York.

Roman, J. 1888. *Répertoire archéologique du département des Hautes-Alpes*. Paris, Hachette.

Salzman, P.C. 2004. *Pastoralists: equality, hierarchy, and the state.* Boulder and Oxford: Westview Press.

Segard, M. 2009. *Les Alpes occidentales à l'époque romaine: développement urbain et exploitation des ressources des régions de montagne (Italie, Gaule Narbonnaise, provinces alpines).* Aix-en-Provence: Editions Errance / Centre Camille Jullian.

Sherratt, A. 1981. Plough and pastoralism: aspects of the secondary products revolution. *Patterns of The Past: studies in honour of David Clarke.* Cambridge: Cambridge University Press, pp. 261–305.

Sherratt, A. 1983. The secondary exploitation of animals in the Old World. *World Archaeology* 15/1:90–104.

Simon, J. 1990. *Un ancien bourg gallo-romain: Forest-Saint-Julien en Champsaur.* Gap: Musée Départemental de Gap.

Voruz, J.L. 1996. Chronologie absolue de l'âge du Bronze Ancien et Moyen. In C. Mordant and O. Gaiffe (eds.) *Cultures et sociétés du Bronze ancien en Europe.* Paris: CTHS (Le comité des travaux historiques et scientifiques), pp. 97–164.

Walsh, K. 2005. Risk and marginality at high altitudes: new interpretations from fieldwork on the Faravel Plateau, Hautes-Alpes. *Antiquity* 79:289–305.

Walsh, K., Mocci, F. and Palet-Martinez, J.M. 2007. Nine thousand years of human/landscape dynamics in a high altitude zone in the southern French Alps (Parc National des Ecrins (Hautes-Alpes). In P. Della-Casa, and K. Walsh (eds.) *Interpretation of Sites and Material Culture from Mid-High Altitude Mountain Environments*. Special Issue of *Preistoria Alpina*, Trentino, pp. 83–93.

Walsh, K., Court-Picon. M., de Beaulieu, J. L., Guiter, F., Mocci, F., Richer, S., Sinet, R., Talon, B. and Tzortzis, S. in press. A historical ecology of the Ecrins (southern French Alps): archaeology and palaeoecology of the Mesolithic to the Medieval Period. *Journal of Quaternary Science*.

Willaume, M. 1991. L'Age du Fer dans les Hautes Alpes. In *Archéologie dans les Hautes-Alpes*. Gap: Musée Départemental de Gap, pp. 133–44.

M. Court-Picon: IMBE (UMR CNRS-IRD 7263), Aix-Marseille Université (AMU), Européôle méditerranéen de l'Arbois, BP 80, F-13545 Aix-en-Provence cedex 04, France. Royal Belgian Institute of Natural Sciences, Department of Palaeontology, Vautier street 29, B-1000 Brussels, Belgium.
Email: mona.courtpicon@naturalsciences.be

J.-L. de Beaulieu: IMBE (UMR CNRS-IRD 7263), Aix-Marseille Université (AMU), Européôle méditerranéen de l'Arbois, BP 80, F-13545 Aix-en-Provence cedex 04, France.
Email: jacques-louis.debeaulieu@imbe.fr

F. Guiter: IMBE (UMR CNRS-IRD 7263), Aix-Marseille Université (AMU), Européôle méditerranéen de l'Arbois, BP 80, F-13545 Aix-en-Provence cedex 04, France.
Email: frederic.guiter@imbe.fr

Florence Mocci: Centre Camille Julian, CNRS, MMSH, Rue Chateau de l'Horloge, Aix-en-Provence.
Email: Mocci@mmsh.univ-aix.fr

S. Richer: Worcestershire Wildlife Trust, Hindlip, Worcestershire, WR3 8SZ, UK.
Email: suzi@worcestershirewildlifetrust.org

B. Talon: IMBE (UMR CNRS-IRD 7263), Aix-Marseille Université (AMU), Européôle méditerranéen de l'Arbois, BP 80, F-13545 Aix-en-Provence cedex 04, France.
Email: brigitte.talon@imbe.fr

Kevin Walsh: Dept of Archaeology, King's Manor, University of York, YORK, Y01 7EP, UK.
Email: Kevin.walsh@york.ac.uk

12. An Archaeological Approach to the *Brañas*: Summer farms in the pastures of the Cantabrian Mountains (northern Spain)

David González Álvarez, Margarita Fernández Mier and Pablo López Gómez

The pastures of the Cantabrian Mountains have constituted an important resource for the subsistence strategies of their inhabitants, from the Neolithic to the present day. Seasonal settlements in these upland areas – so-called brañas *in the case study – were used by shepherds in relation with herding mobility. In this paper we reflect on present-day brañas in order to understand different types of herding settlements in the archaeological record according to different mobility systems or pastoralist strategies. This way, we hope to gain a better understanding on archaeological issues from summer farms. In fact, this kind of approach has been really useful to encourage the discussion around our on-going research on cultural landscapes of the Cantabrian Mountains from a long-term chronological perspective.*

Introduction

The geography of the Cantabrian Mountains is complex, with clearly divided sectors and different biotopes found close together. This has facilitated the development of seasonal pastoralism which uses the mountain pastures at alpine and sub-alpine levels in this cordillera during the summer. This traditional transhumance dating back to at least later prehistory involves the use of seasonal summer farms, called *brañas* in the area under study here.

This study considers the different types of *brañas* or summer steadings found in the area from an ethnoarchaeological perspective, to assess their relevance in the social and productive practices of the contemporary shepherds in the Cantabrian Mountains. This leads to a regressive interpretation of the landscape in dialogue with the historical and archaeological research carried out by our work team. Thus, this research provides material data referring to the use and exploitation of the mountain areas which is then inserted into a long-term view of the herding practices and the use of the *brañas* in this area throughout its history.

Theoretical and methodological framework

Mountain areas are usually considered as marginal or threshold zones of human landscapes. Accordingly, little attention has been paid to them by archaeologists who tend to prefer geographically more hospitable areas. This lack of archaeological studies reinforces the idea that mountain areas in Europe are 'natural' landscapes that have hardly been transformed by human activity. However, these areas are crucial to the subsistence of their human inhabitants. The effect of anthropogenic activity has remodelled their appearance and composition to a considerable extent, as can be seen from palaeo-environmental studies carried out in different mountainous areas from the Iberian Peninsula (Ejarque *et al.* 2009; Gassiot *et al.* 2009; Moreno *et al.* 2011). For this reason, we consider it appropriate to approach these areas from the viewpoint of *cultural* landscapes (Sauer 1925), examining aspects such as the interaction between human communities and their environment, or the social and cultural discourse through which the human beings interpret their perceptions of their environment and make it meaningful.

In more specific terms, pastoralism is fundamental to the economic and cultural contexts of these mountains, where the *brañas* play a central role in the way of life of their inhabitants. These sites act as a base or starting point for the anthropisation processes linked to herding practices. An in-depth diachronic examination of their genealogy will therefore lead to a better understanding of crucial aspects of the landscape biography of the Cantabrian Mountains.

With this aim, it is essential to combine archaeological and historical interpretations with palaeo-environmental studies, calibrating diachronically the incidence of human pressure on the environment. The political and social contexts of each period must also be considered, with an evaluation of the technical possibilities available to the peasant communities at each point in time. Finally, to recognise the historical genealogy of the *brañas* is an essential step towards an understanding of the different livestock-rearing formulas, identifying in each case the models of pastoral mobility, and exam-

Keywords: Landscape archaeology, pastoralism, summer farms, ethnoarchaeology, anthropisation, upland archaeology, agrarian archaeology

Figure 12.1   Ruined hut in the summer braña of Saldepuesto (Cangas del Narcea, Asturias), mentioned in the medieval written sources from the Monastery of Corias as *brannia* de Soldepuesto (Garcia Leal 1998).

ining in the archaeological record the buildings typology and the settlement patterns of these summer farms. These are all fundamental aspects for understanding the processes of social and symbolic construction of the cultural landscapes in the Cantabrian Mountains. They also facilitate an understanding of the anthropisation processes throughout history.

Pastoralist activities form part of an integrated productive system which includes agriculture and sedentary animal husbandry. The types of settlements underpinning this system reflect this pattern, with temporary seasonal sites at higher altitudes and other permanent – or also seasonal – settlements lower down the mountain slopes or in the valleys. Research into the way of life of transhumant groups should therefore integrate their pastoralist activities with other aspects of their subsistence system. It is only by developing projects considering the territories and ways of life peculiar to these communities as a whole that a comprehensive picture can be achieved of these groups. Our work on the *brañas* of the Cantabrian Mountains is therefore considered within a wider research framework, focused on understanding the ways of life developed by the successive peasant communities who have lived in this territory over the course of history (Fernández Mier *et al.* 2013a).

The research carried out by this team of historians and archaeologists began with a preliminary phase combining the morphology of the sub-recent agrarian landscape and micro-toponymy of the area (Fernández Mier 1996). This strategy provided a better understanding of the transformation processes of land division and ownership in modern and contemporary eras, and also allowed us to detect traces of older field systems which in some cases may date back to the Early or High Middle Ages. These regressive readings of the landscape have been combined with the careful examination of written documents (Fernández Conde 2001). Thus, we were able to detect mentions of *brañas* in the written sources as far back as ten centuries ago (Fig. 12.1), and to evaluate the use of summer pastures for herding by medieval peasant communities (Fernández Mier 1999; López Gómez 2012).

Many of the *brañas* were still in use until a few decades ago, although traditional pastoralism is currently in irrevocable decline. The sub-recent modes of animal husbandry have been studied in their final phase as they disappear, and the last shepherds to take part in transhumance in these mountains were interviewed. The information these people shared with us was invaluable when examining the archaeological record of these seasonal settlements as they offered excellent information on the toponymy, the land distribution and tenancy, and the common law regulating how the parish and village were organised. Their accounts also offered an in-depth insight into productive practices and the significance

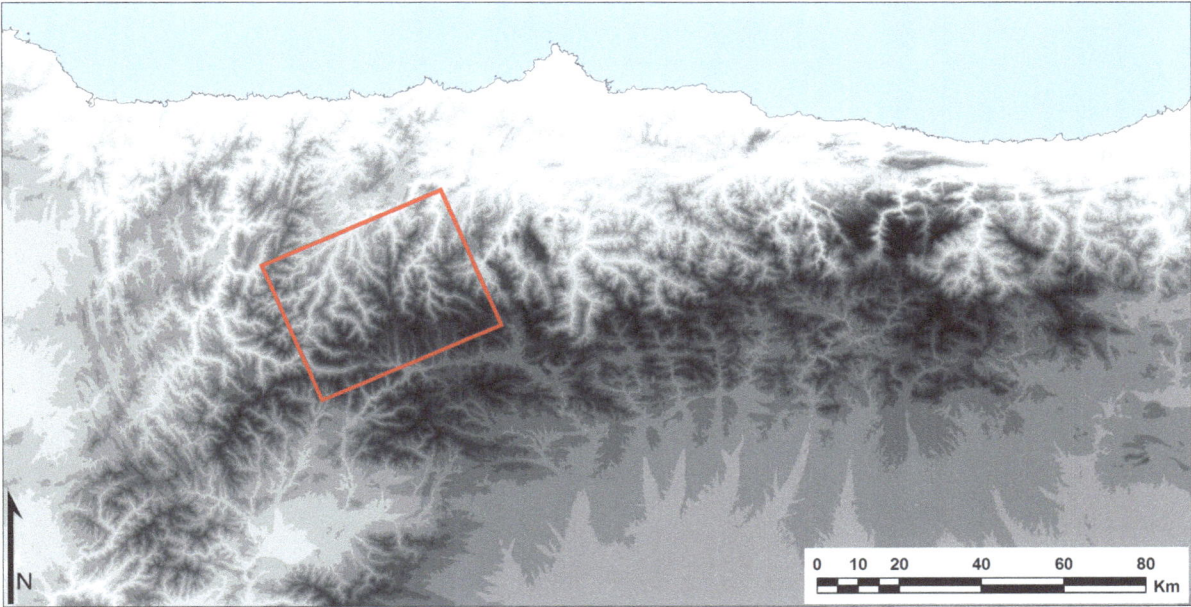

Figure 12.2 The study area.

given to certain landscape features in relation to material remains detected in the field surveys. Here, the ethnographic surveys and ethnoarchaeological studies of the material culture of the herdsmen were important tools. These studies allowed us to reflect on the historical narrative of the culture and subsistence modes in the Cantabrian Mountains, to a degree which archaeological research alone would not have been able to achieve. On the other hand, the ethnographic and anthropological studies carried out by various researchers prior our work (e.g. Barrena 2001; Concepción *et al*. 2008; Corbera 2008; García Martínez 2003; Linares 2004), along with the complementary documentary sources including collections of old photographs from travellers or mountaineers and their accounts (e.g. Lombardía and López Álvarez 2003), completed the ethnoarchaeological analyses recently carried out by this research team (González Álvarez 2013; López Gómez and González Álvarez 2013). In summary, the ethnoarchaeological viewpoint provides us with a better understanding of mountainous cultural landscapes, herding mobility and the pastoralist way of life.

Recent trends in archaeological research led to planning links with other related subject areas. Within our research framework we systematised the data in different formats, using the appropriate tools to relate this wide range of information to the archaeological record itself, with GIS and databases as the central basis for this task. The difficulties of fieldwork in mountainous areas meant applying techniques to maximize the survey results. For example, the use of remote sensing tools or planning surveys using predictive models which take into account features of the relief, the availability of water resources or the ethnoarchaeological stimuli, generate procedures that optimize the available research resources, resulting in substantial improvements in the results obtained (e.g. Carrer 2013). In addition, using GIS along with geographical, geological or environmental information facilitated studies of density, visibility, accessibility or ground quality, etc. Taking these variables into consideration enabled recognition of settlement patterns or subsistence models in the mountains. This led to an enhanced consideration of the historical relevance of these spaces for the communities which used them.

Case study

The study area

The Cantabrian Mountains or *Cordillera Cantábrica* is a mountain range running parallel to the Cantabrian coast in the north of Spain. Its highest peaks reach an altitude of over 2600m with the particular feature that at some points this range runs very close to the Cantabrian Sea, with its northern slopes facing the coast. To the south the Cantabrian Mountains border the central meseta of the Iberian Peninsula, so that these mountains form both a geographical limit and a communicating area between different regions of northwest Iberia.

The steeply sloping relief means deep narrow valleys run between the mountains and areas at high altitudes. The settlement pattern in this mountainous area is currently organised in a network of villages found mid-slope in the narrow mountain valleys (García Fernández 1980). This model dates back to the Late Roman/Early Medieval period (Fernández Mier 2011). The seasonal nature of the Atlantic climate and the close proximity of the different altitudinal zones encourage the use of complementary agrarian uses for the differ-

Figure 12.3 Ruined pastoral hut without a clearly defined chronology found in the summer *braña* of Los Cuartos (Somiedu, Asturias).

ent landscape areas. The agricultural plots are arranged on the gentler slopes around the villages while the extensive pastures at higher altitudes are used during the summer. This has meant that the inhabitants of the Cantabrian Mountains have developed forms of subsistence which combine agriculture with extensive pastoralism based on the seasonal movement of the flocks and herds since the Neolithic (de Blas 2008; Díez Castillo 1996-1997; Marín 2011).

Our research focused particularly on the mountainous municipal areas of Cangas del Narcea and Somiedu (both in Asturias), and the region of L̦laciana (León) (Fig. 12.2). These areas are among the most characteristic pastoralist areas in the western sector of the Cantabrian Mountains, where transhumance has been practised almost up to the present day. In spite of the important transformations since the latter half of the twentieth century, a significant number of *brañas* still in use can be found in the pastures in this area. Many others have been abandoned, although their structures are still recognisable and the last shepherds who used them are still within living memory. The village of Vigaña (Miranda, Asturias), where our research is currently based, has been chosen as a particular microscale case study, with intensive on-going archaeological research in the whole of the surrounding area (Fernández Mier and González Álvarez 2013).

Aims

This paper presents a preliminary overview of our current research. Our basic aim is to understand the social and economic processes involved in the social construction of the cultural landscapes of the Cantabrian Mountains from a diachronic perspective, within the framework of landscape archaeology (Criado 1999).

The archaeological record as well as the documentary, ethnographic and palaeo-environmental ones are used in this research in order to reinforce our understanding of the peasant ways of life in these mountains. At the same time, we are looking to calibrate the impact and rhythms of anthropisation processes over the last millennia, considering the forms of social construction of the cultural landscape in the long term. Starting from the analysis and ethno-archaeological consideration of the traditional uses of the *brañas* by contemporary shepherds, the aim is to reinforce our archaeological research methods. In addition, we will also consider the ways in which the summer farms have been occupied from later prehistory to the present day and we will evaluate the material variability of the different type of *brañas* in order to clarify their typologies in chronological, functional and architectonic terms. We will contemplate too the seasonal nature of their occupation, the productive activities carried out in them and the people involved in their use.

The *brañas* in traditional pastoral systems

The shepherds of the Cantabrian Mountains feed their flocks and herds taking advantage of the ecological niches found close together in the different altitudinal zones in these mountains. To do this, they follow an annual cycle of pastoralist mobility. During the summer they lead their animals up to the mountain pastures where the *brañas* are found. The primary function of these sites is as shelter for the shepherds. They form a habitat with huts of different types built alongside other structures for various herding activities, such as enclosures for keeping, classifying and milking the animals, or hay barns and enclosed fields for cutting fresh fodder. As part of the processing of dairy products there are cheese cellars for curing cheeses and *olleras* or larders designed to store the milk churns in springs or running water to protect the milk and keep it fresh. During the summer season in the *brañas*, the shepherds occasionally take part in other complementary tasks, such as obtaining lime from limestone, producing charcoal or growing certain crops at lower altitudes. These fields are often terraced and are usually for planting corn, potatoes, or cereals suited to higher altitudes such as rye. In other cases they are simply marked out and identified by their system of enclosure, with shallow ditches with a bank of earth on one side to prevent the animals from getting in. The *brañas* are, in fact, multifunctional enclosures of an unexpected and interesting complexity far beyond their basic herding purpose.

An examination of the typology of pastoralist sites mentioned in earlier ethnographic studies shows that there were various formats of the *brañas* related to the different patterns of herding mobility, types of flocks and herds and seasonal use. Thus, in our study area four main types of *brañas* are documented, related to the three main herding mobility patterns. However, defining these generic categories does not imply the existence of a formally defined model for the settlements linked to the use of the pastures in the Cantabrian Mountains. On the contrary, a closer examination of the documented record of our field surveys reveals a much more complex picture, since the categories of settlement shown correspond to conflicting descriptions which conceal a wide range of types of buildings and models of use for these pastures. For example, some of the *brañas* which we identified have been described as hybrid models of the general types, suggesting the co-existence in the same site of different herding patterns. The overall picture becomes even more complicated if the innumerable ruined structures detected in the surveys are taken into consideration (Fig. 12.3). These may be either isolated or linked to the *brañas* and have formal, functional or chronological characteristics which are not clearly defined. Finally, the pastoralist formulas presented here, along with their related settlement patterns, are the result of the agrarian transformations which occurred throughout the nineteenth century, after the confiscation of municipal, communal and waste lands. Thus, we cannot therefore simply take these observations as images derived from the productive forms before that time, but instead we should use them as a starting point for a regressive reading of the landscape through historical and archaeological investigation as we attempt to do.

Valley-transhumance pastoralism

Valley-transhumance is a livestock herding formula involving short-range mobility used by sedentary rural communities. These families live in the mid-slope villages in mountain valleys. From there they make seasonal use of the high pastures located directly above the villages. Normally, each family sends their flocks and herds up to the pastures in the care of one or two family members, while the rest of the family is left in charge of the farm work in the immediate area of the village. The shepherds and cowherds – normally young unmarried men and, to a lesser extent, young women and elderly people – spend the nights in the *brañas*, several hours on foot from the villages.[1] The shepherds often go back down into the valley with the milk, cheese and butter obtained in the summer farms or when they are needed to help with farming work in the village. The usual practice is to relieve those who are looking after the animals in the high pastures from time to time, with other members of the family taking their turn. Several families may share in looking after all their animals, while the better-off families may pay for the services of professional shepherds.

During the summer, the herders in this pastoralist formula lead the animals – mainly cattle, although goats and sheep are also included – up to the mountain pastures following a traditional transhumant model, which at its most complex uses two types of summer farms situated at different altitudes at specific points in the yearly cycle – the equinoctial *brañas* and the summer *brañas* – as well as the village in the valley, the winter quarters for the herders and their animals.

The summer *brañas*

The summer *brañas* are used to take advantage of the highest pastures, at altitudes of around 1500 m.a.s.l. These sites are home to the animals and those who go with them during the three summer months (Fig. 12.4).

The animals are free to graze, with no privately-owned, fenced fields. Cheese and butter-making are

---

1. Although we use here the present tense to describe these herding formulas, in fact most families who currently still engage in extensive pastoralism as a central activity in their way of life have abandoned transhumance. Many of the *brañas* are now abandoned, while others are no longer used to spend the nights and are only used for storage and as reference points for herding tasks. The depleted flocks and herds which still graze these areas are tended by herdsmen who reach them in 4×4 vehicles, and who now only exceptionally continue to make cheese or other dairy products from the milk of their own animals, which in any case would now be made in the villages down in the valley.

Figure 12.4   Summer *braña* of La Mesa (Somiedu, Asturias).

important parallel occupations to herding activities, and so the *brañas* must be near small streams. Although some private *brañas* do exist – which were or still are the private property of a particular landowner – the summer *brañas* are normally used communally by a village or parish. A rota is often drawn up with the villagers taking turns (*veceras*), to look after all the animals of the whole group or sometimes a wage is paid to the person who takes care of the livestock.

The existing structures are small circular or square shielings or huts of around 4m² for the cowherds or shepherds to spend the night in. Normally these shielings are grouped together, close by the pastures, and are built in sunny places, sheltered from the prevailing wind. There are also annexes built on for milking sheds and byres for the animals. All these structures are built with dry-stone walls, except where repairs have been carried out recently. The huts and *beḷḷares* (byres for the animals feeding their young) are roofed with thatch, turf, slates or stone slabs arranged in rows to form a false dome.

The equinoctial brañas

The equinoctial *brañas* (also known as *mayás primaliegas* or *morteras*) are temporary settlements which use medium-altitude zones at around 1000m a.s.l, midway between the highest pastures and the village (Fig. 12.5). These sites taken together with the summer *brañas* form a complementary system to take advantage of all the resources for livestock rearing in the mountain valley. The animals are taken up to these areas in spring and autumn, while in summer they go even further up to the summer *brañas*. With the first snows, the livestock is brought down to their owners' farms in the villages, and is kept semi-stabled until the spring.

The main characteristic of the equinoctial *brañas* is that here the meadows are fenced or walled and/or the hay is gathered and stored to use as winter feed or when the weather is bad or if the pastures for some reason cannot be grazed. For this reason the equinoctial *brañas* have medium-sized or large square buildings with most of the available space intended as a hay barn, normally under cover. Crops of rye or potatoes may also be grown. There is written evidence that these enclosed meadows existed as early as the thirteenth century. As well as being used for the grass which is scythed and stored as hay, they are also used for direct grazing twice a year, in spring and autumn. They are for private or semi-collective use, in contrast to the rest of the pastures which are communal.

The equinoctial *brañas* may form groups of various huts and barns or be separated from the others and located near the enclosed meadows. The buildings are of different types, according to the local tradition in each area, although each settlement also has its own variations in the internal division of the rectangular structures. These are built of stone and the roofs are made of thatch, ceramic tiles, wood or stone slabs

Figure 12.5   Equinoctial *braña* of La Pornacal (Somiedu, Asturias).

(Graña and López Álvarez 2007). The equinoctial *brañas,* found in areas where the mountains are not so high, may also be used in summer, and in this case they have the formal characteristics of the equinoctial type.

The vaqueiros d'alzada

The *vaqueiros d'alzada* form a separate social group within the rural population of Asturias (Cátedra 1989; García Martínez 1988). Their way of life is based on medium-range transhumance, with two houses tens of kilometres from each other. Their yearly cycle involves moving from one house to another with the whole family and all their goods and livestock. This means that they hardly grow any cereal crops. For this reason, these families traditionally supplemented their livestock farming as muleteers, transporting goods between the Cantabrian coast and the inner Iberian Peninsula. These activities declined as the contemporary railway network grew, but provided them with a cash income to buy cereals for flour and other commodities they could not produce themselves in the mountains. These differences from the rest of the peasant communities in Asturias meant that the *vaqueiros d'alzada* tended to be considered as outsiders within their own regional environment and they were even marginalised.

Their summer settlements are the *brañas-pueblo, brañas d'arriba* or summer villages, located at around 1200–1400m a.s.l (Fig. 12.6). They stayed in these for the central 8–9 months of the annual cycle, with their livestock – mainly cows, although they also had sheep, goats and pigs up to the mid-twentieth century, grazing freely on the mountain pastures. These settlements are the nearest thing to permanent homesteads found in all the seasonal sites examined, with large domestic buildings, stables and barns, enclosed fields and even small vegetable gardens. On the other hand, the *vaqueiros d'alzada* spent the winter in the *brañas de invierno, brañas de baxo* or winter villages.

The summer villages show a number of particular features which make them easily recognisable, including the buildings near the enclosed fields, the presence of rectangular or oval dwellings, generally with different rooms for the family, the cattle and tools, as well as annexes used as barns for storing hay. They are built of stone in loosely structured nuclei, with thatched roofs which were gradually replaced by tiles or slates during the twentieth century. Their morphology is similar to that of the equinoctial *brañas,* although the *vaqueiros d'alzada* habitats may be found at higher altitudes, in locations more suited to summer *brañas.* Another characteristic are the terraced areas for growing crops, mainly meadows for hay.

Merino shepherds

The last itinerant herding formula described here took large flocks of sheep and goats from the centre and south of the Iberian Peninsula to the abundant pastures

Figure 12.6   *Braña* of La Peral (Somiedu, Asturias) inhabited by *vaqueiros d'alzada* families.

of the Cantabrian Mountains, mainly on the southern slopes. This pastoral system, the last link with the medieval system of transhumance organised by the *Honrado Concejo de La Mesta* (Klein 1920), finally disappeared in the late twentieth century.

The transhumance route of the merino shepherds started in the *dehesas* or pastures from Castilla and Extremadura, where the flocks wintered. The journey took twenty to thirty days on foot over a distance of hundreds of kilometres, accompanied by mastiffs and pack mules (Fig. 12.7). These shepherds spent the summer in the mountain pastures of the Cantabrian Mountains, from late May to early October, contracted by wealthy landowners who rented the right to pasture from the local population in exchange for monetary payments. The non-local shepherds also paid the local people of the mountains in kind, organising and inviting them to festivals and feasts in the places where they set up their seasonal camps. They lodged in *chozos*, circular shielings or huts made of wood and thatch, built each year for them by local young people. Although they could have been used for several years, these structures were demolished each year when the merino shepherds had left, as rebuilding them was part of the welcome rites which included the invitation to the feasts mentioned above (López Álvarez and Graña 2003).

The huts for the shepherds were built adjoining large pens for their flocks (Fig. 12.8). Many of the structures linked to this transhumance type are nowadays abandoned and they can be detected through archaeological surveying. Their seasonal use determined their simple construction, so it is often difficult to identify them on the upland pastures.

On-going research

The preliminary work of this research team was to carry out ethnographic and ethno-archaeological surveys to assess the productive potential of the mountain landscapes. For example, the closer examination of traditional pastoralism in the study area allowed us to differentiate and locate the altitudinal zones which could accommodate permanent habitats or seasonal camps. Considering the different herding mobility formulas, which may even have been developed in the same area, also allowed us to approach the archaeological record from a non-determinist viewpoint (González Álvarez 2013). Thus, we assume that different productive strategies may have developed in different chronologies, although this does not mean that continuities or unilinear evolutions can be observed in the settlement patterns or subsistence systems.

Written sources from the medieval and modern eras also offer information which varies according to the period but is nevertheless invaluable as it allows us to outline the pressures exercised by different social groups and the taking over of upland pastures. The attitude of these social agents differed depending on their own economic interests, either related to subsistence husbandry activities to complement their agricultural farming or market-oriented livestock rearing practices (Fernández Mier *et al.* 2013b). For this reason, part of

Figure 12.7 Merino shepherds in the upland pastures of Braña Forada (Somiedu, Asturias). Photo taken by J.M. Lueje in 1952 (Lombardía and López Álvarez 2003:202), reproduced with permission from the Muséu del Pueblu d'Asturies.

Figure 12.8 Large pen for the merino shepherds' flocks in Sousas (Ḻḻaciana, León).

Figure 12.9   Braña of L'Estoupieḷḷu, used by the herders from Vigaña (Miranda, Asturias).

our research group has been responsible for the examination and interpretation of documentary sources.

Our archaeological research focusing on the *brañas* consisted basically of surveying the upland pastures searching for pastoralism traces. The preliminary field work campaigns included conventional surface surveys intended to examine some upland areas in the Ḷḷaciana region and the Pigüeña valley. In 2011 these approaches were extended to the mountain area of Cangas del Narcea. During these surveys, the priorities were to locate areas of archaeological interest in the upland pastures, and the reconnaissance and documentation of the *brañas*, in order to identify the different typologies and functions of these settlements from the material record. In addition, we detected many ruined and abandoned buildings of uncertain date. Little attention had been paid to these structures by earlier researchers. Just as in research projects currently underway in other mountain ranges such as the Pyrenees (Gassiot and Jiménez Zamora 2006; Palet *et al.* 2010; Rendu 2003) or the Alps (Walsh *et al.* 2007), these features will focus our attention in the later stages of our research, to attempt to date them and understand their functional characteristics through archaeological excavations.

However, the exuberant vegetation hindered the conventional surveys due to the low or nil visibility of the pastures' surface in the study area. This led us to try out alternative models, such as applied in the equinoctial *braña* of L'Estoupieḷḷu (Fig. 12.9). This grazing area is found at around 900 m.a.s.l., and it is used by the inhabitants of Vigaña. There are five huts, with four of them still in use and in perfect condition. In the highest part there is a large megalithic mound surviving to a maximum height of 4m, and a robber hole in the centre, but there is no evidence of features that would indicate a hypothetical orthostatic chamber. At the extreme southeast of this *braña* there are the remains of an abandoned tilery, mentioned by oral sources. The lowest part of L'Estoupieḷḷu, near the huts, is the setting in mid-August for a *romería* (traditional celebration with feasting, dancing and music) for the people of the neighbouring villages.

The method used consisted of a survey carried out by removing the ground cover at selected points from a random stratified sample. To do this, a 25×25m grid was laid out and random points generated where 1×1m test-pits were dug (Fig. 12.10). At each point we removed the topsoil layer manually to recover material items not visible on the surface. Unfortunately, the results obtained so far are not very promising. From a total 39 test-pits, 13 provided archaeological materials, mainly curved roof tiles from the remains of the tilery documented at the extreme southeast of the surveyed polygon and the huts mentioned.

Worth noting for their singular importance are the lithic materials from test-pits D9, E7 and E12. These are flint flakes – remains of lithic reduction – presumably related to the shepherds who have used these mountain areas since at least the Neolithic period. In fact, these materials are normally found in the vicinity of megaliths (Arias Cabal and Pérez Suárez 1990, 1992; de Blas 1996; Díez Castillo 1996–1997) where,

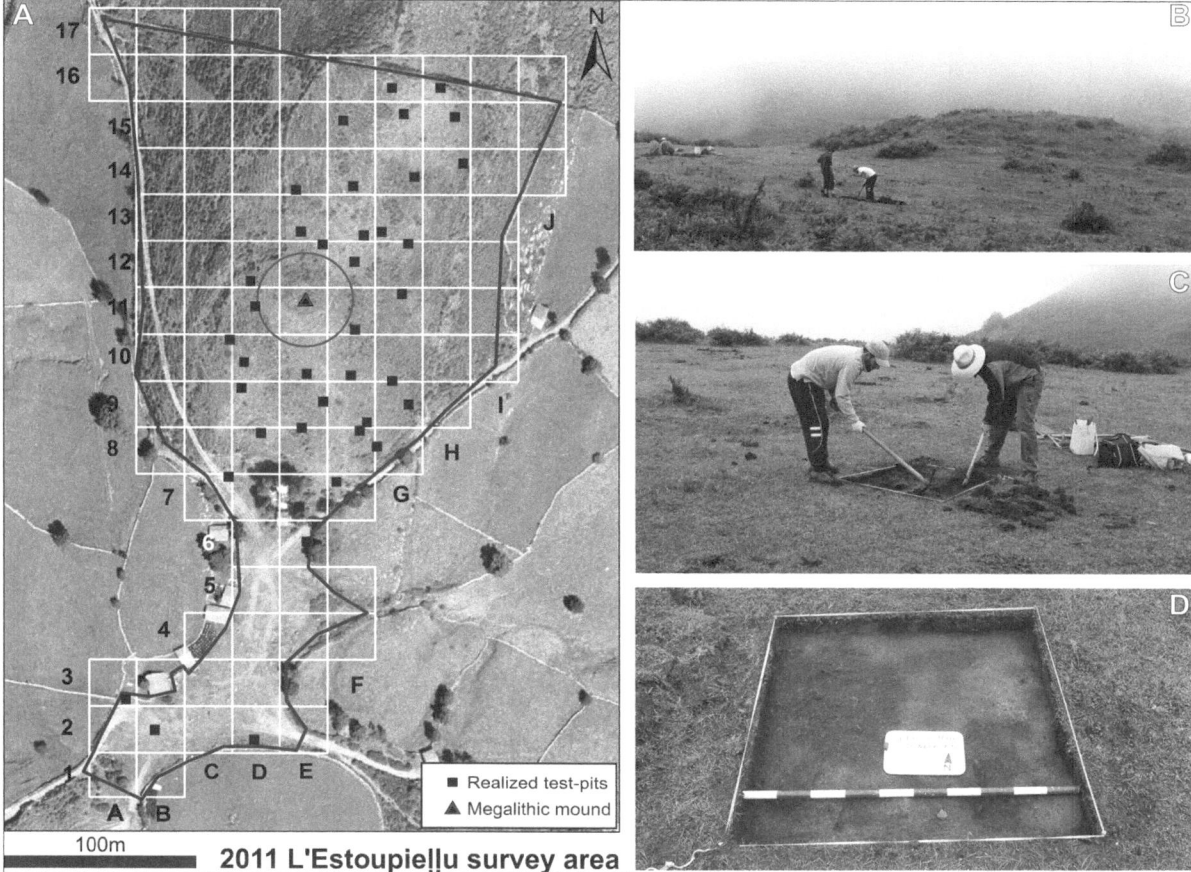

Figure 12.10 Braña of L'Estoupieḷḷu. Using a GPS navigator the test-pits were dug following a random stratified sample within a 25×25m grid (image A shows the test-pits dug during the 2011 survey). The megalithic burial mound is shown in the plan and it is in the back of the image B. The topsoil was manually removed (images B and C) in order to recover archaeological remains related to the pastoralist use of this space. Few materials were recovered, such as the flint flake from the test-pit E12 (image D).

in our opinion, the prehistoric seasonal camps would have been located (González Álvarez 2010). We also collected some contemporary materials, but none of the test-pits provided clear evidence from ancient habitats. However, in some of the test-pits evidence was observed giving information on pasture-clearing practices and how these mountain areas were used, although the studies and analyses to confirm these observations are still pending. Remains of alterations of the clay substratum by high temperatures were recognised in some of the test-pits when the ground cover layer was removed, but unfortunately we could not date them because we did not find enough organic materials.

The experience of the 2011 L'Estoupieḷḷu campaign was positive overall in that it allowed us to calibrate the suitability of the proposed working method, detect its shortcomings and devise possible solutions to maximise the results obtained in future practical field experiences. Basically, we need to establish corrections in the sampling density depending on the presence of certain known anthropogenic features, and also to reject steeply sloping areas or those currently with heavy transit. The results of the case study highlight the strong human presence and anthropogenic activity which contemporary local communities have exerted on this *braña*, presumably altering or erasing the imprint of human presence from the distant past. Nevertheless, the inevitable presence of the prehistoric burial mound and the scarce but significant lithic remains which have been recovered suggest a certain importance for the prehistoric communities in the original use of this *braña*.

On the other hand, the archaeological research carried out around the village of Vigaña (Fig. 12.11) is at a more advanced stage (Fernández Mier and González Álvarez in press.). Starting from the analysis of the field system, toponymy and historical written sources, we reached an acceptable understanding of how the medieval and post-medieval productive spaces were structured. These analyses were matched with the palaeo-environmental studies and interpretative stimulus

Figure 12.11 The village of Vigaña (Miranda, Asturias) in the Pigüeña valley.

provided by the ethnoarchaeological approach, giving us an advanced understanding of the peasant agrarian practices. With these references, we approached the next phase of archaeological fieldwork, which is currently underway, with the excavation of agrarian terraces, cultivated fields, cemeteries, and settlement locations from different periods (Fernández Mier and González Álvarez 2013). These activities provide important data to understand the agrarian system of the human communities throughout history. This has led to the location of inhabited areas with different chronologies in areas which traditionally were not considered to be of archaeological interest. Powerful sedimentation processes had completely concealed these inhabited levels which provide information on historical periods with very little archaeological data previously available on a regional scale, as in the case of the Early Middle Ages or the Neolithic period (Fig. 12.12).

A herding landscape in the *longue durée*

The contemporary peasant landscape in the Cantabrian Mountains is the result of several transformations and human actions on-going since at least the Neolithic period. In order to understand both the changes and the continuities in the formation processes of mountainous landscapes, we are interested in the development of an archaeological genealogy of the peasant landscape. Thus, a long-term view is needed to fully understand the anthropisation process and the cultural relations between the peasant communities and the environment they inhabited. This diachronic approach does not involve the establishment of unwarranted historical continuities. Instead, this approach attempts to evaluate the singularities and diversity of ways of life developed by various human groups in different chronologies and cultural contexts. However, this *longue durée* approach does not mean ignoring the short term, as Ian Morris says (2000:5).

A careful examination of the palaeo-environmental record is particularly relevant when studying the incidence of the local peasant communities in mountain areas. Thus, the use of the alpine and sub-alpine levels in the Cantabrian Mountains generated changes reflected in the data from palaeo-environmental studies which must be distinguished from natural changes. The earliest archaeological evidence which shows continued use for livestock rearing in the Cantabrian Mountains is related to Neolithic groups. Previous to this, these areas had already been used by Mesolithic hunter-gatherers, but their anthropogenic impact was really small. An interpretation of the pollen records shows a decreasing evolution of the arboreal mass compared to a greater importance of grasses and shrubs since the Neolithic and, above all, the Bronze Age (López Merino 2009:228–230; Moreno *et al.* 2011:344). This can be interpreted as the result of increasing deforestation linked to the gradual clearing of new grazing areas throughout later prehistory.

The earlier use of the upland summer pastures is related to the Neolithic and Bronze Age communities which were developing nomadic ways of life (Marín 2011). These groups left a fragile archaeological record, difficult to recognise in archaeological surveys. The identification of their seasonal settlements has hardly begun (Camino and Estrada 2012), and their formal and functional characteristics may have been assimilated into itinerant camps established in locations equivalent to the sub-recent *brañas* (Criado *et al.* 2000). Other remains which can be used to identify their presence in the mountains are lithic materi-

Figure 12.12 Las Corvas site, in the agrarian surroundings of the village of Vigaña. Under two metres of sub-recent agrarian layers, a later prehistoric hut was found. It was dated back to Neolithic from a radiocarbon analysis sample taken from the hearth fire place: DSH3620 4770 ± 31 BP; 3640–3516 cal BC (96.2%), 3421–3419 cal BC (0.4%), 3410–3405 cal BC (0.6%), and 3398–3384 cal BC (2.7%) (2 sigma, after Reimer et al. 2009).

als on the surface, Chalcolithic and Bronze Age votive metal hoards or Bronze Age schematic rock art. The megalithic monuments are the exception to this near invisibility, arranged at visually dominant points in the mountain ranges, in mountain passes or in the areas where sub-recent *brañas* are located. The distribution of these funerary monuments shows the high mobility of these mainly nomadic herders (Criado and Mañana 2003), who did not have stable settlements in the valleys either, as they practised itinerant slash-and-burn agriculture. The function of these monuments may be related to the topographical writing mechanisms which were used by these groups to code the landscape for symbolically adopting the usufruct of the pastures in connection with the cult of their ancestors (*sensu* Santos Granero 1998).

From the Iron Age onwards, for the first time in history, successive generations of the inhabitants of these mountains were born and died in the same settlement. The *castros* or hillforts – fortified and monumentalised sedentary villages – served as the central hub in a territorialised cultural landscape in which the livestock rearing activities followed a model of specialised herding with valley-transhumance (González Álvarez 2011). In this context, Iron Age communities probably had secondary settlements up in the mountains available for seasonal occupation, similar to the *brañas*, to support the herding activities carried out by a few members of each group. Unfortunately, little attention has been paid beyond the outer walls of the *castros* in northwest Iberia (e.g. López Sáez *et al.* 2009; Parcero 2006). For this reason, further research is needed to relate these settlements to their productive environments. In this scenario, our micro-scale research in Vigaña provides a good opportunity for remedying this. The off-site study of the area is combined with the excavation of the hillfort of *El Castru*. On this site we have recovered an interesting collection of animal remains which on detailed examination should offer a clearer picture of the husbandry activities of their inhabitants.

After the Roman conquest, the *castros* were gradually abandoned and the population became based in small hamlets where little archaeological information is available (Marín and González Álvarez 2011). However, presumably pastoralism continued in a similar way to the preceding period. The main difference arose from the integration of these peasant communities into a state-wide political and administrative system which most probably made changes to the regulation of the seasonal use of pastures and introduced taxation systems. The continuation of these pastoral practices is easily seen in the palaeo-environmental records, which show a high level of deforestation in the mountain areas, as well as the extension of pastures and shrubs. Thus, the impact of the anthropisation processes was even more marked in the Roman period (López Merino *et al.* 2009).

In the transition from the Late Roman period to the Early Middle Ages the network of villages was firmly established, the forerunner of the traditional system of settlement in the Cantabrian Mountains (Fernández Mier 2011). In the Middle Ages, the transhumant model of livestock herding can be seen more clearly as a result of the careful examination of written sources. Within this framework, the herding activities carried out in the upland pastures were included in an annual cycle marked by short-range transhumant movements in the mountain valleys. The impact of deforestation was much stronger in the Early Middle Ages (López Merino *et al.* 2011:2751–2752), when the mountainous grazing areas became valuable assets for the feudal lords and ecclesiastical authorities of the new feudal system (Fernández Mier *et al.* 2013b). In the High and Late Middle Ages, the written sources referring to pastoralism and the *brañas* multiplied. At this time, the image of transhumant pastoralism in these mountains was formed, with the establishing of *brañas* at different altitudes and the overlapping in the same space of different herding regimes, including the *vaqueiros d'alzada* or the shepherds of La Mesta.

Final remarks

Since the Neolithic period, mountainous areas have constituted significant spaces for the subsistence of peasant communities, with an increasing presence of hunters, shepherds, miners and metal workers. The extension of productive activities such as seasonal herding led to the establishment of important interactions between the human groups and the mountain environment. This active role of human beings initiated a marked anthropogenic process at alpine and sub-alpine altitudes in European mountains with the special incidence of practices such as burning and deforestation to clear the land for new pastures. Parallel processes of symbolic appropriation of these spaces were initiated through political and religious strategies which imbued the human experience and perception of these spaces with cultural significance. We should therefore not view mountains as static natural landscapes but instead consider them as dynamic cultural ones.

The *brañas* are 'summer farms' used for herding activities in the high pastures during the summer season by the shepherds in the western Cantabrian Mountains. These sites are a local version of the many different examples of seasonal pastoralist settlements in mountain regions, such as the *mayadas* nearby in the eastern part of Asturias (Barrena 2001), and the *seles* in Cantabria (Corbera 2008), and further away, the *malghe* in the Italian Alps (Kezich and Viazzo 2004), the *alpeggi* in the Apennines (Barker and Grant 1991) or the *yayla* in the mountains of Anatolia (Yakar 2000).

The *brañas* are a central piece in the anthropisation processes which have occurred in high mountain areas such as the Cantabrian Mountains. Because of their relevance in the social construction of the cultural landscapes of these mountains, we need to develop strategies which will encourage the archaeological study of these summer farms. The archaeological characterisation of different models of the *brañas* which have existed from later prehistory to the present day will enrich the understanding of the herding mobility patterns, types of settlements and subsistence models developed by the peasant communities in these mountains. Moreover, the development of an archaeological approach to the *brañas* may help us to comprehend the genealogy of these 'summer farms' in a broad chronological view.

Some of the *brañas* in the study area are still currently in use and many others were studied by geographers and ethnographers during the twentieth century. Their diversity will lead us to consider the multiplicity of forms that summer farms have adopted over the last millennia, related to types of transhumance, the predominant productive scale, the available agrarian technology, the animal species involved in pastoralism and other identity and cultural questions. For this reason, an ethno-archaeological approach to these realities may offer interpretive stimuli which enrich the archaeological research about the *brañas*. The ethno-archaeological observations may encourage further archaeological studies, offering a deeper reflection from an experiential viewpoint, and contributing to intensify the archaeological interpretations.

The scenario of the irrevocable disappearance of the transhumant herding way of life is fast approaching in the study area. The *brañas* are being abandoned and falling into ruins while the memory and example of the elderly shepherds are disappearing with them. This is why it is a matter of some urgency to carry out various types of studies on these seasonal settlements, before the landscape itself of the *brañas* in the Cantabrian Mountains becomes merely a fossilised reality.

The *brañas* and seasonal herding practices have to be considered as part of the integral system of peasant economy. It is therefore essential to integrate the research on this type of 'summer farms' into an overall study of the territoriality and subsistence of these peasant communities. If we only study the *brañas* themselves or the seasonal use of the mountain pastures we will ignore the inhabited villages with which they were related, located in the lower parts of the mountains. Inversely, the historical and archaeological studies of mountain communities cannot be complete without a consideration of the higher areas used in summer.

Mountains are not marginal or threshold areas. They are important areas for local subsistence, as well as key points for understanding aspects of trade, inter-community relationships and marriage practices. In fact, mountains are a meeting point for people of different provenance within the framework of the summer pastures use (Fig. 12.13). In this context, there are

Figure 12.13 Annual cattle fair celebrated in El Puertu (Somiedu, Asturias). Image of the 1952 edition by J.M. Lueje (Lombardía and López Álvarez 2003:202, reproduced with permission from the Muséu del Pueblu d'Asturies).

frequent exchanges of ideas or products, cattle fairs, feasts and marriage arrangements between families from different villages. Mountains are also focal points of tensions and problems such as the rights to the usufruct of pastures.

Looking ahead to our future research, the development of archaeological excavations in the *brañas* should lead to precise chronologies for the occupation of these settlements. The chronological depth emerging from these archaeological studies, along with the analyses derived from the interdisciplinary dialogue with other historic, ethnographic, geographic or palaeo-environmental sources, will allow us to question current evolutionary interpretations of the genealogy of the *brañas* in the Cantabrian Mountains (García Martínez 1988, 2003). In fact, we believe that this variety is not related to successive and unilinear chronological phases, but rather reproduces a complex cultural social and productive diversity in the different subsistence strategies used by the human groups present in these upland areas throughout history (González Álvarez 2013; López Gómez and González Álvarez 2013). The evidence that different types of transhumance, linked to different settlement patterns, may have come together in the same area, leads us to consider that the geographical factors are important, but never determinant in their development. Over and above these factors there are always culture and identity-based influences, which shape and develop the different productive possibilities of the mountain areas. Thus, the archaeological study of the anthropisation processes and the social construction of the landscapes must be approached in context to be able to put forward historically informed interpretations.

Bibliography

Arias Cabal, P. and Pérez Suárez, C. 1990. Investigaciones prehistóricas en la Sierra Plana de La Borbolla (1979–1986). *Excavaciones Arqueológicas en Asturias* 1:143–151.

Arias Cabal, P. and Pérez Suárez, C. 1992. Los yacimientos al aire libre del Llano de Los Carriles en el concejo de Llanes (Asturias). *Boletín del Instituto de Estudios Asturianos* 140:513–558.

Barker, G. and Grant, A. 1991. Ancient and modern pastoralism in central Italy: an interdisciplinary study in the Cicolano mountains. *Papers of the British School at Rome* 59:15–88.

Barrena, G. 2001. El hábitat de los pastores de los Picos de Europa. In F. Rodríguez (ed.) *Paisajes y paisanajes de Asturias. Organización del espacio y vida cotidiana tradicional.* Pp. 65–84. Gijón: Trea.

Camino, J. and Estrada, R. 2012. El mayéu Busián (Ḷḷena): orixe d'una braña na Edá del Bronce. *Asturies: memoria encesa d'un país* 32:4–11.

Carrer, F. 2013. An ethnoarchaeological inductive model for predicting archaeological site location: a case-study of pastoral settlement patterns in the Val di Fiemme and Val di Sole (Trentino, Italian Alps). *Journal of Anthropological Archaeology* 32/1:54–62.

Cátedra, M. 1989. *La vida y el mundo de los vaqueiros de alzada*. Madrid: Centro de Investigaciones Sociológicas.

Corbera, M. 2008. El proceso de colonización y la construcción del paisaje en los Montes de Pas. *Ería* 77:293–314.

Concepción Suárez, J. García Martínez, A and Mayor López, M. 2008. *Las brañas asturianas: un estudio etnográfico, etnobotánico y toponímico*. Oviedo: Real Instituto de Estudios Asturianos.

Criado, F. 1999. *Del terreno al espacio: planteamientos y perspectivas para la Arqueología del Paisaje*. Santiago de Compostela: Universidade de Santiago de Compostela (CAPA 6).

Criado, F., Gianotti, C. and Villoch, V. 2000. Los túmulos como asentamientos. In P. Arias Cabal, P. Bueno, D. Cruz, J.X. Enríquez, J. Oliveira and M.J. Sanches (eds.) *Actas do 3º Congresso de Arqueologia Peninsular, Vol. 3: Neolitizaçao e Megalitismo da Peninsula Ibérica*. Pp. 289–302. Porto: Associação para o Desenvolvimento da Cooperação em Arqueologia Peninsular.

Criado, F. and Mañana, P. 2003. Arquitectura como materialización de un concepto: la espacialidad Megalítica. *Arqueología de la Arquitectura* 2:103–111.

De Blas, M.A. 1996. Espacio funerario – Espacio económico: las sugerencias del registro arqueológico en el entorno de un dolmen de montaña. In A.A. Rodríguez Casal (ed.) *Humanitas: estudios en homenaxe ó Prof. Dr. Carlos Alonso del Real*. Pp. 125–150. Santiago de Compostela: Universidade de Santiago de Compostela.

De Blas, M.A. 2008. La Prehistoria reciente: el brumoso inicio de las sociedades neolíticas en Asturias. In J. Rodríguez (ed.) *La Prehistoria en Asturias: un legado artístico único en el mundo*. Pp. 489–566. Oviedo: Editorial Prensa Asturiana.

Díez Castillo, A. 1996–1997. *Utilización de los recursos en la Marina y Montaña cantábricas: una prehistoria ecológica de los valles del Deva y Nansa*. Gernika: AGIRI Arkeologia Kultur Elkartea.

Ejarque, A., Julià, R., Riera, S., Palet, J.M., Orengo, H.A., Miras, Y. and Gascón, C. 2009. Tracing the history of highland human management in the eastern Pre-Pyrenees: an interdisciplinary palaeoenvironmental study at the Pradell fen, Spain. *The Holocene* 19/8:1241–1255.

Fernández Conde, F.J. 2001. Ganadería en Asturias en la Primera Edad Media: algunas características de la economía castreña y romana. In J.L. Gómez-Pantoja (ed.) *Los rebaños de Gerión: pastores y trashumancia en Iberia antigua y medieval*. Pp. 139–158. Madrid: Casa de Velázquez.

Fernández Mier, M. 1996. Análisis arqueológico de la configuración del espacio agrario medieval asturiano. *Mélanges de la Casa de Velázquez* 33:287–318.

Fernández Mier, M. 1999. *Génesis del territorio en la Edad Media: arqueología del paisaje y evolución histórica en la montaña asturiana*. Oviedo: Universidad de Oviedo.

Fernández Mier, M. 2010. Campos de cultivo en la Cordillera Cantábrica. La Agricultura en zonas de montaña. In H. Kirchner (ed.) *Por una arqueología agraria: perspectivas de investigación sobre espacios de cultivo en las sociedades medievales hispánicas*: 41–59. Oxford: Archaeopress (BAR International Series 2062).

Fernández Mier, M. 2011. Changing scales of local power in the Early Medieval Iberian North-West. In J. Escalona and A. Reynolds (eds.) *Scale and Scale changes in the Early Middle Ages: exploring landscape, local society and the World Beyond*. Pp. 87–117. Turnhout: Brepols (The Medieval Countryside 6).

Fernández Mier, M., Aparicio, P., González Álvarez, D., Fernández Fernández, J. and Alonso González, P. 2013a. Proyecto de Investigación: la formación de los paisajes agrarios del Noroeste peninsular durante la Edad Media (siglos V al XII). *Debates de Arqueología Medieval* 3:359–374.

Fernández Mier, M. and D. González Álvarez 2013. Más allá de la aldea: Estudio diacrónico del paisaje en el entorno de Vigaña (Belmonte de Miranda). *Excavaciones Arqueológicas en Asturias 2007-2012. En el centenario del descubrimiento de la caverna de La Peña de Candamo*. Pp. 353–365. Oviedo: Consejería de Educación, Cultura y Deporte del Principado de Asturias. Dirección General de Patrimonio Cultural.

Fernández Mier, M., López Gómez, P. and González Álvarez, D. 2013b. Prácticas ganaderas en la Cordillera Cantábrica: aproximación multidisciplinar al estudio de las áreas de pasto en la Edad Media. *Debates de Arqueología Medieval* 3:167–219.

García Fernández, J., 1980. *Sociedad y organización tradicional del espacio en Asturias*. Gijón: Silverio Cañada Editor, Biblioteca Julio Somoza.

García Leal, A. 1998. *Colección diplomatica del Monasterio de San Juan Bautista de Corías*. Oviedo: Universidad de Oviedo.

García Martínez, A. 1988. *Los vaqueiros de alzada de Asturias: un estudio histórico-antropológico*. Oviedo: Principado de Asturias.

García Martínez, A. 2003. La trashumancia en Asturias. In L. Elías and F. Novoa (eds.) *Un camino de ida y vuelta: la trashumancia en España*. Pp. 95–107. Barcelona: Lunwerg.

Gassiot, E. and Jiménez Zamora, J. 2006. El poblament prefeudal de l'alta muntanya dels Pirineus occidentals catalans (Pallars Sobirà i Alta Ribagorça). *Tribuna d'Arqueologia* 2004–2005:89–122.

Gassiot, E., Pèlachs, A., Bal, M.-C., García, V., Julià, R., Rodríguez Antón, D. and Astrou, A.-C. 2009. Dynamiques des activités anthropiques sur un milieu montagnard dans les píreneénne occidentales catalanes pendant la période de la préhistoire: une approche multidisciplinaire. *Bibliothèque d'Archéologie Méditerranéenne et Africaine* 4:33–43.

González Álvarez, D. 2010. El Parque Eólico Sierra de Carondio: una oportunidad perdida para el conocimiento de la Prehistoria reciente cantábrica. *Estrat Crític* 4:75–88.

González Álvarez, D. 2011. Movilidad ganadera entre las comunidades castreñas cantábricas: el valle del Pigüeña (Asturias) como caso de estudio. In OrJIA (eds.), *Actas de las II Jornadas de Jóvenes en Investigación Arqueológica (JIA 2009)*. Tomo I:147–156. Zaragoza: Pórtico.

González Álvarez, D. 2013. Traditional pastoralism in the Asturian Mountains: an ethnoarchaeological view on mobility and settlement patterns. In F. Lugli, A.A. Stoppiello and S. Biagetti (eds.) *Ethnoarchaeology: current research and field methods. Conference Proceedings, Rome, Italy, 13th–14th May 2010*: 202–208. Oxford: Archaeopress (BAR International Series 2472).

Graña, A. and López Álvarez, J. 2007. *Los teitos en Asturias: un estudio sobre la arquitectura con cubierta vegetal*. Gijón: Red de Museos Etnográficos de Asturias.

Kezich, G. and Viazzo, P.P. (eds.) 2004. *Il destino delle malghe: transformazioni nello spazio alpino e scenari futuribili di un sistema di consuetudini d'alpeggio*. San

Michelle all'Adige: Museo delgli Usi e Costumi della Gente Trentina (SM Annali di San Michele 17).

Klein, J. 1920. *The Mesta: a study in Spanish Economy History 1273–1836*. Cambridge: Harvard University Press.

Linares, F. 2004. *La arquitectura de las brañas somedanas*. Valladolid: Universidad de Valladolid.

Lombardía, C. and López Álvarez, J. (eds.) 2003. *José Ramón Lueje: la montaña fotografiada (1936–1975)*. Gijón: Ayuntamiento de Gijón.

López Álvarez, J. and Graña, A. 2003. Noticias sobre pastores y vaqueros. In C. Lombardía and J. López Álvarez (eds.) *José Ramón Lueje: la montaña fotografiada (1936-1975)*: 103–122. Gijón: Ayuntamiento de Gijón.

López Gómez, P. 2012. Ganadería de alta montaña en la Edad Media: el caso de Cangas del Narcea, Asturias. *@rqueología y Territorio* 9:185–199.

López Gómez, P. and González Álvarez, D. 2013. Etnoarqueología de los asentamientos pastoriles en la Cordillera Cantábrica: las brañas de Somiedu y Cangas del Narcea (Asturias). In G. Compañy, J. Fonte, B. Gómez-Arribas, L. Moragón and J.M. Señorán (eds.) *Actas de las V Jornadas de Jóvenes en Investigación Arqueológica – JIA 2012*: 362–366. Madrid: JAS Arqueología.

López Merino, L. 2009. *Paleoambiente y Antropización en Asturias durante el Holoceno*. Departamento de Ecología, Universidad Autónoma de Madrid. Unpublished PhD dissertation.

López Merino, L., López Sáez, J.A., Sánchez-Palencia, F.J., Reher, G.S. and Pérez Díaz, S. 2009. Castaños, nogales y cereales: la antropización de los paisajes de Asturias y León en época romana. *Cuadernos de la Sociedad Española de Ciencias Forestales* 30:93–99.

López Merino, L., Martínez Cortizas, A. and López Sáez, J.A. 2011. Human-induced changes on wetlands: a study case from NW Iberia. *Quaternary Science Reviews* 30/19–20:2745–2754.

López Sáez, J.A., López Merino, L., Pérez Díaz, S., Parcero, C. and Criado, F. 2009. Contribución a la caracterización de los espacios agrarios castreños: documentación y análisis palinológico de una posible terraza de cultivo en el castro de Follente (Caldas de Reis, Pontevedra). *Trabajos de Prehistoria* 66/2:171–182.

Marín, C. 2011. Las montañas cantábricas en el II y I milenio a.C.: un espacio de encuentro entre los grupos cantábricos y meseteños. In OrJIA (eds.), *Actas de las II Jornadas de Jóvenes en Investigación Arqueológica (Madrid, 6, 7 y 8 de mayo de 2009). JIA 2009, Tomo I*: 137–145. Zaragoza: Pórtico.

Marín, C. and González Álvarez, D. 2011. La romanización del Occidente Cantábrico: de la violencia física a la violencia simbólica. *Férvedes* 7:197–206.

Moreno, A., López Merino, L., Leira, M., Marco-Barba, J., González-Sampériz, P., Valero-Garcés, B.L., López Sáez, J.A., Santos, L., Mata, P. and Ito, E. 2011. Revealing the last 13,500 years of environmental history from the multiproxy record of a mountain lake (Lago Enol, northern Iberian Peninsula). *Journal of Paleolimnology* 46/3:327–349.

Morris, I. 2000. *Archaeology as Cultural History: words and things in Iron Age Greece*. Oxford: Blackwell.

Palet, J.M., Orengo, H.A., Ejarque, A., Euba, I., Miras, Y. and Riera, S. 2010. Formas de paisaje de montaña y ocupación del territorio en los Pirineos orientales en época romana: estudios pluridisciplinares en el valle del Madriu-Perafita-Claror (Andorra) y en la Sierra del Cadí (Cataluña). *Bollettino di Archeologia Online, Volume speciale A/A8/5. Edizione speciale–Congreso di Archeologia A.I.A.C. 2008*:67–79.

Parcero, C. 2006. Los paisajes agrarios castreños: modelos de construcción del espacio agrario a lo largo de la Edad del Hierro del noroeste. *Arqueología Espacial* 26:57–85.

Reimer, P.J. et al. 2009. IntCal09 and Marine09 radiocarbon age calibration curves, 0–50,000 years cal BP. *Radiocarbon* 51/4:1111–1150.

Rendu, C. 2003. *La Montagne d'Enveig: une estive pyrénéenne dans la longue durée*. Canet: Trabucaire.

Santos Granero, F. 1998. Writing history into the landscape: space, myth, and ritual in contemporany Amazonia. *American Ethnologist* 25(2):128–148.

Sauer, C.O. 1925. *The Morphology of Landscape*. University of California Publications in Geography 2(2):19–53.

Walsh, K., Mocci, F. and Palet, J.M. 2007. Nine thousand years of human/landscape dynamics in a high altitude zone in the southern French Alps (Parc National des Ecrins, Hautes-Alpes). *Preistoria Alpina* 42:9–22.

Yakar, J. 2000. *Ethnoarchaeology of Anatolia: rural socio-economy in the Bronze and Iron Ages*. Jerusalem: Emery and Claire Yass Publications in Archaeology.

David González Álvarez: FPU Researcher. Department of Prehistory, Universidad Complutense de Madrid.
Email: davidgon@ucm.es

Margarita Fernández Mier: Lecturer in Medieval History. Department of History, Universidad de León.
Email: margarita.mier@unileon.es

Pablo López Gómez: Postgraduate student. Department of Prehistory and Archaeology, Universidad de Granada.
Email: pirilopez@correo.ugr.es

13. Elusive *sel* sites: The geoarchaeological quest for Icelandic shielings and the case of Þorvaldsstaðasel, in northeast Iceland

Patrycja Kupiec, Karen Milek, Guðrún Alda Gísladóttir and James Woollett

The seasonal movement of grazing livestock to upland pastures (shielings) is believed to have played an important role in the subsistence economies of small farms in Scandinavia and the North Atlantic region during the Viking Age, Medieval and Post-Medieval periods. Historical sources, saga literature and place-name evidence strongly suggest that transhumance had been practised in Iceland since the settlement period in the ninth century, and that it formed an important part of a decentralised farming economy. However, since only eight putative shieling sites have been subjected to archaeological investigation in Iceland, little is known about the size and character of these sites, the full range of activities that took place in them, or the degree to which they were materially distinct from upland farms. This paper examines how microscopic analysis of floor surfaces can aid the detection of seasonally occupied sites and improve our understanding of livestock management in the North Atlantic region. The potential of micromorphological analysis of floor deposits to distinguish between periodically and permanently occupied sites is illustrated by a case study of a putative shieling site at Þorvaldsstaðasel, in northeast Iceland. The analysis of two thin sections taken from this site showed that its floor deposits exhibited a pattern of thin, periodic occupation surfaces, separated by thicker and less compacted accumulations of aeolian silt and fine sand. This pattern reflects the periods of intermittent occupation, separated by the periods of abandonment, and therefore it is consistent with the interpretation of the site as a periodically occupied shieling. The thin sections also captured a phase when the site was permanently occupied (characterised by thick and compacted floor deposits), which suggest that the occupation history of Icelandic shieling sites may have been more variable and complex than is normally recognised by archaeologists.

Introduction

The earliest historical sources, saga literature, place-name evidence, and small-scale archaeological investigations suggest that upland pastures were important economic resources in Iceland since the beginning of the settlement period in the 9th century (Sveinbjarnardóttir 1991). The practice of keeping livestock at summer farms (shielings, or *sel* in Icelandic) served two interconnected purposes: to utilise upland areas for grazing by dairy animals; and to remove livestock from the infield in the growing season. The latter was one of the strategies that Icelandic farmers employed to minimise land degradation, prevent overgrazing of home pastures, and ensure sufficient supply of hay to feed the animals during long winter months (Brown *et al.* 2012). Additional winter fodder could also be collected at shieling sites, and, by using the dung accumulated during the shieling season, certain unproductive areas could be fertilised, thus increasing their grazing potential and the future harvest of fodder. Summer transhumance of milking livestock was also a common practice elsewhere in the North Atlantic region, and archaeological evidence for the use of shielings has been found in Norway, where it dates from at least the Iron Age (Magnus 1986; Skrede 2005), the British Isles (Gelling 1964), the Faroe Islands (Mahler 1995), and Greenland (Keller and Albrethsen 1986).

The study of transhumant landscapes in Iceland has gained attention in recent years, with numerous suspected shieling sites located during area-based surveys undertaken in different parts of the country (Gunnarsdóttir 2002; Guðmundsson 2008; Lárusdóttir 2006; Pálsdóttir 2005; Sveinbjarnardóttir 1991). To date eight putative shielings have been the subject of archaeological investigation in Iceland: the Viking Age site of Kot in southern Iceland (Hallmundsdóttir 2009), the Viking Age site of Pálstóftir in eastern Iceland (Lucas 2008), the medieval sites of Faxadalur, Norðtungusel, Reykholtssel, Þórsárdalssel and Gautsstaðagrófarsel in western Iceland (Ellertsson Csillag and Hermannsdóttir 2013; Sveinbjarnardóttir *et al.* 2011), and the late medieval site of Þorvaldsstaðasel, in northeast Iceland (Gísladóttir *et al.* 2011). Of these, only Pálstóftir and Kot have been subjected to full-scale excavation. The identification of putative

Keywords: shieling, Iceland, micromorphology, seasonal occupation, Icelandic sagas, Íslendingasögur

shieling sites has often been based only on documentary sources, such as the 1703 land register (Árni Magnússon and Pál Vídalín 1703), and place-names (names with *sel* suffix or prefix). Many suspected shieling sites identified in this way have been classified as multi-period sites and could have served different functions at different times, including that of small marginal farms (Sveinbjarnardóttir *et al.* 2011). Distinguishing shielings from upland farms has proved problematic due to their potentially similar physical characteristics and the periodic use of some shielings as permanently occupied farms and *vice versa* (Hitzler 1979). In light of this, the identification of shieling sites should not be based only on the location of the site in upland/marginal areas, the size and the layout of the upstanding ruins, and *sel* place-names.

The main aim of this paper is to explore the potential of micromorphological analysis (the study of undisturbed soils and sediments in thin section) to detect punctuated signatures of occupation, which is the main characteristic that distinguishes shieling sites from perennially occupied farms. Since the microstratigraphy of the occupation deposits is captured in thin section, the method makes it possible to study changes in a site's main function and continuity of occupation, from permanent to periodic (possibly seasonal) occupation and *vice versa*. The method also allows the study and semi-quantification of anthropogenic inclusions, such as organic matter, ash, bone, charcoal, and various artefacts, and it can offer clues about the use of different structures at shieling sites and spatial organisation of activities within the structures, which may not be always visible during macroscale investigations.

A suspected shieling (Fig. 13.1) site of at Þorvaldsstaðasel, northeast Iceland, identified on the basis of its place-name (*sel* suffix), upland location, and the small size of its ruins, was subjected to test trenching and small-scale micromorphological sampling. The analysis of the samples taken at Þorvaldsstaðasel provides an ideal opportunity to test the potential of microscopic analysis for prospecting for shieling sites in Iceland. The data generated by this research may also help to detect shielings in other areas of the North Atlantic region, thereby providing a more nuanced understanding of pastoral practices and adaptations to local environments.

Historical sources

In contrast to scarce archaeological evidence for shieling sites in Iceland, the documentary record attesting to the importance of summer pastures for the Icelandic economy is considerable. The significance that Icelandic farmers attached to the shieling economy is reflected by the presence of laws that permitted travelling to shielings on Sundays in both the oldest Icelandic lawbook, *Grágás* (written down in 1117–1118, but thought to incorporate some earlier, Viking Age, laws), and the late thirteenth-century *Jónsbok*, which served as a legal framework in the country as late as the 19th century (Eggertsson 1992). This stands in contrast to the earliest Norwegian laws, *Gulaþing* and *Frostaþing* laws (written down in the mid-13th century, but incorporating laws dating back to the 11th century), which forbade most travels during the major Christian celebrations and on the holy festival of Sunday (that is, from Saturday afternoon to Monday morning), with the exception of fishing excursions and journeys that started on an ordinary work day and could not be stopped on a holy day due to the lack of appropriate shelter (*The Gulathing Law and the Frostathing Law* 2008:45, translated by L.M. Larson). In Iceland, we know from *Grágás* that men were permitted to drive livestock to and from pastures, and that women were permitted to milk animals, to carry the milk to a place where it could be stored or processed (even if the distance between a milking station and a dairy was significant), and to make dairy products on Sundays (Grágás 2006:40, translated by A. Dennis, P. Foote and R. Perkins).. The laws concerning work permissible on Sundays highlight the main difference between Icelandic and Norwegian laws regulating their respective shieling systems. The absence of laws permitting haymaking and travel to upland pastures on holy days in the *Gulaþing* and *Frostaþing* codices suggests that the need to gather enough animal fodder and produce enough stored dairy product for the winter was more pronounced in Iceland than in the Scandinavian homelands of the Norsemen. This can be explained by a heavier reliance on animal husbandry in Iceland, where the harsher climate and shorter growing season were not favourable for horticulture. Highly erodible volcanic soils (andosols) also meant that the effects of overgrazing and vegetation cover removal were noticeable in Iceland shortly after the 9th-century *landnám* (Old Norse meaning 'land-taking'), and summer pastures could have been perceived as a way of stopping the encroachment of erosion by putting less grazing pressure on already affected areas.

Icelandic and Norwegian laws also stipulated when livestock had to be brought to summer pastures, and driven back to home pastures. According to the earliest Icelandic and Norwegian laws, the duration of the shieling season was about two months, from mid-June to the latter half of August at the latest (*Jónsbok* 2010:237, translated by K. Schulman; *The Gulathing Law and the Frostathing Law* 2008:94, translated by L.M. Larson). However, this clause in Icelandic law seems to be derived directly from Norwegian law and it has been argued that the shieling season in Iceland was in fact shorter (Benediktsson 1978). The existence of the month-name *selmánuður*, 'the month of shielings', in the old Nordic calendar, may indicate that the period spent at summer pastures was actually closer to one month.

Laws also stipulated the preferred location of shielings, which had to be within the boundaries of the holding and could not be located on communal grazing land. Both

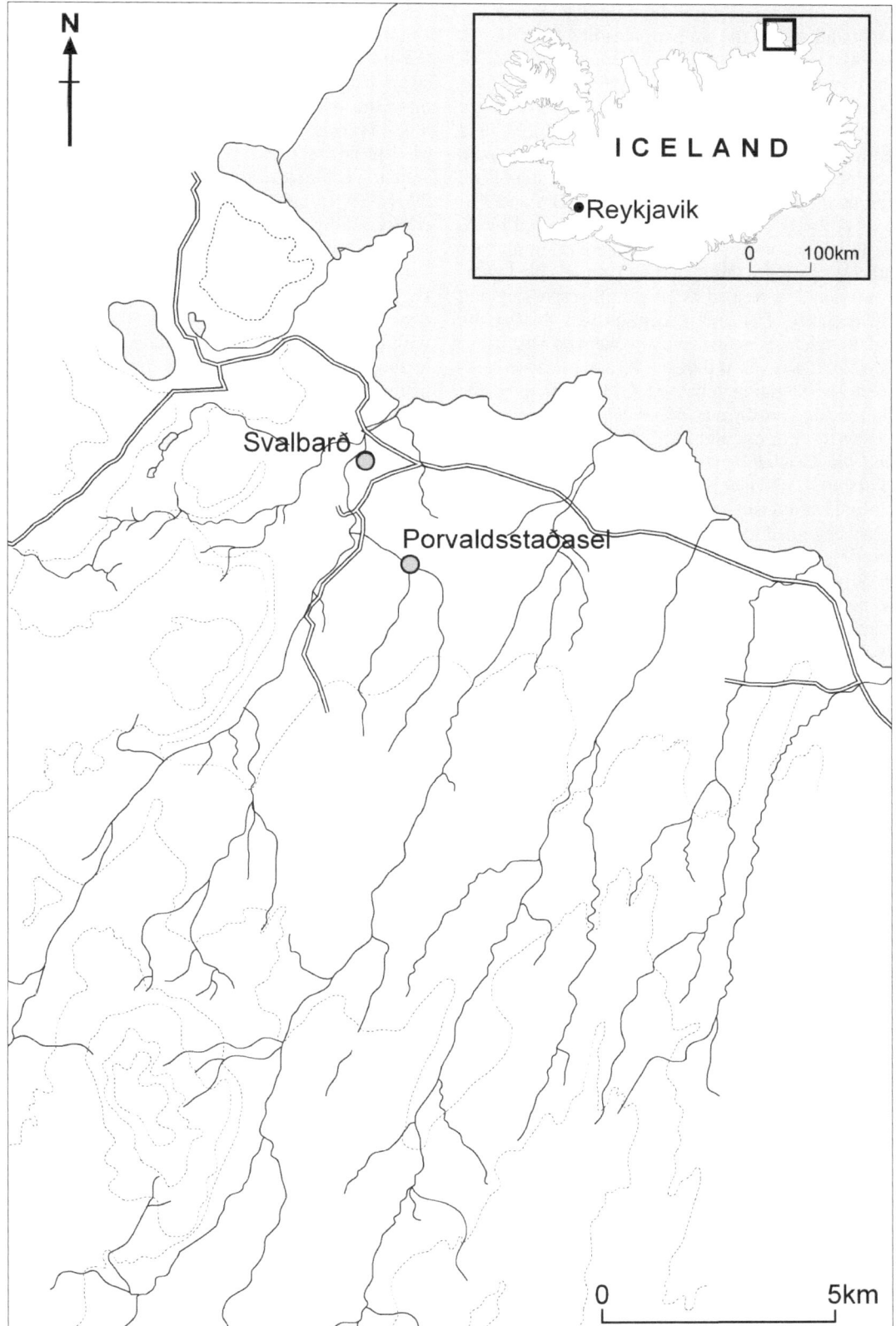

Figure 13.1 Location of the Þorvaldsstaðasel site, northeast Iceland.

Norwegian and Icelandic laws did not allow changes to shieling boundaries or the pathways leading to shielings, and they were to 'be kept as of old' (*Jónsbok* 2010:239, translated by K. Schulman; *The Gulathing Law and the Frostathing Law* 2008:96, translated by L.M. Larson). Those who did not respect shieling boundaries and grazed their animals on pastures belonging to other men could be punished by 'lesser outlawry', i.e. expulsion from the country for the period of three years and forfeiture of property (*Grágás* 2006:130, translated by A. Dennis, P. Foote and R. Perkins). Moreover, in all four law codices, travelling through other people's land without permission was treated as an act of trespassing and fined accordingly. The only exception was driving the livestock to upland pastures, and anyone who obstructed paths to shielings (even if they cut through private land) was fined for the damage caused to the farmer by the delay in reaching pastures (*Grágás* 2006:111, translated by A. Dennis, P. Foote and R. Perkins; *The Gulathing Law and the Frostathing Law* 2008:101, translated by L.M. Larson). This law suggests that shielings were often located a considerable distance away from the farms they belonged to.

Interestingly, the sagas of Icelanders, which were written down in the 13th and 14th centuries, paint a more diverse picture of the possible locations of shielings. For example, the shieling owned by Auðunn in *Grettis Saga* was located within a day's walk of the main farm and he was able to regularly fetch dairy products from it (*Grettis Saga* 1974:61, translated by D. Fox and H. Pálsson) while in *Laxdæla Saga* the shieling belonging to Bolli Þorleiksson's farm in Sælingsdalr was located both further in and higher up in the valley, but still within reasonable distance to the main farm (*Laxdæla Saga* 2001:379, translated by K. Kunz). The shieling belonging to Þorstein in *Heiðarvíga Saga* was also located within a day's walk from the main farm, and Þorstein's wife and daughters brought clean laundry from the shieling to the farm on Sundays (*Heiðarvíga Saga* 1998:90, translated by K. Kunz). Shielings could also be located further away from the home farm, in the next valley, or in the highlands between two valleys. When Einarr Þorbjarnarson went searching for lost sheep in *Hrafnkels Saga Freysgoða*, he rode from the head of the Hrafnkelsdalr valley south as far as the glaciers, across the marshy area between Þuríðarstaðalur and Jökuldalur, and then along the Jökulsá River (*Hrafnkels Saga Freysgoða* 2001:441-2, translated by T. Gunnel). During his journey he stopped at several shielings along the way to inquire after the lost sheep. Einarr's route was many kilometres long, suggesting that the shielings were located a considerable distance away from the farms in the valley. The shieling that belonged to Guðrún and Þord in *Laxdæla Saga* was also located at a great distance from their farm, and as a consequence the couple spent their summers apart (*Laxdæla Saga* 2001:335, translated by K. Kunz).

Some sagas also give an indication about how the shielings were constructed and how they appeared: the information that is often very difficult to reconstruct from the archaeological record only. In *Laxdæla Saga* the shieling where Bolli Þorleiksson was staying when he was subjected to a violent assault that ended his life is described as having two dairy buildings, a sleeping cabin, a storehouse and a byre (*Laxdæla Saga* 2001:380, translated by K. Kunz), while the shieling where Hallfreðr met with his old lover consisted of several huts, each suitable as sleeping quarters (*Hallfreðar saga vandræðaskálds* 1998:244, translated by D. Whaley). The shieling hut where Helgi was murdered in *Laxdæla Saga* is described as being "built with a single roof beam, which reached from one gable to the other and protruded at the ends, with a thatch of turf only a year old, which had not yet fully taken root" (*Laxdæla Saga* 2001:396-7, translated by K. Kunz). The construction of this building is depicted as rather weak, and the assailants trying to get into it managed to break the roof beam without much trouble. In *Þorsteins Saga Hvíta*, men attacking the shieling also manage to remove its roof to get inside (*Þorsteins Saga Hvíta* 1999:308, translated by A. Maxwell).

In addition to *Jónsbok*, which framed Icelandic law until the 20th century, a distinct 'household law' was first introduced in the early 15th century, which was subsequently reproduced and amended until the 18th century. This *Búalög* stipulated duties and payments for daily work on the farm. According to the earlier versions of *Búalög*, only three women were required to stay at a shieling site, in addition to *selráðskona* (the housekeeper of the shieling), in order to milk eighty ewes and twelve cows a day. In the afternoon, when their milking work was finished, the milkmaids were to return to work on the main farm, while the *selráðskona* stayed behind to cook (presumably for farmhands who were attending to haymaking) and to keep the site in good order (*Búalög* 1915:22, 34, 61). In the 17th century the work rate seems to have gone up, and three milkmaids were supposed to manage ninety sheep and fifteen cows, besides attending to *heyvinnu* (haymaking) and *heimannu* (housework) (*Búalög* 1915:191). The introduction of these new 'household laws' has at least two interesting implications for the functioning of the shieling economy in the Post-Medieval period. First, the practice of transhumance became a feature of large-scale farming, as it is very unlikely that more than a few landowners would own this number of cattle in the 17th century. This notion is supported by the descriptions of large farms (including church farms), which frequently mention shielings as important economic assets in documents related to legal disputes and inheritance (e.g. the *máldagi* for Reykjaholt church from 1185, DI 1:279-80). Second, the fact that the *sel* women were able to travel to upland pastures every day suggests that in the Post-Medieval period these sites were located within walking distance from farms they belonged to.

It seems that in spite of the recurrent references to shielings in the 17th century versions of *Búalög*, shieling sites gradually fell out of use from the early 18th century onwards, and by the late 19th century the practice of transhumance was already a feature of the distant past. The decline of the shieling system coincided with a period of widespread economic and social crisis. The 18th and 19th centuries in Iceland were characterised by recurrent famines, worsening climatic conditions, and a shortage of land for the increasing population, which pushed some farmers to try to occupy upland pastures formerly used as shielings (Hastrup 1989; Lamb 2011). By the 18th century, erosion had also begun to seriously affect farmland all over the country. The land register of 1703 (Árni Magnússon and Páll Vídalín 1703) and the annual sheriff's letters of the 17th and 18th centuries paint a grim picture of pastures and entire farmsteads being lost to rapid wind erosion, destabilisation of slopes, and sudden hydrological changes in river and stream regimes (Ogilvie 1984:2001). *Jarðabók* also contains frequent references to abandoned shieling sites, such as the shieling Fornusel, in Sýrholt, and the shieling Péturssel, in Eyjafjallasveit. In the 18th century Icelandic scholars blamed the abandonment of summer pastures on the destitute state of the Icelandic economy and they called for the re-introduction of shielings as an invaluable feature of the traditional Icelandic economy (Eggers 1786; Olavius 1780; Ólafsson 1772). In 1754 the Danish king issued a decree that reinforced medieval laws regulating the use of upland pastures. The decree encouraged the practice of transhumance as a means of protecting infields from overgrazing in the summer. Despite the joint efforts of the king and the agricultural reformers of the late 18th century, the shieling system was never re-introduced in Iceland. The occupation of the Þorvaldsstaðasel site in northeast Iceland spans the last phase in the history of Icelandic shielings, from the intensification and professionalisation of shieling activities in the 15th century to the decline of the shieling system in the 18th century (Gísladóttir *et al.* 2011).

Methods

The activities at Þorvaldsstaðasel started before AD 1300, as tephra from the Hekla eruption of this date capped a layer of peat ash exposed in test trench B (Gísladóttir *et al.* 2011). However, the structures exposed in the trench at Þorvaldsstaðasel were built after the Veiðivötn 1477 AD eruption, since it was captured *in situ* and clearly it was deposited on open ground. Based on the typology of a distinctive clay pipe bowl found in the deposit belonging to the last phase of occupation, the site was completely abandoned in the eighteenth century. The ruins at Þorvaldsstaðasel comprised a dwelling mound and an enclosed animal pen (Gísladóttir *et al.* 2011). The main structure appeared to have been divided into three rooms (R1, R2 and R3) and test trenches A and B were opened to expose their occupation deposits (Fig. 13.2). The most substantial floor deposits were uncovered in test trench B, which was placed across the width of R1, and therefore two soil blocks (samples 44B and 45B) were taken from the east section exposed by this test trench (Fig. 13.3). The blocks were thin sectioned at the McBurney Geoarchaeology Laboratory at the University of Cambridge.

Thin sections from Þorvaldsstaðasel were first photographed with reflected light and then examined at a scale of 1:1 using a light box, which helped to highlight colour differences that were used to identify layers and

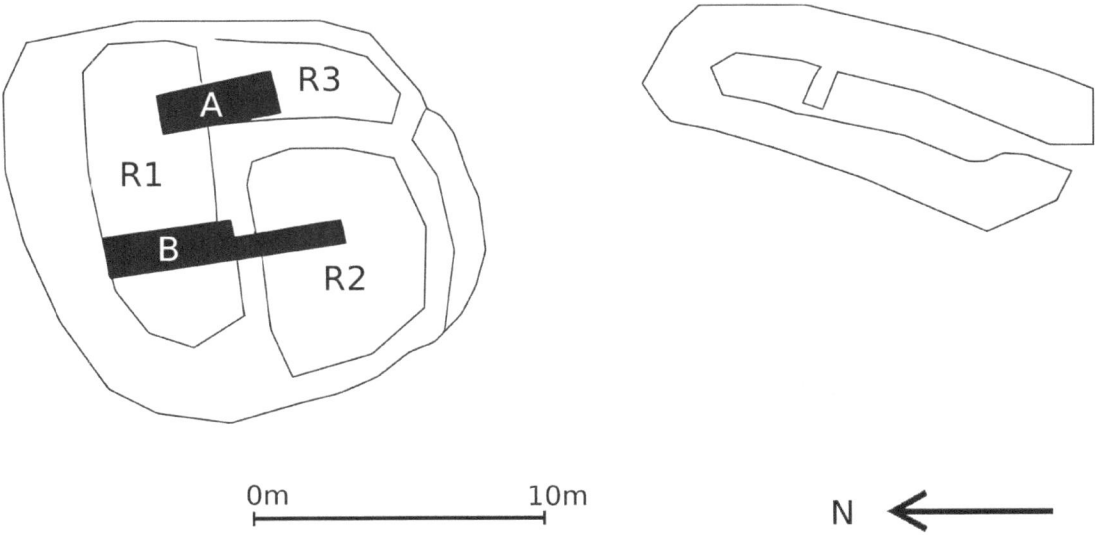

Figure 13.2 Plan of Þorvaldsstaðasel site showing the location of test trench B from which micromorphology samples were taken (adapted from Gísladóttir *et al.* 2011).

Table 13.1 Summary of features observed in thin sections 45B and 44B at the Þorvaldsstaðasel site.

Sample	Layer and microstratigraphic unit	Maximum thickness in thin section (mm)	Micromass: Nature of fine material (PPL)	Structure: Porosity	Mineral components: Degree of sorting	Mineral components: Microstructure	Mineral components: Soil texture	Groundmass: C/F$_{(100\,\mu m)}$ related distribution	Groundmass: C/F$_{(50\,\mu m)}$ ratio	Organic and Anthropogenic components: Charred organic matter	Charred wood	Charred plant	Fungal sclerotia	Wood tissues	Plant tissues	Diatoms	Phytoliths	Bone	Burnt bone	Amorphous organic matter	Excremental pedofeatures
45B 071.1	48	B-DB, DT	⁞⁞⁞⁞	PO	G, C, LN	SSL	CE, PR	40:60	⁞⁞				+	+	+	⁞⁞	+			⁞⁞	⁞
45B 071.2	10	B, DT	⁞⁞	W	I	LS	EE-CE	15:85	.	.	+		.	+	⁞	.			⁞⁞	.
45B 055.1	16	B-DB, DT	⁞⁞⁞⁞	MD-PO	LN	SSL	PR	25:75	.	+	⁞	+	+	+	+	+	+		⁞⁞	.
45B 055.2	13	DB-BL, DT	⁞⁞⁞	PO	LN	SIL	PR	20:80	⁞	+	+	.	.	+	.	+	+		⁞⁞	+
45B 055.3	4	DB, DT	⁞⁞	MD	LN, C	SSL	PR	25:75	⁞	+	⁞	+	.	+	.	+	+		⁞⁞	.
45B 055.4	28	DB, DT	⁞⁞⁞⁞	PO	LN, C	SIL	PR	20:80	+	⁞	⁞	+	+	+	+	⁞	.		⁞⁞⁞⁞	+
44B 036.1	38	YB, DT	⁞⁞⁞⁞	PO	LN, C	LS	PR	15:85	+	⁞⁞	⁞	+	+	+	⁞	⁞	⁞		⁞⁞⁞	⁞
44B 048.1	15	B-DB, DT	⁞⁞⁞⁞	MD-PO	LN	SSL	PR	30:70	⁞	⁞	⁞		+	.	+	+			⁞⁞⁞	+
44B 038.1	5	YB, DT	⁞⁞⁞	MD	LN	LS	PR	15:85	⁞	⁞		+	+	+	+	⁞		.	⁞⁞⁞	.
44B 038.2	10	B, DT	⁞⁞⁞	MD	LN	SIL	PR	20:80	+	⁞	+	+	.	+	+	.	+		⁞⁞⁞	.
44B 038.3	37	YB, DT	⁞⁞⁞	PO	LN	SSL	EE, PR	40:60	+	.	+	.	+	+	.	+			⁞⁞⁞	⁞
44B 038.4	3	B-DB, DT	⁞⁞⁞⁞⁞	MD-PO	G, C	SSL	PR, EE	40:60	+					.	.	⁞⁞	⁞⁞⁞		⁞⁞⁞	⁞⁞

Microstructure:
G: granular
V: vughy
PL: platy
L: lenticular
I: intergrain microaggregate
SG: single grain
AB: angular blocky
SB: subangular blocky

Degree of sorting:
P: poorly sorted
MSR: moderately sorted
W: well sorted

Soil texture:
SD: sand
LS: loamy sand
SL: sandy loam
SSL: sandy silt loam
SI: silt
SIL: silt loam

C/F related distribution:
SPR: single spaced porphyric
DPR: double spaced porphyric
OPR: open porphyric
CE: coarse enaulic
EE: equal enaulic
FE: fine enaulic

Fine material:
YB: yellowish brown
Y: yellow
LY: light yellow
B: brown
DB: dark brown
GR: grey
BL: black

Limpidity:
S: speckled
D: dotted

Birefringence:
U: undifferentiated

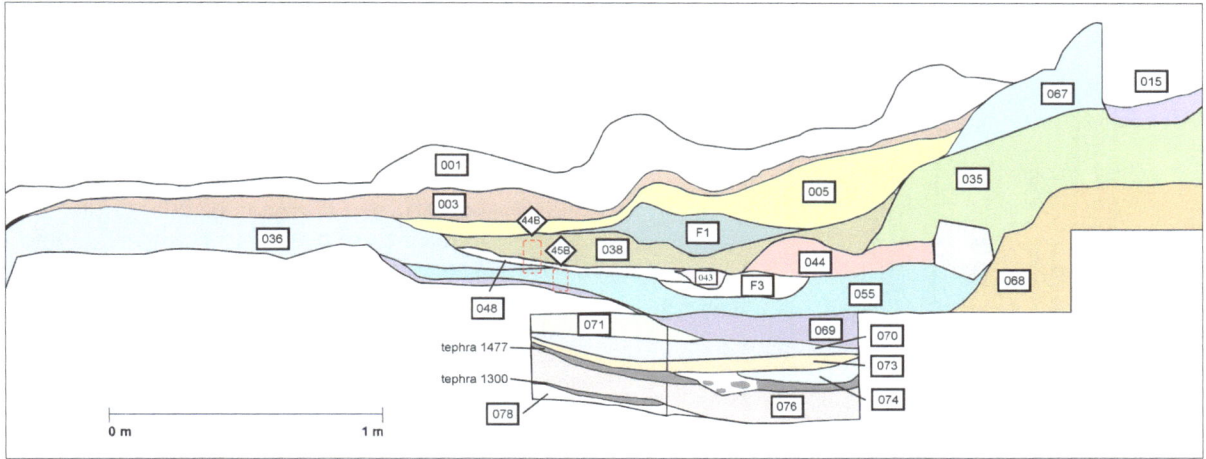

Figure 13.3　Section drawing showing the contexts visible in the test trench B, R1 at Þorvaldsstaðasel, and the location of the samples 44B and 45B.

measure their thicknesses. The thin sections were then analysed with petrographic microscopes at magnifications ranging from ×40 to ×400 with plane-polarised light (PPL), cross-polarised light (XPL), and oblique incident light, in which light is reflected on the surface of the thin section at an angle (OIL). Micromorphological descriptions followed the international standards in Bullock *et al.* (1985) and Stoops (2003), and utilised additional reference works such as FitzPatrick (1993) and Canti (2003). Summary descriptions are provided in Table 13.1, and illustrations of key micromorphological features are provided in Figs. 13.4–6.

Micromorphological analysis was aimed at the provision of microscopic evidence for activities associated with shielings and for evidence of periodic occupation. Micromorphological descriptions there-fore focused on the characteristics considered to be the most diagnostic: the thickness of the layers, the microstructure and porosity of the sediment, which can provide an indication of compaction under a vertical load (e.g. trampling); the nature of pedofeatures such as soil fauna excrement, which is indicative of bioturbation; the degree of sorting (including anthropogenic inclusions), the average particle size, soil texture and the ratio of coarse to fine material, which are indicative of the agent of deposition and subsequent anthropogenic and pedogenic influences, and, of course, all organic matter (in various stages of decomposition), biomineral inclusions (phytoliths and diatoms) and anthropogenic components (charred wood and plant, and burnt and unburnt bone). Each component or feature observed in thin section was quantified on the basis of the percentage of the visible area it represented, averaged over ten fields of view or, in the case of fine layers and lenses, averaged over the maximum number of fields of view possible. Visual percentage charts were used for this procedure, and the data should therefore be considered semi-quantitative. While quantifying features in thin section, care was taken to avoid areas obviously affected by bioturbation, such as worm channels, which are likely to contain intrusive material.

Results

General overview

The basic mineral composition of the occupation deposits captured in thin sections taken at Þorvaldsstaðasel was dominated by very fine silt and sand. The dominant size and the mineralogy of these deposits suggest that they originated in the andosols that surrounded the site. Due to their sandiness, high water retention, low cohesion, the presence of sand-sized silty aggregates, and low bulk density of soil grains, Icelandic andosols are very susceptible to wind and water erosion (Arnalds 2004; Wada *et al.* 1992), which can explain accumulations of aeolian sand in R1. The floor layers contained varying quantities of charred wood (2–10%), charred plant tissues (up to 3%), unidentifiable charred organic matter lacking visible cell structure (1–5%), and very low quantities of uncharred plant tissues (less than 1%). All layers captured in thin section were enriched in amorphous organic matter, which ranged from 5–20%. Their microstructure (a characterisation of the size, shape and arrangement of grains, aggregates and voids) was mainly lenticular, and stacks of elongated lenticular aggregates were separated by (sub)horizontal planar voids (Fig. 13.4a). The presence of silt cappings on the lenticular aggregates and a skeletal fabric of loose sand within the voids strongly suggest that the lenticular microstructure formed as a result of repeated freezing and thawing. A granular microstructure, in which granular shaped aggregates are separated by compound packing voids, was also present. This type of microstructure can be associated with fabric alteration by repeated freeze-thaw cycles (Coutard and Moucher 1985), and it is also universally observed in andosol topsoils due to the types of clay that are present (Stoops *et al.* 2008).

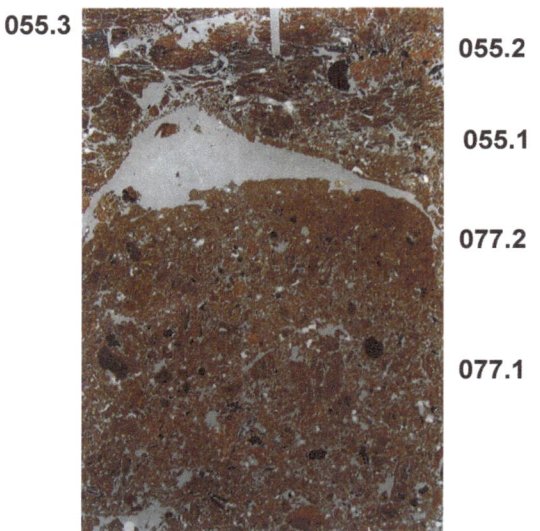

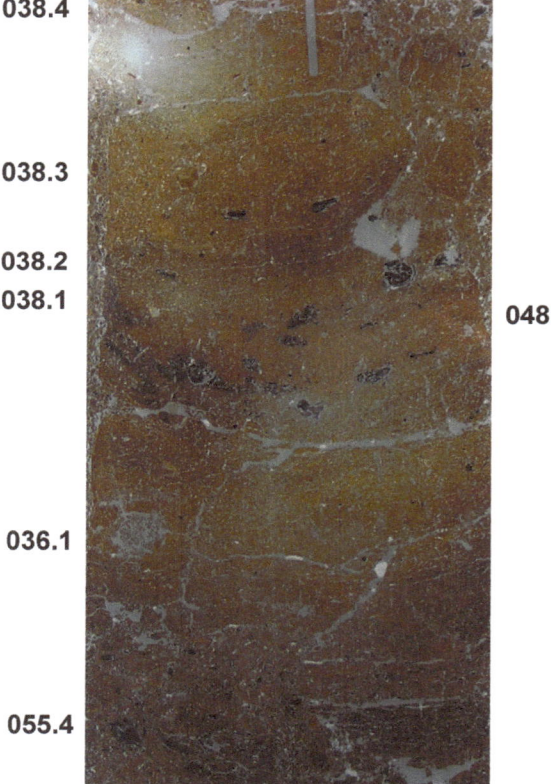

Figure 13.4 Micromorphology samples from test trench B in R1 at Þorvaldsstaðasel site (samples 45 and 44B) showing the layers that were observed in thin section.

All occupation deposits were subjected to intense reworking by soil fauna, which resulted in relatively high quantities of granular soil fauna excrement in the form of minute aggregates of fine material and numerous channels and irregularly shaped voids. Nevertheless, with the exception of heavily reworked context 0.71, the boundaries between the layers were clearly visible in thin section. In thin section it is possible to distinguish between the original sediment fabric and areas that had been disturbed by soil fauna, allowing a more accurate characterisation of the original sediments. Many organic and anthropogenic components in these thin sections, for example, appeared to have been dislocated by soil fauna, resulting in their movement, vertical orientation, and presence in earthworm channels that cut across sediment boundaries (Fig. 13.5i). As components disturbed by soil fauna could have originally derived from any of the layers, they were not included in visual quantifications presented in Table 13.1. During micromorphological analysis, care was taken to record the composition, compaction, and internal organisation of the preserved aggregates of original sediment. In case of significantly bioturbated context 069, it was possible to separate two microfabrics that related to different phases of the site's occupation: accumulations of windblown sand (indicative of periods when the site was abandoned) punctuated by thin floor surfaces (indicative of periods when the site was periodically occupied). These thin floor layers have similar composition to deposits from other seasonally occupied sites, including the shieling at Pálstóftir, in eastern Iceland (Milek 2007; Kupiec 2010), and the seasonal trading site at Gásir, northern Iceland (Guðmundsdóttir Beck 2011).

Since the archaeological deposits at Þorvaldsstaðasel had been significantly altered by freeze-thaw processes and bioturbation, the floor deposits could not be recognised only on the basis of the characteristics considered typical for trampled floors: increased compaction, dominant platy microstructure, or the presence of planar voids. Instead they were categorised on the basis of the horizontal/subhorizontal orientation of their components, which indicate that they accumulated on a gradually accruing occupation surface, a low degree of sorting compared to windblown layers or turf layers derived from andosols, and increased quantities of anthropogenic inclusions. The thickness of the best preserved floor surfaces varied from 4mm to 28mm but it was not possible to measure the maximum thickness of the heavily disturbed floor layers captured in context 071.1. Four thick floor layers (all belonging to context 055, as recorded in the field) were similar to thick floor deposits studied micromorphologically in the residential buildings of Icelandic farms (e.g. Milek 2012a; 2012b; Milek and Roberts 2013), and they are likely to have formed when the site was permanently occupied. After this phase of occupation, the structure was abandoned and the

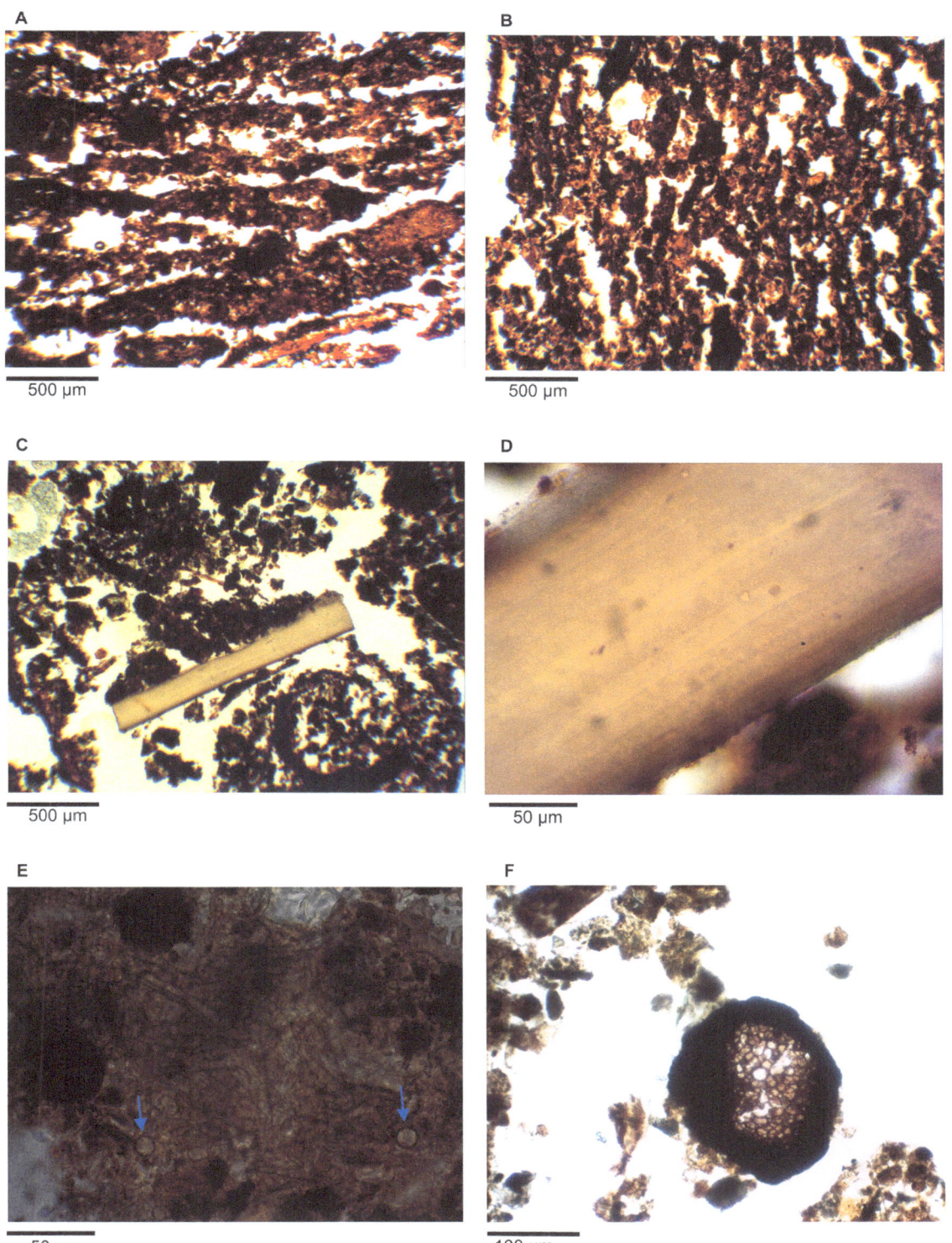

Figure 13.5 Photographs of micromorphology samples as seen through a petrographic microscope in PPL. A: lenticular microstructure with planar voids in sample 45B; B: vertically oriented planar voids resulting from soil fauna disturbance, sample 45B; C: fish bone in sample 45B; D: close up of fish bone, horizontal banding visible; E: aggregate of phytoliths and decomposed organic matter, grass pollen grains marked with blue arrows, 44B; F: fungal sclerotium in sample 44B.

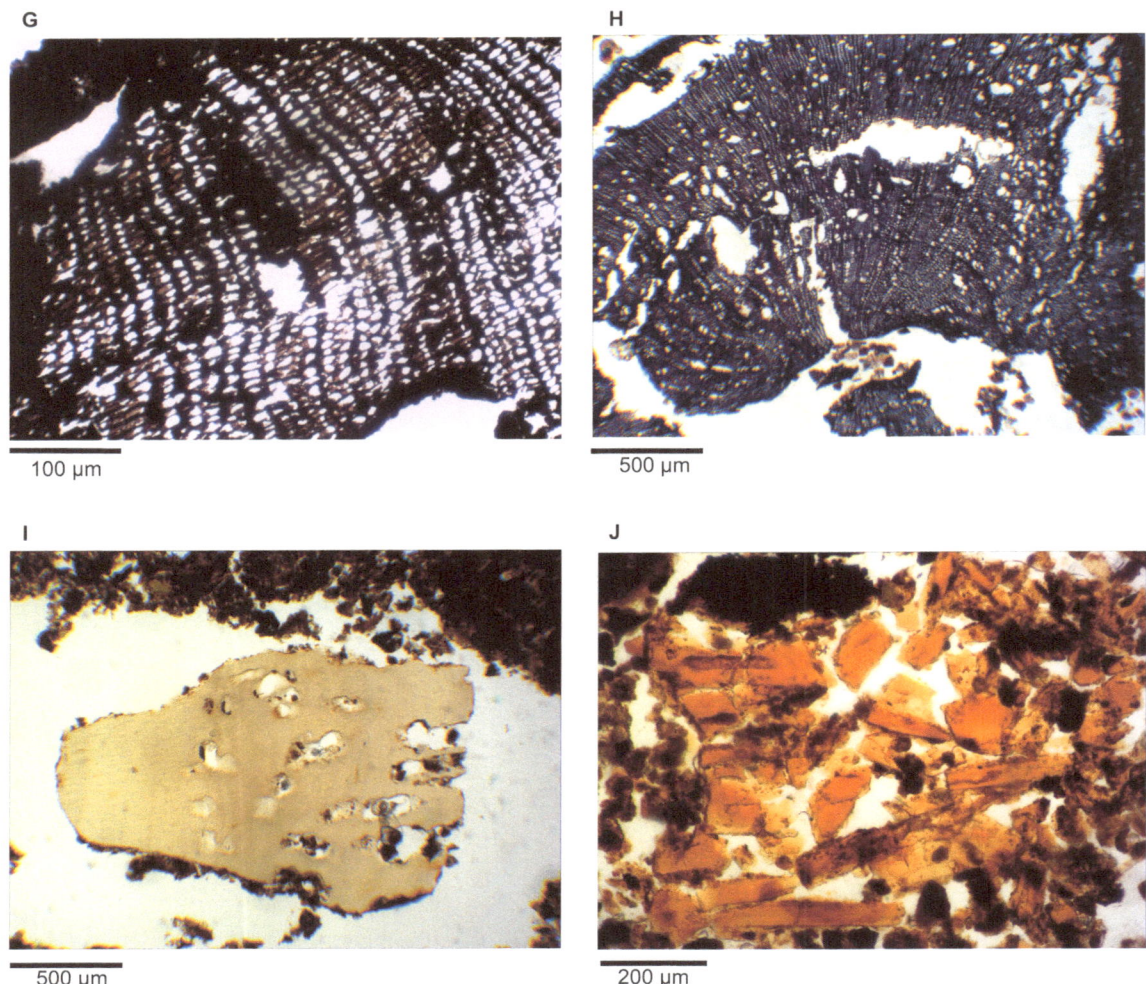

Figure 13.6 Photographs of micromorphology samples as seen through a petrographic microscope in PPL. G: charcoal fragment identified as coniferous wood in sample 44B; H: charcoal fragment identified as decidous wood in sample 45B; I: bone fragment in earthworm channel, sample 44B; J: fragmented and weathered bones in sample 44B.

collapse of its turf built structural remains began. The samples taken at Þorvaldsstaðasel, therefore, captured a shift in the structure's use, from periodic occupation to permanent occupation, and final abandonment.

Sample THS 45B

Sample THS 45B captured the most significant change in the function of structure R1 at Þorvaldsstaðasel, from periodically occupied settlement (possibly a shieling) to permanently occupied farm. The lowermost context captured in this sample, context 071 was associated with periodic occupation and periodic abandonment phases. The context, which was described in the field as a light brown silt deposit with charcoal fragments, was not a homogenous layer and it consisted of two disturbed fabrics: 071.1, a bioturbated layer consisting of two disturbed fabrics; and 069.2, an accumulation of well sorted aeolian sand (Fig. 13.4). 069.1 was penetrated by numerous channels filled with intrusive material, it had an average porosity of 25%, and it contained 5–10% excremental pedofeatures, which are typical indicators of bioturbation by soil fauna. The origin of its dominant granular microstructure is also most likely due to intensive activity by soil fauna, which produced minute aggregates of fine material. Many of the inclusions captured in the layer were either present in earthworm channels or were vertically oriented, including small aggregates of floor deposits with vertically oriented planar voids (Fig. 13.5b). Since the layer formed as a result of intense reworking by soil fauna its original composition was difficult to assess. However, after excluding intrusive and bioturbated material it became clear that the layer was composed of two different fabrics. The first fabric consisted of fine sand grains, which, based on their size and degree of sorting were most likely aeolian in origin. This fabric is likely to date to periods when the structure was not used, when the aeolian sediments, which originated from local andosols, could accumulate on the floor of the structure. The second fabric was preserved as compacted aggregates with dominant lenticular structure enriched with amorphous organic matter and horizontally bedded charred wood and organic matter. In comparison to the windblown sediments, these aggregates were poorly sorted. The distribution of individual particles in relation to finer material (the c.f. related distribution) in these aggregates was such that the finer material filled all the spaces between the coarser constituents (a so-called porphyric, or embedded distribution), which implies that the layer had been subjected to some compaction. However, since these aggregates were altered by freeze-thaw processes it is not possible to say whether this compaction was caused by downward pressure of trampling or by expanding ice lenses. Their composition, degree of sorting, and the horizontal orientation of anthropogenic inclusions suggest that these aggregates are remnants of thin floor layers. Due to bioturbation by soil fauna, many of these aggregates were vertically or sub-vertically oriented. However, since their lenticular microstructure was not disturbed, it is unlikely that the aggregates were moved a significant distance from their original position. The aggregates clustered in three distinctive lenses, strongly suggesting that prior to intensive reworking by soil fauna, 071.1 consisted of at least three periodic occupation and abandonment episodes. Interestingly, 071.1 contained the only fishbone captured in thin section (Figs. 13.5c and 13.5d), which suggests that during its periodic occupation phase, either the local rivers (such as Þorvaldsstaðaá, 'river of Þorvaldstaðir') were utilised by farmhands working at the site, or fish was brought to this upland site from lowland areas – perhaps indicating provisioning of the shieling with dried fish products.

Layer 071.1 was capped by a layer of fine, well sorted sand, most likely of aeolian origin. The layer contained 5–10% of amorphous organic matter and trace quantities of plant tissues, and it seems likely that it accumulated as a result of the weathering of the turf-built structural elements. When the roof and wall turves started to gradually disintegrate, mineral and organic matter contained within them, and aeolian sands, which originated from eroded local andosols, were able to rain for some time on the floors. The layer also contained 1–2% of minute fragments of charred organic matter, which could easily be transported by wind. Its dominant intergrain microaggregate microstructure, with its mixture of single particles, such as sand grains, and small aggregates of finer material (also fine-sand sized), also supports the interpretation of this layer as a windblown deposit, in which the finer material acted as a pseudosand. The layer had low porosity (5–10%) and low percentage of excremental pedofeatures (less than 2%), which suggests that its reworking by soil fauna was minimal. It is possible that reworking of underlying substrates by soil fauna began during this abandonment phase, and since organic-rich occupation surfaces in 070.1 were more attractive to soil fauna, 069.1 was not affected by bioturbation. Layer 069.1 and the windblown sand component of the underlying layer were the only deposits captured in thin section that were not affected by freeze-thaw processes, which could be explained by the better drainage of these well sorted windblown sands.

After this abandonment phase, sample 45 B captured a significant occupation deposit, context 055, which was described in the field as a compact, dark brown layer, rich in organic matter and interpreted as a floor deposit. The layer could be divided into three sub-phases. The earliest sub-phase was associated with a layer enriched in amorphous organic matter (10–20%), with less than 2% charred wood, plant tissues and organic matter and trace quantities of plant tissues, including plant tissues that contained phlobaphene, and that may be interpreted as bark. When cell structures were preserved or visible in thin section the source of the wood could be identified as deciduous wood species (Fig. 7.6h). Context 055.3 displayed all the characteristics that are generally considered to be typical of trampled floor deposits: a compact structure with a porphyric c.f. related distribution; planar voids; and horizontally bedded occupation debris. However, its dominant lenticular structure with localised pockets of granular microstructure attests to the post-depositional processes that affected this floor deposit. The relatively high porosity of the layer (20–30%) was due to earthworm channels and wide planar voids associated with freeze-thaw processes. With only 1–2% excremental pedofeatures, bioturbation of the layer was less significant than that of context 071. It is possible that the compaction of the original layer made it difficult for soil fauna to penetrate it. A layer composed almost entirely of ash was deposited above 051.1. Layer 051.3 contained 5–10% deciduous wood charcoal, 5–10% charred amorphous organic matter, and 2–5% charred plant tissues, suggesting that peat, rather than wood, might have been the main source of fuel at Þorvaldsstaðasel. The layer also captured several aggregates of diatoms, single celled photosynthetic organisms that produce siliceous cell walls and live in moist soils or wet environments. It is difficult to interpret the source of these diatoms without background reference samples from local rivers, bogs and lakes, but the diatomaceous aggregates are very likely to be remnants of ash created by the burning of peat or turf taken from a wet area. An experimental study of various fuel resources available in Iceland in the Viking Age, showed that diatoms in peat burnt at temperatures lower than 800 °C do not lose their structure (Simpson *et al.* 2003), was and this likely to be the case with the diatomaceous aggregates captured in sample 45B. Low temperature burning, which preserves the siliceous structures of both diatoms and phytoliths, is consistent with a temperature range of a cooking fire (McDonnell 2001). The layer also contained three minute fragments of unburnt bone. Layer 055.2 may represent either accidental or intentional spreading of ash from the hearth and its subsequent trampling. The intentional spreading of ash from the hearth on to the floors has been historically recorded in Iceland and it was practiced to create even floor surfaces and to diminish dampness or smell in wet conditions (Milek 2012b). There was no calcareous ash present in the layer, so this component must have been dissolved and washed down the profile by percolating rainwater. The final sub-phase, layer 055.1, was similar in composition and structure to layer 055.3, suggesting that the nature of occupation and the function of the structure did not change after the ash dumping episode. Finally, it is worth noting that due to the reworking of all three sub-phases by soil fauna, some bigger fragments of charcoal, which were moved and redeposited in earthworm channels, could not be quantified, since they might have come from the layers above or below 055.

Sample THS 44B

Sample THS 44B also captured the main occupation deposit, context 055. Layer 055.4 was very similar to layers 055.1 and 055.3 in terms of composition and structure, with the main difference being a higher amount of charred components: deciduous wood charcoal (2–5%); charred plant tissues (2–5%); and charred amorphous organic matter (2–5%). The layer also had higher percentage of excremental pedofeatures (2–5%), which can be explained by lack of unpalatable ash in context 055 captured in sample 44B. The layer also contained two fragments of weathered unburnt bone, which may relate to the last phase before the structure's abandonment, when the bones left on the floor could have been exposed to percolating rain water.

The layer that accumulated above the last phase of permanent occupation of R1 could be interpreted as the first stage of a complex abandonment sequence. Layer 036 can be categorised as turf collapse (most probably from the collapse of the roof) based on its composition, which was dominated by fine sands, amorphous organic matter (10–20%), horizontally bedded plant and woody tissues in trace quantities, and horizontally bedded aggregates of decomposed organic matter with grass phytoliths and grass pollen (see Fig. 13.5e). The layer also contained trace amounts of fungal sclerotia, fungal resting bodies that are commonly found in soils and are indicative of conditions adverse to fungal germination (usually a drop in temperature and moisture content of soil) (Fig. 13.5f). Grass phytoliths, grass pollen, plant tissues and amorphous organic matter are indicative of decomposed grass, while fungal sclerotia and the sandy component of layer 036 are most likely to be from the soil held by grass roots. Trace quantities of woody tissues captured in the layer may be the remains of wooden structural elements that collapsed with the roof, but, since the quantities are not substantial, they could also be derived from the woody plants commonly found on Icelandic heathland soils, juniper, crowberry, or blueberry, for example. The layer also contained trace amounts of charred organic matter, which could be indicative of soot accumulation in the roof turf.

After the structure at Þorvaldsstaðasel was abandoned and its roof collapsed, 048.1, a layer of ash, 5–10% charcoal, 2–5% charred amorphous organic matter, and 2–5% amorphous organic matter was deposited in R1. The random orientation of its constituents suggests that, in contrast to floor layer 055, this ash rich deposit did not gradually accumulate on a flat floor surface, and instead it was most likely dumped in R1. The layer contained the only identifiable fragment of coniferous wood charcoal (Fig. 13.6g). Except for juniper, coniferous wood in Iceland was derived from spruce, pine, or larch driftwood, and was commonly used for artefacts and as a building material, but was rarely used as a source of fuel (Dawn Mooney, pers. comm.). Its presence in this post-abandonment layer can be explained if collapsed or otherwise unused wooden structural elements of R1 had been used for fuel. The layer also captured 1–2% moderately burnt bone with browned (slightly burnt) edges. The small size of these bone fragments and their presence in an ash layer suggest that they had been intentionally thrown into the fire once the meal was consumed as a convenient and sanitary method of waste disposal. More burnt bone had also been redeposited in an earthworm channel directly underneath the layer, and burnt bone was also present in channels cutting through aeolian silt layer above it, suggesting that 048 could have originally contained higher quantities of burnt bone.

It is possible to say that even after R1 was abandoned, some activities were still taking place at Þorvaldsstaðasel, perhaps seasonal in nature and performed outside the structure, or in the neighbouring structure, which has not been subjected to test trenching yet. The collapse sequence that followed could be divided into three sub-phases. The first, layer 038.1 was a wind-blown deposit of very fine sand mixed with amorphous organic matter that probably derived from decomposing turf walls. This deposit also captured minute fragments of charred amorphous matter, which could have been transported by wind from nearby structures that were still in use, or it could have derived from soot covered walls. The next phase saw a collapse of the turf wall, context [039], with its soot-covered face represented by layer 39.2. The composition of this deposit was very similar to that of a likely turf roof collapse captured below. Context [039] was recognized as a turf layer in the field, but it was interpreted as a leveling deposit, rather than a collapse layer. The final phase captured in sample THS 44B showed that the partially collapsed R1 was still occasionally used as a short term shelter or a rubbish dump for nearby structures or outdoor functional areas. Layer 005.1 contained an accumulation of horizontally bedded unburnt bone. The bones were weathered, with frequent cracks and yellowish-brown rims on their surfaces (so-called pellicular alteration) and around internal cracks (irregular alteration). The physical alteration of bones suggests that they were subject to dissolution by slightly acidic rainwater and/or microbial attack, and the breakdown of collagen (Hedges 2002). The alteration of the bones could be a result of exposure to cold and damp conditions in an abandoned and partially collapsed structure.

Discussion

Although the earliest occupation deposits captured in the thin sections taken from the Þorvaldsstaðasel site were significantly affected by bioturbation, their microscopic analysis showed that they had originally contained a pattern of thin, short-term, periodic occupation surfaces, between which there were thicker, cleaner, and less compacted accumulations of fine wind-blown sand. When the roof and wall turves started to slowly disintegrate, these aeolian sediments, which originated from local andosols (soils derived from wind-blown, recycled materials of volcanic origin), 'entered' the floors of the structure. Therefore, amorphous organic matter visible under the microscope in the aeolian sand layers probably derived from organic matter from the turf roof, while the sand grains themselves could be a product of either roof weathering or wind erosion of local soils. This type of floor deposit could form if the site experienced periods of intermittent occupation alternating with periods of abandonment, perhaps lasting a year (if the site was occupied only during one particular season), or even longer. It is not possible to know the length of time that passed between periods of occupation. Estimates for the duration of turf roofs and walls before they require maintenance vary from three to twenty years, depending on the local climate (Ólafsson and Ágústsson n.d.); however, the process of disintegration intensifies when a turf structure is not heated during the cold seasons, and, once this process starts, it can progress at an ever increasing rate until the turves have to be completely replaced (Beresford 1988; Stell and Beaton 1984). Also, it is possible that less effort was put into the construction of periodically used structures, such as shielings, and therefore, after abandonment, they deteriorated at a faster rate than permanently occupied farm buildings. This notion is supported by the references to shielings in saga literature, which suggest that these structures were not always built to last. The poor quality of turves used to construct the buildings at Þorvaldsstaðasel was also noted by the site excavators (Gísladóttir *et al.* 2011).

The pattern of periodic occupation and abandonment phases evident at the Þorvaldsstaðasel site suggests that it could have functioned as a seasonal pastoral settlement. However, due to disturbed nature of the deposit that captured its periodic floor surfaces and short-term abandonment layers it is not clear for how many seasons Þorvaldsstaðasel was used as a shieling. At least three occupation-abandonment sequences are likely. According to the Icelandic sagas and medieval law codices, people and their milking animals travelled to upland pastures in June and stayed at shieling sites until August–September. However, the historical sources do not specify whether all shieling sites belonging to one farm were visited on an annual basis. Since transhumance in Iceland was one of the strategies employed to slow down the process of land degradation and to counteract the effects of overgrazing (Brown *et al.* 2012), it is possible that Icelandic farmers alternated between different upland pastures in order to put less grazing pressure on them. The relatively high organic content of the wind-blown deposits at Þorvaldsstaðasel, which suggests that the turf construction materials were degrading during periods of abandonment, indicate that this site probably witnessed abandonment phases longer than twelve months. Moreover, it is possible that the thin and ephemeral floor surfaces associated with the earliest occupation phases at Þorvaldsstaðasel did not only form as a result of annual transhumant activities, but also the periodic use of the shieling structures as a shelter for other upland activities such as hunting.

It is interesting to note that the main function of Þorvaldsstaðasel shieling changed at least once throughout its occupation sequence, and the site went through a period of permanent, long-term occupation after its use as a shieling, before it was abandoned and subsequently experienced large-scale collapse of its turf roof and walls. Since only part of its occupation sequence was sampled for micromorphological analysis, at present it is impossible to say how many times this structure changed its main function and the character of occupation. After the period of permanent occupation, structure R1 at Þorvaldsstaðasel was left to collapse and it was only occasionally used to dump rubbish, such as hearth deposits and bones. The fact that dumping of waste continued to occur in this phase shows that the site was still being visited, but activities in this phase were restricted to other buildings or to the outdoors.

It is likely that the two samples taken from Þorvaldsstaðasel's occupation deposits captured its earliest phase of use shortly after the eruption of Veiðivötn that has been dated to AD 1477; however, without micromorphological sampling of its entire occupational sequence or full excavation of the site, the possibility cannot be excluded that the site began with a short period of permanent occupation. The results of more intensive micromorphological sampling will provide further insight into the site's depositional history, and the question of the character of its initial occupation. If the site was first used as a shieling, the change of its function to a permanent farm several seasons after it was established could be indicative of land pressure and the expansion of settlement into marginal areas. However, the 15th and 16th centuries witnessed a sharp decline in the Icelandic population, which was caused by two plague epidemics. This resulted in the desertion of a great number of farms and the lowering of land rents on the remaining farms (Júlíusson 2007). The final abandonment of the site after a period of permanent occupation could also have been due to worsening climatic conditions, which made winter occupation impossible.

Conclusion

The results of this study showed that micromorphological analysis can be used to identify shieling sites (and other seasonally/periodically occupied sites) prior to excavation. The technique can also be used when there are no other indicators of seasonality, which could include, for example, the lack of synanthropic insect species in occupation deposits, the presence of bones from migratory animals and birds, and faunal assemblages dominated by marine mollusc shells, fish otoliths or mammalian teeth from the animals that died in the same season, based on the oxygen isotopes ratios and/or the study of annual incremental growth structures. The small geoarchaeological reference collection created by this research project can provide a basis for future studies of transhumant landscapes in the North Atlantic region. Moreover, it is notable that even this limited micromorphological sampling programme at Þorvaldsstaðasel captured the change of the building's function from a periodically occupied shieling to permanently occupied dwelling, a shift which was not detected during the trial excavation of this site.

Documentary sources and the results of archaeological investigations of Norse shielings indicate that the shieling economy included a broad spectrum of subsidiary activities, from iron smelting to hunting and textile production, which were organised around the seasonal movement of the people and their livestock to summer pastures (Amudsen 2007; Emanuelsson et al. 2000; Keller and Albrethsen 1986; Lucas 2008; Mahler 1991; Skrede 2005). With the great diversity currently displayed in the size and number of structures in these upland sites, it is possible that shielings with different secondary functions also had different typologies. The only way to explore all the complexities of the shieling economy in Scandinavia and the North Atlantic region is to identify and excavate more shielings. The micromorphological analysis of occupation deposits can facilitate our quest for these elusive sites.

Further geoarchaeological investigations of putative shieling sites in Scandinavia and its North Atlantic colonies are also needed to determine how the Norse adapted their traditional pastoral practices to new environments and social contexts. Due to the paucity of excavated shieling sites we do not know how (and if) the shieling economy varied between regions and through time. Inter-regional studies of transhumant landscapes are therefore needed to evaluate the uniformity of the Norse shieling economy that is currently suggested by historical sources, but which so far does not seem to be supported by archaeological evidence (as limited as it is). The ideal methodology would utilise several methods simultaneously, combining geoarchaeological analysis of putative shielings with other archaeological methods that can contribute to the interpretation of punctuated or seasonal occupation of a site, as well as contributing to an understanding of the full range of activities that might have taken place at shielings.

Acknowledgments

The authors would like to thank Dr Charles French for his support, advice in micromorphology, and enthusiasm for this research. We are also grateful to the team at Þorvaldsstaðasel site for taking micromorphological samples for this research. Special thanks are also due to Tonko Rajkovaca for manufacturing Þorvaldsstaðasel thin sections. This project has been supported by grants from Bill & Melinda Gates Foundation and from the Carnegie Trust for the Universities of Scotland.

Bibliography

Albrethsen, S.E. and Keller, C. 1986. The use of the *saeter* in medieval Norse farming in Greenland. *Arctic Anthropology* 23:91–107.

Amudsen, T. (ed.) 2007. *Elgfangst og Bosetning i Gråfjellområdet*. Oslo, Kulturhistorisk Museum.

Arnalds, O. 2004. Volcanic soils of Iceland. *Catena* 56:3–20.

Benediktsson, J. 1978. Some problems in the history of the settlement of Iceland. In T. Andersson (ed.) *The Vikings: Proceedings of the Symposium of the Faculty of Arts of Uppsala University, June 6–9, 1977*, pp. 161–65. Uppsala, Almqvist and Wiksell.

Beresford, G. 1988. Three deserted medieval settlements on Dartmoor: a comment on David Austin's reinterpretations. *Medieval Archaeology* 32:175–83.

Brown, J.L., Simpson, I.A., Morrison, S.J.L., Adderley, W.P., Tisdall, E. and Vésteinsson, O. 2012. Shieling areas: historical grazing pressures and landscape responses in northern Iceland. *Human Ecology* 40:81–99.

Bullock, P., Fedoroff, N., Jongerius, A., Stoops, G. and Tursina, T., 1985. *Handbook for Soil Thin Section Description*. Wolverhampton, Waine Research Publications.

Búalög, 1915–22. Reykjavik, Sögufélagið.

Canti, M.G. 2003. Aspects of chemical and microscopic characteristics of plant ashes found in archaeological soils. *Catena* 54/3:339–361.

Coutard, J.P. and Mucher, J. 1985. Deformation of laminated silt due to repeated freezing and thawing cycles. *Earth Surface Processes and Landforms* 10:309–319.

Dennis, A., Foote, P., and Perkins, R., trans. 2006. *Laws of early Iceland: Grágás, the Codex Regius of Grágás, with material from other manuscripts*. Winnipeg, University of Manitoba Press.

DI (1857–1972). *Diplomataricum Islandicum Íslenzkt Fornbréfasafn*. Volumes 1–16. Copenhagen and Reykjavik.

Eggers, C.U.D. 1786. *Philosophische Schilderung der Gegenwärtigen Verfassung von Island*. Altona, J.D.V. Eckhardt.

Ellertsson Csillag, Hermannsdóttir, S. and Hermannsdóttir, Á. 2013. *Seljabúskapur á norðanverðu Snæfellsnesi.* Skýrsla II. Reykjavík: Fornleifafræðistofan.

Emanuelsson, M., Berquist, U., Segerström, U., Svensson, E. and von Steding, H. 2000. Shieling or something else? An Iron Age and medieval forest settlement at Gammelvallen in Ångersjö, central Sweden. *Lund Archaeological Review* 6:123–138.

Fitzpatrick, E.A. 1993. *Soil microscopy and micromorphology.* New York, John Wiley and Sons.

Fox, D. and Pálsson, H. (trans.) 1974. *Grettir's Saga (Grettis Saga).* Toronto, University of Toronto Press.

Gelling, P.S., 1964. Medieval shielings in the Isle of Man. *Medieval Archaeology* 6–7:156–172.

Gísladóttir Alda, G., Ævarsson, U. and Wollett, J. 2011. *Interim Report of Archaeological Fieldwork on Svalbarð in summer of 2010 and winter 2011.* Reykjavik, Fornleifastofnun Íslands.

Guðmundsdóttir Beck, S. 2011. The micromorphology of the Gásir trading booths. In Sólveig Guðmundsdóttir Beck and M. Hayeur Smith. *Gásir Post-Excavation Reports* 3:6–69. Reykjavik, Fornleifastofnun Íslands.

Gunnarsdóttir, S. 2002. *The transhumant landscape of Saurbæjarhreppur: a study of shielings, dependent farms and their locations in connection to the mother settlements.* Unpublished MA thesis, School of Archaeological Studies, University of Leicester.

Gunnel, T. trans. 2001. The saga of Hrafnkel Frey's Godi (Hrafnkels Saga Freysgoda). In Ö. Thorsson (ed.). *The Sagas of the Icelanders*, pp. 436–463. New York, Penguin Books.

Guðmundsson, B. 2008. *Sel og Selstöður við Dýrafjörð.* Unpublished manuscript.

Jorgensen, P. A. trans., 1998 Hen-Thorir's Saga (Hænsna-Þóris Saga). In *The Complete Sagas of Icelanders 5.* Reykjavík, Leifur Eiríksson Publishing

Halmundsdóttir, M. H. 2009. *Upphaf og þróun byggðar við Hekluraetur: Fornleifarannsókn á rúst í landi Kots í Rangárvallasýslu.* Unpublished report.

Hastrup, K. 1989. Saeters in Iceland 900–1600. An anthropological analysis of economy and cosmology. *Acta Borealia* 6:72–85.

Hastrup, K. 1990. *Island of Anthropology: Studies in past and present Iceland.* Odense, Odense University Press.

Hedges, R.E.M. 2002. Bone diagenesis: an overview of processes. *Archaeometry* 44:319–28.

Heinemann, F.J. trans., 1998. The saga of the people of Svarfadardal (Svarfdæla saga). In V. Hreinsson (ed.). *The Complete Sagas of Icelanders* 4. Reykjavík, Leifur Eiríksson Publishing

Júlíusson, A.D. 2007. Peasants, aristocracy, and state power in Iceland 1400–1650. *CAHD papers* 2:1–9.

Kunz, K. trans., 2001. The saga of the people of Laxardal (Laxdæla saga). In Ö. Thorsson (ed.) *The Sagas of the Icelanders*, pp. 270–422. New York, Penguin Books.

Kunz, K. trans., 1998. The saga of the slayings on the heath (Heidarvíga saga). In V. Hreinsson (ed.). *The Complete Sagas of Icelanders* 4. Reykjavík, Leifur Eiríksson Publishing.

Kupiec, P.M. 2010. *Viking Age Shielings: finding evidence for seasonal occupation.* Unpublished undergraduate dissertation. Department of Archaeology, University of Aberdeen.

Lamb, H.H. 2011. *Climate: present, past and future.* London, Routledge.

Larson, L.M., trans., 2008. *The earliest Norwegian laws: being the Gulathing law and the Frostathing law.* Clark, New Jersey, The Lawbook Exchange Ltd.

Lárusdóttir, B. 2006. Settlement, organization and farm abandonment: the curious landscape of Reykjahverfi, North-East Iceland. In W. Davies, G. Halsall and A. Reynolds (eds.) *People and Space in Middle Ages, 300–1300*:45–64. Brepols, Netherlands, Turnhout.

Lucas, G. 2007. *Fornleifauppgröftur á Pálstóftirum við Kárahnjúka 2005.* Unpublished report. Landsvirkjun, Iceland.

Lucas, G. 2008. Pálstóftir: a Viking Age shieling in Iceland. *Norwegian Archaeological review* 41/1:85–101.

Magnus, B. 1986. Iron Age exploitation of high mountain resources in Sogn. *Norwegian Archaeological Review* 19/1:44–50.

Magnússonar, Á. and Vídalín, P. 1924–1947. *Manntal á Íslandi Árið 1703 Tekið að Tilhlutun Árna Magnússonar og Páls Vídalín Ásamt Manntali 1729 í Þrem Sýslum (Jarðabók 1703).* Reykjavík, Hagstofu Íslands.

Mahler, D.L.D. 1991. Agrisbekka: new evidence of shielings in the Faroe Islands. *Acta Archaeologica* 61:60–72.

Mahler, D.L.D. 1995. Shielings and their role in the Viking-Age economy: new evidence from the Faroe Islands. In C.E. Batey, J. Jesch and C.D. Morris (eds.). *The Viking Age in Caithness and the North Atlantic*, pp. 487–506. Edinburgh, Edinburgh University Press.

Matras, C. 1956. Gammelfærøsk ærgi, n., og dermed beslægtede ord. *Namn och Bygd* 44:51–67.

Maxwell, A. 1998. The saga of Thorstein the White (Þorsteins Saga Hvíta). In V. Hreinsson (ed.) *The Complete Sagas of Icelanders* 4. Reykjavík, Leifur Eiríksson Publishing.

Milek, K. 2007. Micromorphological analysis of the floor sediments from Pálstóftir: evidence for seasonal activity and the functions of the buildings. In G. Lucas (ed.) *Fornleifauppgröftur á Pálstóftum við Kárahnjúka* pp. 83–92. Reykjavik, Fornleifastofnun Íslands.

Milek, K.B. 2012a. The roles of pit houses and gendered spaces on Viking-Age farmsteads in Iceland. *Medieval Archaeology* 56:85–130.

Milek, K.B. 2012b. Floor formation processes and the interpretation of activity areas: an ethnoarchaeological study of turf buildings at Thverá, northeast Iceland. *Journal of Anthropological Archaeology* 31:119–137.

Milek, K.B. and Roberts, H.M. 2013. Integrated geoarchaeological methods for the determination of site activity areas: a study of a Viking Age house in Reykjavik, Iceland. *Journal of Archaeological Science* 40:1845–1865.

Olavius, O. 1780. *Oeconomisk Reise igiennem Island.* Kjøbenhavn, Gyldendals Forlag.

Ólafsson, E. 1832. *Kvædi Eggerts Ólafssonar.* Kaupmannahöfn, S.L. Møller.

Ólafsson, G. and Ágústsson, H. (n.d.). *The Reconstructed Medieval Farm in Þjórsárdalur and the Development of the Icelandic Turf house.* Reykjavik, National Museum of Iceland.

Pálsdóttir, A.H. 2005. *Segðu Mér Sögu af Seli: Fornleifafræðileg Úttekt á Íslenskum Seljum.* Unpublished BA thesis, University of Iceland.

Scudder, B. trans., 2001. Egil's saga (Egils saga). In Ö. Thorsson (ed.) *The Sagas of the Icelanders*, pp. 3–185. New York, Penguin Books.

Schulman, J.K. trans. 2010. Jónsbók, the laws of later Iceland: *the Icelandic text according to MS AM 351 fol. Skalholtsbok eldri*. Saarbrücken, AQ-Verlag.

Sigurjónsson, A. (ed.) 1966. *Búalög: verðlag á Íslandi á 12–19 öld*. Reykjavik, Framleiðsluráð Landbúnaðarins.

Skaptason, J. and Pulsiano, P. eds. and trans., 1984. *Bárðar Saga*. Garland Library of Medieval Literature 8A. New York, Garland.

Skrede, M.A. 2005. Shielings and landscape in western Norway: research traditions and recent traditions. In I. Holm, S. Innselset and I. Øye (eds.) *Utmark: the outfield as industry and ideology in the Iron Age and the Middle Ages*, pp. 31–43. Bergen, University of Bergen.

Stell, G. and Beaton, E. 1984. Local building traditions. In D. Omond (ed.) *The Ross and Cromarty Book*, pp. 207–18. Golspie, Northern Times.

Stoops, G. 2003. *Guidelines for the Analysis and Description of Soil and Regolith Thin Sections*. Madison, SSSA.

Stoops, G., Gérard, M. and Arnalds, O. 2008. A micromorphological study of andosol genesis in Iceland. In S. Kapur, A. Mermut and G. Stoops (eds.) *New Trends in Micromorphology*, pp67–90. Heidelberg, Springer.

Sveinbjarnardóttir, G. 1991. Shielings in Iceland: an archaeological and historical survey, *Acta Archaeologica* 61:76–91.

Sveinbjarnardóttir, G., Dahle, K., Erlendsson, E., Gísladóttir, G. and Vickers, K. 2011. The shielings of Reykholt: some preliminary results. In S. Sigmundsson (ed.), *Viking Settlements and Viking Society. Papers from the Proceedings of the Sixteenth Viking Congress, Iceland, 17–23 August 2009*, pp. 164–77. Reykjavík, Háskólaútgáfan; Hið íslenzka fornleifafélag.

Tucker, J. trans. 1998. The saga of the people of Vopnafjord (Vopnfirdinga Saga). In V. Hreinsson (ed.) *The Complete Sagas of Icelanders*, 4. Reykjavík, Leifur Eiríksson Publishing.

Wada, K., Arnalds, O., Kakuto, Y., Wilding, L.P. and Hallmark, C.T. 1992. Clay minerals in four soils formed in eolian and tephra materials in Iceland. *Geoderma* 52:351–365.

Whaley, D. trans. 1998. The saga of Hallfred the troublesome poet (Hallfredar Saga Vandrædaskálds). In V. Hreinsson (ed.) *The Complete Sagas of Icelanders* 1. Reykjavík, Leifur Eiríksson Publishing.

Guðrún Alda Gísladóttir: Fornleifastofnun Íslands, Bárugötu 3, 101 Reykjavík, Iceland.
Email: fsi@instarch.is

Patrycja Kupiec: Department of Archaeology, School of Geosciences, University of Aberdeen, St. Mary's, Elphinstone Road, Aberdeen, AB24 3UF, UK.
Email: r01pmk12@abdn.ac.uk

Karen Milek: Department of Archaeology, School of Geosciences, University of Aberdeen, St. Mary's, Elphinstone Road, Aberdeen, AB24 3UF, UK.
Email: k.milek@abdn.ac.uk

James Woollett: Université Laval, Département d'histoire, 1030, avenue des Sciences-Humaines, Bureau 5309, Université Laval, Québec, G1V 0A6, Canada.
Email: james.woollett@hst.ulaval.ca

Index

Aare-Gotthard Massif 156
Aare River 156, 158
Abies alba 125
Abtei 52
Acorns 127
Acqua Fredda 48, 53
Acta Murensia 161
Adige River 51, 59
Agno-Leogra ridge 70
Agno Valley 59, 65
Albanbühel 48, 52
ALGM 112
Allier Valley 1
Alm 1, 16
Almhütte 35
Almwirtschaft 37
Alpeggio 198
Alpes-de-Haute-Provence 197
Alpilles 194
Alpine Last Glacial Maximum 112
Alpine Pasture Inspection Report of 1902 175
Alps 7, 47, 53, 140, 153, 156, 161–2, 170, 173, 179, 188, 198, 212
Alp-system 33
Alpwirtschaft 19
Altipiano dei Sette Comuni 53
Alto Adige 51, 52
Álvarez Sanchís, Jesús 1
Ämpächli 171
Ancien régime 75
Andres, Brigitte 12–3, 15
animal husbandry 71, 147
animal teeth, analysis of 22
Apennines 47, 75
aqueduct 81
archaeobotany 109
Argentières-la-Bessée 195
arrowheads 128
Asiago plateau 62
Astico Valley 66
Asturias 97, 99, 206
Augustus 66
Austria 1, 19, 39
Auvergne 7, 11

Avanzini, Marco 12, 13
Avena 41
Aveto Valley 76–7, 90, 92–3
Ávila 1
Avisio 99
Axalp Chüemad 171, 173
Axalp Litschentellti 176
axe vi, 37, 39, 40, 64, 65, 161
Azov Sea 25, 29, 30

Bacchiglione 59
Badia 52
Bagós 49
baiti (*baito*) 63, 142–3, 146–7
Baito dei Ciocchi 101
Baito della Bassa 102–3
Balkans 9, 97
Ballenberg 175
Bargone 77–8
Barker, Graeme 1, 52
Bauden 35
Baudenwirtschaft 33
Baumgarten Alp 166
Bavaria 35, 58
Bayesian modelling 53
Beata Maria Vergine mine 69
Bedollo 53
beer bottles 36
Beḷḷares 208
Bell Beakers 189
'Belles Aigues', Laveissière, Cantal 4, 15
belt buckles 147
Belverde di Cetona 49
Ben Lawers Historic Landscape Project 17
Beresford, Maurice 11
Bergeten ob Braunwald 155, 170
Berne 158
Bernese Alps 155–6, 174, 178
Bernese Oberland 162, 174, 178
Beskydy Mountains 38–9, 43
Bílé Karpaty Mountains 36, 43
Bivio del Pidocchio 65
Bjørge (Vestfold, Norway) 4, 6, 11, 14
Black Forest, Germany 38

Black Sea 25, 29–30
Blockbau 61
Bocca Lorenza 68
Bodmin Moor 18
Bohemia 33, 34–5, 39–41, 43–4
Bohemian and Moravian lowlands 37
Bohemian Forest 35
Böhmerwald 35
Bolzano 3
bone collagen 24
Bosnia-Hercegovina 34, 38
  Eastern Bosnia 40
Brachypodio 43
Braña Forada 211
Braña of L'Estoupieḷḷu 212
brañas 203–4, 206–9, 212, 214–7
Brañas de baxo 209
Brañas de invierno, 209
Brannia 204
Brenta River 57, 59
Brentonico 51
Bressanone 52
Briançon 194
Brienz 158, 171
Brienzer Rothorn 161
Brigantium 194
Brixen 52
Bronze Age 1, 2, 4, 21, 23–4, 30, 37–40, 43, 47, 49, 50–1, 64–5, 70–1, 113, 127, 131, 161, 183, 186–9, 196–7, 199, 214–5
Brünig 155
Brünig Pass 158, 162
Búalög 224
buron 1, 4, 8, 12, 15
Busimo 71
Busimo Mountain 64–5, 71
byres 208

Cairo Massif 17
Calécc 16
Camargue 194
Camera del foco 142
Camera del latte 142–3
Campaniform (Bell Beaker) 189
Campo Pericoli 48, 50
Campo Rotondo 62–3
Cangas del Narcea 206, 212
Cantabrian Mountains 9, 203–5, 207, 210, 214, 216–7
Cantabrian Sea 205
Cantal 1, 4–5, 8, 13
Canton Bern 156
Canton Glarus 155
Canton Nidwalden 157

Canton Uri 156
Canton Valais 156
Caratata 81
Careggi (hamlet) 79
Caroso (hamlet) 79
Carpathian Mountains 34, 39
Carpathian region 36
Carpathians 9, 33, 97
Carpological remains 124
Carrer, Francesco 3, 9
Casa Carletti 49
Casanova di Rovegno 92
Casanova (hamlet) 78
casara 63
casaria (mobile wooden huts) 146
Case delle Barche 88
casera 143
Casone del Giazzo 73, 76, 90, 92
Casoni della Pietra 76, 77, 84, 86
Casoni detti li Fei 82
Casoni di Bargone 73, 76, 86, 88
Casoni di Perlezzi 78
Casoni Lagorara 73–4, 90
casoni (wooden huts) 4, 63, 73, 75–7, 79, 81–2, 85–6, 89, 91–2, 142
Caspian region 24
Caspian Sea 25
Caspian Steppes 21, 23, 25
Castelir di Bellamonte 48, 51
Castellieri 71
Castilla 210
Castilverio 65
Castion di Erbè 70
Castros 215
Catacomb Culture 21, 25
  catacomb groups 24
Catholic Church 194
cattle
  cattle-shed 177
  cow-shed 175
  dung 19
  manure 91
Caucasus 21
cauldron hanger 147
Cayolar 1
Cellars 36
Cerealia 41
cereals
  barley 41, 126
  einkorn wheat 41, 126
  emmer wheat 41, 126
  oats 41
  rye 41, 43
  Secale cereale 42, 43

Index

spelt 127
Triticum 41, 43
Triticum dicoccum 126
Triticum monococcum 126
wheat 41
Českomoravská vrchovina 39
Chalcolithic 189, 215
Chamois 19
Champsaur 185–6
Champsaur Valley 195
Charcoal-burners 53
Chiampo 66
Chichin 39, 187
Chichin Valley 187
Chiese 109
Chilguir 25
Chozos 210
Cima Marana 65
Circe 50
Civillina 70
Civillina Mountain 68
Clarke, David 1
Clark, Grahame 1
clay pipe 225
clearance cairn 163
climate
  climatic change 23
  climatic variation 140
Coe Veronesi Mountains 65, 71
Cogola of Giazzera 140
coins 36
cold store 36
collagen 233
Col Perpetue Mountains 112
Columella 47
Conelle di Arcevia 49
copper 53, 185
  Copper Age 19
  copper mine 53
Cornwall 2
Costoni pasture 144, 146
Coulanjou 8
Crau 194
cultural landscapes 109
'culture historical' paradigm 11
Czech Republic 33, 39

Dachstein 39
daggers 64
Dahle, Kristoffer 11
dairying 103–4
  butter 156
  cheese 3, 156
    cheese graters 51

  cheese making 4, 15, 51
    gorgonzola 49
    hard cheese 49
    rennet 52
  cream 156
  milk 36
    milk boiler 49, 50
    milk churns 207
    milking huts 175
    milkmaids 224
    milk parlour 146
    milk processing 147, 156, 178
Dalton minimum 140
Dartmoor 1
  Reaves 2
deer 197
Deforestation 216
Dehesas 210
dentine 22
diatoms 227
diet 21
Digital Terrain Model 144
Dinarides 34
Dolomites 34
Domenico Carbonara 81
Domodossola 158, 162
Don River region 25
Don Steppes 21, 24
Doria family 90
Dos Grum di Cadine 48, 52
Dosso Rotondo 48, 51, 109
Dreslerová, Dagmar 9, 13
driftwood 232
drought 3
Durance River 185
Durance Valley 194
Dzhurak-Sal River 29

Early Bronze Age 47, 51, 131, 187, 189
Earthworm 230
East Manych 21
East Manych Catacomb culture 23, 25
Economic Society of Bern 173, 175
Ecrins 39, 187
Ecrins National Park 184
eggshell 22
Einzelsennerei 170
El Castru 215
Elm 171
El Puertu 217
Eneolithic 197
England 140
Engstlen Alp 157, 159
Engstlen Lake 163

epidemics 147
Equinoctial *brañas* 207
Erbezzo 64, 65
Erzgebirge 34
Eschen Valley 158
Espinasse Soubro 5
ethnoarchaeology 11, 57
Euboea 51
Eurasian Steppes 21
exchange 21
Extremadura 210

Fabriano 49
Falconi 58
Fangeas 195
Faravel 187, 189, 194
Faroe Islands 221
Färrich 171
Faxadalur 221
fertilizer 91
Fettziger 174
feudal manor farms 34
Fiavé 47–8
Fiavé Carera 125–6, 131
Fior, Dalla 127
First World War 70, 139
fishbone 231
fishing 21, 222
fish otoliths 234
Fittanze Pass 65
flash-flooding 197
Fleming, Andrew 2
flint tools 128
Foehn 156
Folesani 70
Fontanabuona Valley 76
forestry 4
Forest-Saint-Julien 194
fortified hill top settlement 58
Fournel Valley 185, 198
France 1–2, 140, 162, 197
Frasassi cave 49
Freissinières 185
Freissinières Valley 186, 189, 194–5
French Alps 34, 38, 109, 183, 185, 196
Frostaþing law 222
Fungal sclerotia 232
furnace 53

Gadmen 159, 170
Gadmen, Gries 170
Gadmen, Mettlenberg 177
Gadmen Valley 156–7, 159, 162, 163–4, 170–2
Gap 194

Gashun-Sala 24
Gásir 228
Gasparine di Mezzo 65
Gaul 197
Gautsstaðagrófarsel 221
Gazzolo (hamlet) 82
Gebsen 174
Gen Valley 156–7, 159, 163–4, 166
Georgenfelder Hochmoor 43
Germany 38, 162
Giant Mountains 35
Giazzo 78
Giuseppe Ferrretto 81
Glasinac Plateau 40
glass 36
 glass making 35
goats 21
 goat bell clappers 147
gold 185
González-Álvarez, David 9, 13
Gotthard Pass 158
Grágás 222
Gran Sasso 50
grasses 127
Graveglia Valley 76
Great Scheidegg Pass 174
Great St Bernard Pass 158, 161
Greece 198
Greenland 221
Grettis Saga 224
Grevena 198
Gries 1 172
Gries Pass 155, 158, 162, 179
Grimsel 155
Grimsel area 178
Grimsel Pass 158, 161–2, 179
Grindel Alp 161
Grindelwald 158, 161, 175
grindstone 36
Grisons 174
Grisons Pass 161
Grosse Scheidegg Pass 158
Grotta a Male 48, 50
Grotta di Pertosa 51
Gsteigwiler, Breitlaunen 176
Guaite 70
Guelder rose 125
Guizza-Faedo Mountain 59
Gulaþing 222
Gwavos 18

Habsburgic Cadastre 142, 144, 146
Hafod 1
hafod names 2

Index

Hallfreðar saga vandræðaskálds 224
Hallstatt 40–3, 52
Hardanger Plateau 8
Hasliberg 169, 177
Hasliberg area 163
Hasliberg Basin 156, 157, 164
Hasliberg Mountain 160
Hasli Valley 158
Hat 36
Haute Fournel 186
Havos 18
hay 36
  hay-barn 73
Heiðarvíga Saga 224
Heidenhüttchen 155
Hekla eruption 225
Hendre 18
Herring, Peter 2
Higgs, Eric 1
hillforts 39
Hinder Tschuggi, Hasliberg 9, 171 177
historical ecology 75
hoards 215
Holocene 42, 185, 186
Honrado Concejo de La Mesta 210
Hordeum vulgare 41, 126
horses 21
  horse harness 147
  horse nails 147
  horseshoe 150
horticulture 222
Hoskins, W.G.
  The Making of the English Landscape 11
Hrafnkelsdalr 224
Hrafnkels Saga Freysgoða 224
Hrubý Jeseník Mountains 43
human teeth, analysis of 22
hunting 7

Iberia 215
Iberian Peninsula 97, 203, 209
ibex 19
Iceland 17, 221, 222, 223, 228, 233
  Icelandic lawbook 222
  Icelandic sagas 221
Iliad 50
Illasi 58
Indo-Europeans 11
Innertkirchen 157, 158, 165
Innsbruck 40
inscriptions 74
Interlaken 158, 174
Iron Age 1, 17, 24, 37–8, 40–1, 43, 47, 64, 66, 70–1, 183, 186, 189, 197–9, 215, 221
  Sub-Atlantic 37, 41, 43
iron fittings 36
iron smelting 4
Isergebirge 35
Isle of Man 19
isotopes 21
Italian Alps 99, 102, 185, 198
Italy 1, 9, 47, 140, 158, 162, 174

Jahren Johansen, Astri 11
Jarðabók 225
Jizera Mountains 35
Jizerské Hory 35
Joch Pass 157, 158
Jökuldalur 224
Jökulsá River 224
Jónsbok 222
Juglans 194

Kalaus 24
Kalmykia Steppes 23
Karst limestone 34
Khar-Zukha 22, 25
Kleine Osser, Bavaria 35
Klisura-Kadića Brdo 40
knee-buckles 147
knife 36
Koliba 37
Kot 221
Krkonoše forests 35
Krkonoše Mountains 33, 35–6, 42
Krušné Hory Mountains 34, 39, 43
Kupiec, Patrycja 12
kurgans 21–4, 29
Kyniaf-vod 18

La Courbatière 3
La Croix Blanche 4
La Croix Blanche (Cantal, France) 11
Lagorara 78
Lagorara Valley 76
Laiteries 8
Lake Brienz 161
Lake Garda 57
Lake Idro 110
Lake Lucerne 158
Lake Zurich 155
Landnám 222
landscape archaeology 11
Langquaid 161
Larionova, Yuri 9, 12
Las Corvas 215
Late Bronze Age 43, 50, 188, 197
La Tène 41–43

Late Sub-Boreal 38
Lauterbrunnen 174
Lauzes 2
Lavaggi di Chiappozzo 76, 92
Lavarone 53
Laveissière 4, 11
Laxdæla Saga 224
LBK Culture 38
lead 185
leather 70
Lech Sant 48, 52
Lefkandi 50–1
Le Guaite 58
Leinton, Lars 11
Leno Valley 140
Leogra Valley 59, 65
León 206
Les Sagnes 197
Lessinia 139
Lessini Highlands 57, 60–3, 65, 71
Lessini Plateau 70
L'Estoupieḷḷu 212
LiDAR 11, 143
Liguria 84
Ligurian Apennines 73, 75, 78, 93
Ligurian Mountains 75
Linear B 51
lintel 84
Little Ice Age 140, 196, 198, 199
Ḷḷaciana 206, 212
Loch Tay 17
logging 35
Lola 23
Lombard Southern Alps 112
Lombardy 16, 162
Longue durée 214
Louis XVI 7
Lower Champsaur 194
Lower Don River region 29, 30
Lower Engadin (Switzerland) 110
Lower Tatra Mountains (Slovakia) 37
Lucerne 158
Luco Culture 70
Lusatian Mountains 35, 43
Luserna 53
Lusia Pass 101
Lužické Hory 35

Machaon 50
Mägisalp, Hasliberg 169, 175
Magnifica Comunità di Fiemme 99
Magrè 68
Maison de Tailleur 3

Maissana 86
Maissana (hamlet) 86
Majkop 25
Malga (*malghe*) 1, 58
Malga Agnelezza 101
Malga Cadinello Alta 101
Malga Dosso Rotondo 112
Malga Lagorai 102
Malga Principi 48, 53
Malga Vacil 48, 51, 110, 112
Mandjikiny 22, 25–6, 29
Mandrom de Camp 48, 51
Manfron 68
Manfron Pass 59
Manych 24
Maquis 7
Marine model 27
marine mollusc 234
market economy 54
marriage 216
Massif Central 1, 4, 7, 9, 13
Maunder 139
Maunder minimum 140
Mayás primaliegas 208
meadows 208
Medieval period 18, 37, 43, 74, 97, 185
　Early Medieval 40
　Medieval Climate Optimum 140
　Medieval Warm Period (MWP) 140
　Post-Medieval period 74, 185
Mediterranean 51
megalithic tombs 1
Meiringen 156, 157, 158
Melkgang 175
Mercanti Valley 59
Merino 210
Mesolithic 1, 38, 185
Mesta (Spain) 3, 9
Mettlenberg 170
Meyer, Werner 155
Middle Ages 49, 53, 75, 156, 161, 174, 178, 183, 216
Middle Bronze Age 3, 47, 64, 111, 130–1, 161, 189
Migliavacca, Mara 13
migration 24
mineral deposits 11
mining 7
Miranda 206
mobility 21
　mobile huts 146
　mobile pen 101
Modetto Mountains 71
Modo Mountains 71
Mokrá žába 43
Molinietum 43

Index

Molise 9
Monastery of Corias 204
Montafon Valley 110
Montagne de la Mouche, Cantal 12
Montebello Vicentino 70
Monte Cengio 68
Monte Croce 70
Montefalcone 65
Monte Loffa 58, 70
Monte Loreto 48, 53
Monte Purga 58, 71
Monte Purga di Velo 64
Monte San Giovanni 58
Monte Santa Maria 86
Monte Tonolo 112
Monte Tregin 88
Monticazione 76
Monts Dore 9
Morava River 36
Moravia 33–4, 36
Moreno, Diego 75
Morris, Ian 214
Morteras 208
Moscow 25
mountain pastures 75
Mount Cornon 102
Mount Porcile 86
Mount Titlis 157, 159
Mraznica 38
Mühlebach Valley 155
mule-track 158
Muotatal SZ 171
Muri Abbey 161
Muséu del Pueblu d'Asturies 211
Mu-Sharet-4 25
Mutatio(nes) 194, 198

nails 36
Napoleonic cadastre 7
Neolithic 16, 37–41, 43, 117, 161, 183, 185–9, 196, 198, 206, 214–5
 Neolithic stone tools 38
Nestor 50
'New Archaeology' 11
Nicolis, Franco 12–3
Niercombe 2
Norðtungusel 221
North Atlantic 17
North Caucasus culture 25
North Caucasus region 24
Norway 4, 221
 Norwegian law 222

Oberhasli 155, 156
Oberhasli region 156–8, 160–2, 165, 171, 174–5, 177–8
Oberland Tour 174
Obermad 157
Obří Hrad u Studence 39
Obušek 37
Obwalden 158, 162
Odyssey 50
 Odysseus 50
Oeggl, Klaus 3, 12
Olleras 207
Onde Valley 185
O'Neill, Sue 2
Open-air museum 175
Oppeano 70
Orcières 189
Ore Mountains 34
Orobic Alps 16
Orry 1
osteological analyses 22
Ötzi the Iceman 3, 12, 19, 110
Ötz valley 19

Padua 70
palaeoclimate 23
Pálstóftir 221
Pasekarska 36
Passo Mucchione 70
pasture walls 163
Pasubio 140, 142
Pasubio Massif 140, 145
Pasubio Plateau 139
Pasubio territory 141
pathways 36
Paznaun Valley (Austria) 110
Pearce, Mark 4
Pennines 1
Perlezzi 73, 77–9, 81–2, 85, 89, 92
Petronio Valley 76, 89
Pferch 171
phytoliths 21, 227
Pierrefort 1
Pievebelvicino 59
pig 4
Pigüeña valley 212
pile-dwellings 155
pipe 36
plague 234
plants
 Sparganium (reed) 189
 crowberry 232
 blueberry 232
 Juncaceae (rush) 127–8
 laburnum (Laburnum sp.) 125

Plantago 41
   Plantago lanceolata 43
Poaceae 41
rhododendron 112
Rosaceae tipo 127
Viburnum 125
see also cereals, hay, phytoliths, pollen
podzols 113
Poland 35
polished stone axes 1
pollen 18, 43
   pollen analysis 21, 22, 38
Po Plain 53
porcelain 36
Posina River Basin 59
potatoes 208
pottery 36
Prášily 42
Prato di Pozzo 88, 89
Prato (hamlet) 79
Predazzo 51
primary products 104
Protovillanovan Culture 70
Provence 194, 198
Puglisi, Salvatore 4, 49
Pylos 51
Pyrenees 7, 92, 97, 212

Radhošť 38
Raeti 70
Rama 194
razors 65
reaves 2
Recent Bronze Age 64–5
Recoaro-Schio mining district 57
Redebus 53
Reichenbach Valley 161
Repubblica di Genova 90
Reuss Valley 158
Reykholtssel 221
Reykjaholt 224
Rezzoaglio 52
Riesengebirge 35
Rio Bucato 87
Rio S. Barbara 112
River Aare 156
River Chiese 131
River Leogra 59
River Triftwasser 157
Roccopiano 71
Roc, Jean-Claude 4, 11, 12
rock art 215
rock engravings 74
rock-salt 52

rock shelters 74, 163
Roman campaigns 11
Roman era 9, 11, 40, 43, 97, 183, 194, 198–9, 216
Romania 9
Romería 212
Rosso ammonitico 58, 63
round houses 18
Royal Commission on Historical Monuments 11
Ruiz Zapatero, Gonzalo 1
Rumex acetosella 43
rural archaeology 75

Saint-Laurent du Cros 194
salaš (farmhouses) 1, 37
Salers 7
salt 52
salt marshes 2
Salto 76
Saltwort Desert 23
Salvador, Isabella 12
Salzkammergut 52
Sanchorreja 1
San Fortunato 49
Sankt Christina in Gröden 52
Santa Cristina Valgardena 52
Santorso 68
San Vitale 58
Sardinia 93
Sarpa 24
Sassi Neri 68
Savoie 189
Saxony 35
Scaglia rossa 58
Scandinavia 1, 4, 7, 11, 15, 221
Scharfs Baude 36
Scheuchzer, Johann Jakob 155, 173
Schio-Recoaro mining district 69–71
Schnidejoch 110
Schöllenen Gorge 162
Schwabian Jura 34
scissors 147
Scotland 4
Scottish Highlands 17
scrapers 130
scythe 36
seasonal hunting 3
   versus transhumance 3
seasonality 22
secondary products revolution 3
Second World War 7–8, 36, 65, 102
Sedlo u Albrechtic 39
Sel 222
Sella pass 51
Selva di Val Gardena / Wolkenstein 48, 52

Sennen 162
Sennerei 156
Sentino Gorge 48, 50
Serre de l'Homme 188–9, 198
seter 1, 4, 7, 11
settlement patterns 75
sheep 21
   shearing 70
   sheep fold 57
   sheep/goat bell 36
   sheep husbandry 34
   shepherds 36
shell middens 1
Sherratt, Andrew 3
shielings 1, 4
Shishlina, Natalia 9, 12
shoe 36
sickles
  bronze sickle fragment 52
  flint sickle blades 128
Silesia 9, 36
silver 185
Silvretta 110
Silvretta range 19
slash-and-burn 215
slates 208
Slovakia 34, 36
smelting 7
  smelters 53
solar activity 140
Soldepuesto 204
Somiedu 206, 211
Sotćiastel 48, 52
Sottosengia 58
Spain 9
spearheads 64
Spindler, Konrad 3
Spitz 59
Square Mouth Jar Culture 68
Stafel 161
Stäfelti 4 174
Stagno, Anna 7
Star Carr 1
Steenstrup, Axel 1
steppe 21
Steppe model 27
Steppe North Caucasus culture 23
St Nectaire 7
stone fences 36
stone foundations 36
Storo Dosso Rotondo 109–13, 115–6, 118, 120–2,
    128–9, 130, 132–3
stove tiles 36
strainers 49

Strontium isotope 25, 27
Stubai Alps 40
Sturla Valley 76, 78, 90
Sub-Atlantic 38
Sub-Boreal 38
Südtirol 51, 52
Sukhaya Termista 22, 26, 29
Šumava foothills 42
Šumava Forests 35
Šumava Mountains 33, 38–9, 42
Summer brañas 207
Susten Pass 157, 158
Sweden 4, 11
Swiss Alps 171, 185
Switzerland 1, 19, 38, 155, 157–8, 161–2, 168, 170
   Swiss Plateau 158, 174

Tecchiati, Umberto 52
Temrta 22, 25
Terek 24
terraces 81
Terragnolo Valley 140
test-pits 212
textile 34
thatch 208
thin section 227
Thun 158
tiles 208
Tilia 125
Tipo geometrico 81
toponymy 204
Torque 189
Torrebelvicino 59
trading site 228
transhumance 1, 2, 7, 9, 15, 33, 34, 38–9, 97, 103, 233
  long, horizontal or Mediterranean 99
  Long transhumance 101
  short, vertical or Alpine 99, 104
Trebbia Valley 75, 92
tree fruits
  chestnuts 89
  Cornelian cherry berries (*Cornus mas*) 126
  hazelnuts (*Corylus avellana*) 126
  juniper berries (cf. *Juniperus sp.*) 127
  various 232
tree species
  alder
    green alder (*Alnus viridis*) 125
  apple (*Malus*) 125
  beech
    European beech (*Fagus sylvatica*) 84, 125
  birch (*Betula*) 41
  chestnut (*Castanea sativa*) 194

conifers 117, 125
Cornelian cherry (*Cornus*) 125
elder (*Alnus sp.*) 125
fir
   silver fir (*Abies alba*) 125
hazel
   European/common hazel (*Corylus avellana*) 125
hornbeam (*Carpinus*) 42
larch (*Larix decidua*) 140, 232
lime (*Tilia sp.*) 125
maple (*Acer*) 125
   Norway maple (*Acer platanoides*) 125
   sycamore (*Acer pseudoplatanus*) 125
mountain ash (*S. aucuparia*) 127
pear (*Pyrus*) 125
pine (*Pinus*) 41
   Pinus sp. 189
   Scots pine/mountain pine (*Pinus sylvestris/mugo*) 125
   Swiss stone pine (*Pinus cembra*) 125, 140, 185
Service tree (*Sorbus domestica*) 128
spruce
   Norway spruce (*Picea excelsa*) 125, 232
walnut (*Juglans*) 194
whitebeam (*Sorbus sp.*) 125
   common whitebeam (*Sorbus aria*) 127
   dwarf whitebeam (*Sorbus chamaemespilus*) 127
Trentino 51
Trift Glacier 157, 163
Trisa 59
Tschudi, Johann Heinrich 155
turf 233
turves 233
Tyrol 19

Ukraine 36
Ulan 22, 26, 28, 29
Upper Palaeolithic 38, 185
U Puzzu di Ertola 48, 52
Urtica 43
use-wear analysis 109

Vaccinium myrtillus 112
Váh River 36
Vaihingen 38
Valahians (*see* Walachians) 36
Valais 174
Valaška 37
Valbella 68
Val Cadino 101
Val d'Aveto 90
Val de Bagnes 156
Val di Fassa 101
Val di Fiemme 97–100, 103–4

Val d'Ossola 158
Val Lagorara 84
Valle del Caffaro 110
Valle del Chiese 110, 125
Valli del Bitto 16
Val Mercanti 68
Val Morta 53
Val Petronio 86
Val Riolo 68
Val Senales/Schnals (Alto Adige) 110
Val Sugana 53
Val Travignolo 51
Val Venosta/Vinschgau 110
Vapincum 194
Vaqueiros d'alzada 209
Var 198
Vara Valley 75–6, 84
Varolo Mountains 59
Varro 52
veceras 208
Vegetable garden 92
Veiðivötn 233
Věnec u Lčovic 39
Veneti 70
Veneto 51
Ventarola (hamlet) 77–8, 92
Verona 57–8
Verruga mountains 75
Vestfold 7
Vézzena 49
Vigaña 206, 212–5
Viking Age 22–2
villa sites 198
Vinschgau 19
Vlachs (*see also* Walachians) 9
Volga River 21
Volga River region 25
Vołosi 36
Volti 142
Vorarlberg, Austria 110

Wales 2
Wallachians (Vlachs) 9, 36, 37
Walsh, Kevin 12–3
Wassen 158
weapons 70
Wendenboden 170–1
Wenden Glacier 157
Wendenläger 170, 172
Wenden Valley 157, 162, 170
Western Caspian Plain 24
Western Isles 17
whetstones 36
Wildbach 157

Index

window-panes 36
winter
  winter sports 162
  winter villages 209
Wolkenstein in Gröden BZ 52
wood
  woodcutters 53
  wooden vessels 4, 174
  woodland 11
  woodworking 131
wool 34
Wörgetal 40
Wyss, Johann Rudolf 174

Yamnaya
  Yamnaya burials 24
  Yamnaya culture 23–4

Yamnaya kurgans 24
Yamnaya people 189
Yamnaya period 21, 25
Yamnaya sites 24
Yergueni Hills 25, 27

Zatta 75
Zum See, Innertkirchen 166, 169, 171
Zürich 19

Þórsárdalssel 221
Þorsteins Saga Hvíta 224
Þorvaldsstaðasel 221, 223, 225, 227, 230–1, 233
Þuríðarstaðalur 224

Printed in the USA
CPSIA information can be obtained
at www.ICGtesting.com
JSHW041025261223
54199JS00005B/63